Will Wilson
Zoology Department
Duke University
Durham, NC 27708-0325

AN INTRODUCTION TO COMPUTATIONAL PHYSICS

D0706174

AN INTRODUCTION TO COMPUTATIONAL PHYSICS

TAO PANG

University of Nevada, Las Vegas

CAMBRIDGE
UNIVERSITY PRESS

PUBLISHED BY THE PRESS SYNDICATE OF THE UNIVERSITY OF CAMBRIDGE
The Pitt Building, Trumpington Street, Cambridge CB2 1RP, United Kingdom

CAMBRIDGE UNIVERSITY PRESS
The Edinburgh Building, Cambridge CB2 2RU, United Kingdom
40 West 20th Street, New York, NY 10011-4211, USA
10 Stamford Road, Oakleigh, Melbourne 3166, Australia

First published 1997

Printed in the United States of America

Typeset in Times Roman

Library of Congress Cataloging-in-Publication Data
Pang, Tao, 1959–
Introduction to computational physics / Tao Pang.
p.   cm.
Includes bibliographical references and index.
ISBN 0-521-48143-0 (hardback). – ISBN 0-521-48592-4 (pbk.)
1. Physics–Data processing.   2. Physics–Methodology.
3. Numerical calculations.   4. Mathematical physics.   I. Title.
QC20.7.E4P36   1997
530′.0285–dc21   96-45574
                                                     CIP

A catalog record for this book is available from
the British Library.

ISBN 0 521 48143 0   hardback
ISBN 0 521 48592 4   paperback

To my parents, Shuru and Bifu, my wife, Yunhua,
and my children, Joyce and Deric

CONTENTS

PREFACE

The beauty of Nature is in its detail. If we are to understand different layers of scientific phenomena, tedious computations are inevitable. In the last half-century, computational approaches to many problems in science and engineering have clearly evolved into a new branch of science, *computational science*. With the increasing computing power of modern computers and the availability of new numerical techniques, scientists in different disciplines have started to unfold the mysteries of the so-called *grand challenges,* which are identified as scientific problems that will remain significant for years to come and may require teraflop computing power. These problems include, but are not limited to, global environmental modeling, virus vaccine design, and new electronic materials simulation.

Computational physics, in my view, is the foundation of computational science. It deals with basic computational problems in physics, which are closely related to the equations and computational problems in other scientific and engineering fields. For example, numerical schemes for Newton's equation can be implemented in the study of the dynamics of large molecules in chemistry and biology; algorithms for solving the Schrödinger equation are necessary in the study of electronic structures in materials science; the techniques used to solve the diffusion equation can be applied to air pollution control problems; and numerical simulations of hydrodynamic equations are needed in weather prediction and oceanic dynamics.

Important as computational physics is, it has not yet become a standard course in the curricula of many universities. But clearly its importance will increase with the further development of computational science. Almost every college or university now has some networked workstations available to students. Probably many of them will have some closely linked parallel or distributed computing systems in the near future. Students from many disciplines within science and engineering now demand the basic knowledge of scientific computing, which will certainly be important in their future careers. This book is written to fulfill this need.

Some of the materials in this book come from my lecture notes for a computational physics course I have been teaching at the University of Nevada, Las Vegas. I usually have a combination of graduate and undergraduate students from physics, engineering, and other majors. All of them have some access to the workstations or supercomputers on campus. The purpose of my lectures is to provide the students with some basic materials and necessary guidance so they can work out the assigned problems and selected projects on the computers available to them and in a programming language of their choice.

This book is made up of two parts. The first part (Chapter 1 through Chapter 6) deals with the basics of computational physics. Enough detail is provided so that a well-prepared upper division undergraduate student in science or engineering will have no difficulty in following the material. The second part of the book (Chapter 7 through Chapter 12) introduces some currently used simulation techniques and some of the newest developments in the field. The choice of subjects in the second part is based on my judgment of the importance of the subjects in the future. This part is specifically written for students or beginning researchers who want to know the new directions in computational physics or plan to enter the research areas of scientific computing. Many references are given there to help in further studies.

In order to make the course easy to digest and also to show some practical aspects of the materials introduced in the text, I have selected quite a few exercises. The exercises have different levels of difficulty and can be grouped into three categories. Those in the first category are simple, short problems; a student with little preparation can still work them out with some effort at filling in the gaps they have in both physics and numerical analysis. The exercises in the second category are more involved and aimed at well-prepared students. Those in the third category are mostly selected from current research topics, which will certainly benefit those students who are going to do research in computational science.

Programs for the examples discussed in the text are all written in standard Fortran 77, with a few exceptions that are available on almost all Fortran compilers. Some more advanced programming languages for data parallel or distributed computing are also discussed in Chapter 12. I have tried to keep all programs in the book structured and transparent, and I hope that anyone with knowledge of any programming language will be able to understand the content without extra effort. As a convention, all statements are written in upper case and all comments are given in lower case. From my experience, this is the best way of presenting a clear and concise Fortran program. Many sample programs in the text are explained in sufficient detail with commentary statements. I find that the most efficient approach to learning computational physics is to study well-prepared programs. Related programs used in the book can be accessed via the World Wide Web at the URL http://www.physics.unlv.edu/~pang/cp.html. Corresponding programs in C and Fortran 90 and other related materials will also be available at this site in the future.

This book can be used as a textbook for a computational physics course. If it is a one-semester course, my recommendation is to select materials from Chapters 1 through 7 and Chapter 11. Some sections, such as 4.6 through 4.8, 5.6, and 7.8, are good for graduate students or beginning researchers but may pose some challenges to most undergraduate students.

Tao Pang
Las Vegas, Nevada

DISCLAIMER OF WARRANTIES

The author and Cambridge University Press make no warranties, express or implied, that the materials contained in this book are free of error or are consistent with any particular standard of merchantability, or that they will meet the requirements of any particular application. They should not be relied on for such a usage whose incorrect result could cause injury to a person or loss of property. If one does use the materials in such a manner, it is at one's own risk. The author and Cambridge University Press disclaim any liability for direct or consequential damages resulting from any use of the materials contained in this book.

ACKNOWLEDGMENT

Many chapters of this book have been strongly influenced by my research work in the last ten years, and I am extremely grateful to the University of Minnesota; Miller Institute for Basic Research in Science at the University of California, Berkeley; National Science Foundation; Department of Energy; and W.M. Keck Foundation for their generous support of my research work. Many people have contributed to this book directly or indirectly. Here I would like to thank Dr. René Fournier, Mr. John Kilburg, Dr. Eunja Kim, Dr. Kathleen Robins, and Dr. Lon Spight for carefully reviewing part of the manuscript and for their comments. I would also like to make a special acknowledgment of the excellent work of Dr. Philip Meyler, Physical Sciences Editor at Cambridge University Press, who has reviewed the entire manuscript and has made many constructive suggestions.

Several trademarks appear in the text of this book. IBM is a trademark of the International Business Machines Corporation, UNIX is a trademark of AT&T, *Mathematica* is a trademark of Wolfram Research Inc., MACSYMA is a trademark of MACSYMA Inc., Maple is a trademark of Waterloo Maple Inc., MATLAB is a trademark of The MathWorks Inc., Mathcad is a trademark of MathSoft Inc., Theorist is a trademark of Prescience Corporation, REDUCE is a trademark of RAND Corporation, and CM-2 and CM-5 are trademarks of Thinking Machines Corporation.

1

Introduction

Computing has become a necessary means of scientific analysis. Even in ancient times, quantification of gained knowledge played an essential role for the further development of mankind.

In this chapter, we will discuss the role played by computation in advancing scientific knowledge and outline the current status of computational science. We will give only a very short discussion of the subject here. A detailed discussion of the development of computational science and computers can be found in Nash (1990) and Moreau (1984). Recent progress in high-performance computing is elucidated in Loshin (1994).

1.1 Computation and science

Modern societies are not the only ones that rely on computation. Ancient societies also had to deal with quantifying their knowledge and events. For example, the Mayans recorded multiplication tables in bars and dots and used the tables to figure out the period of lunar eclipses.

More than two thousand years ago, the ancient Chinese were able to estimate the ratio of the circumference to the diameter of a circle, the π value, by measuring to a very high accuracy the ratio of the length of the side of a polygon to its diameter. Let us take the ancient Chinese method as a simple example here. If the length of the side of a regular polygon with n sides is denoted as ℓ_n and the diameter is taken as the unit of length, then the approximation of π is given by

$$\pi_n = n\ell_n. \tag{1.1}$$

The exact π value is the limit of π_n as $n \to \infty$. The values of π_n obtained from the measurements on the polygons can be formally written as

$$\pi_n = \pi_\infty + \frac{a_1}{n} + \frac{a_2}{n^2} + \frac{a_3}{n^3} + \cdots, \tag{1.2}$$

1

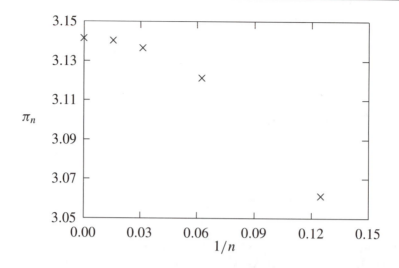

Fig. 1.1. π_n with $n = 8, 16, 32, 64$ plotted together with the extrapolated value π_∞.

where $\pi_\infty = \pi$ and a_i $(i = 1, 2, \ldots, \infty)$ are the coefficients to be determined. The expansion in Eq. (1.2) is truncated in practice when one wants to obtain an approximate value of π. For example, if one has measured $\pi_8 = 3.061467$, $\pi_{16} = 3.121445$, $\pi_{32} = 3.136548$, and $\pi_{64} = 3.140331$, one can truncate the expansion at the third order and then solve the equation set (see Exercise 1.1) to obtain π_∞, a_1, a_2, and a_3 from the given π_n. The approximation of $\pi \simeq \pi_\infty$ from the given π_8, π_{16}, π_{32}, and π_{64} is 3.141583, which has five digits of accuracy in comparison with the exact value of $\pi = 3.1415926\ldots$. The values of π_n for $n = 8, 16, 32, 64$ and the extrapolated value of π_∞ are all plotted in Fig. 1.1. The extrapolated value can be further improved if one can accurately measure π_n with higher values of n.

In a modern society, we need to deal with a lot more computations daily. Almost every event in science or technology requires quantification of the data involved. For example, before a jet aircraft can be actually manufactured, extensive computer simulations of different flight conditions are now performed to check whether there is any design deficiency. This is not only necessary economically but may help avoid loss of life. A related use of the computer is in the reconstruction of an unexplained flight accident. This is extremely important in preventing the same accident from happening again. It is fair to say that computing has become part of our lives, permanently.

1.2 The emergence of modern computers

The advantages of computing machinery were realized a long time ago. The abacus was invented by the Babylonians about five thousand years ago, and it was a very efficient device for addition, subtraction, multiplication, and division for several thousand years.

The concept of an all-purpose, automatic, programmable computing machine was introduced by British mathematician and astronomer Charles Babbage in the early nineteenth century. Babbage proposed the construction of a computing machine, called an *analytical engine*, which could be programmed to perform any type of computation. Unfortunately, the technology at that time was not advanced enough to provide Babbage with the necessary machinery to actually build the analytical engine. In the late nineteenth century, Spanish engineer Leonardo Torres y Quevedo showed that with the availability of electromechanical technology, which had just been developed, one might be able to construct the machine conceived earlier by Babbage. However, he could not actually build the whole machine either, due to lack of funds. American engineer and inventor Herman Hollerith built the very first electromechanical counting machine with punch cards, which was used to sort the populations in the 1890 American census. Hollerith used the profit obtained from selling this machine to set up a company, the Tabulating Machine Company, which was the predecessor of the International Business Machines Corporation (IBM). These developments continued in the early twentieth century. In the 1930s, scientists and engineers at IBM built the first difference tabulator and researchers at Bell Laboratories built the first relay calculator, which were among the very first electromechanical calculators built during that time.

The real beginning of the computer era started with the advent of electronic digital computers. John Vincent Atanasoff, a theoretical physicist at Iowa State University at Ames, invented the electronic digital computer between 1937 and 1939. The history regarding Atanasoff's accomplishment is described in Mackintosh (1987), Burks and Burks (1988), and Mollenhoff (1988). Atanasoff introduced vacuum tubes (instead of electromechanical devices used earlier by other people) as basic elements, a separated memory unit, and a scheme to keep the memory updated in his computer. With the assistance of Clifford E. Berry, a graduate assistant, Atanasoff built the first electronic computer in 1939. Most computer history books have cited ENIAC (Electronic Numerical Integrator and Computer), built by John W. Mauchly and J. Presper Eckert at the Moore School of the University of Pennsylvania with their colleagues in 1945, as the first electronic computer. ENIAC could complete about 5,000 additions or 400 multiplications in one second. Some very impressive scientific computations were performed on ENIAC as soon as it appeared, including the study of nuclear fission with the liquid drop model by Metropolis and Frankel (1947). In the early 1950s, scientists at Los Alamos built another electronic digital computer, called MANIAC I (Mathematical Analyzer, Numerator, Integrator, and Computer), which was very similar to ENIAC. Many important numerical studies, including Monte Carlo simulations of classical liquids (Metropolis et al., 1953), were completed on MANIAC I.

All these computationally intensive research projects accomplished in the 1950s showed that computation was no longer just a supporting tool for scientific research but rather an actual means of probing scientific problems and predicting new scientific phenomena. A new branch of science, *computational science*, was born. Since then, the field of scientific computing has developed and grown rapidly.

The computational power of new computers has been increasing exponentially. To be specific, the computing power of a single computer system has doubled almost every two years in the last fifty years. Computers with transistors replaced those with vacuum tubes in the late 1950s and early 1960s, and computers with very-large-scale integrated circuits were built in the 1970s. Microprocessors also became available in the 1970s. In the mid-1970s, vector processor supercomputers were built to set the stage for supercomputing.

In the 1980s, microprocessor-based personal computers and workstations appeared. Now they have penetrated into all aspects of our lives, as well as all scientific disciplines, because of their affordability and low maintenance cost. With technological breakthroughs in the RISC (Reduced Instruction Set Computer) architecture, cache memory, and multiple instruction units, the speed of each microprocessor is now faster than that of the first generation of supercomputers. In the last few years, these fast microprocessors have been combined to form parallel or distributed computers, which can easily deliver a computing power of a few tens of gigaflops (10^9 floating-point operations per second).

Teraflop (10^{12} floating-point operations per second) computers are now emerging. With the availability of teraflop computers, scientists can start unfolding the mysteries of the grand challenges, such as the dynamics of the global environment; the mechanism of DNA sequencing; computer design of drugs deadly to viruses; and computer simulation of future electronic materials, structures, and devices.

1.3 Computer algorithms and languages

Before one can use a computer to solve a specific problem, one must instruct the computer to follow specific procedures and to carry out the desired computational task. The process involves two steps. First, one needs to transform the problem, typically in the form of an equation, into a set of logical steps that a computer can follow, and second, one needs to inform the computer to follow and complete these logical steps.

Computer algorithms

The complete set of the logical steps for a specific computational problem is called a *numerical algorithm*. Some popular numerical algorithms can be traced back over a hundred years. For example, Carl Friedrich Gauss (1866) published an article on the fast Fourier transform (FFT) algorithm (Goldstine, 1977, pp. 249–53).

Let us here use a very simple and familiar example in physics to illustrate how a typical numerical algorithm is constructed. Assume that a particle of mass m is confined to move along the x-axis under a force $f(x)$. If we describe its motion through Newton's equation, we have

$$f = ma, \tag{1.3}$$

where a is the acceleration of the particle. If we divide the time into small equal intervals τ, we know from introductory physics that the average velocity during the time interval $[t_n, t_{n+1}]$ is

$$v_n = \frac{x_{n+1} - x_n}{\tau}, \tag{1.4}$$

whereas the average acceleration in the same time interval is

$$a_n = \frac{v_{n+1} - v_n}{\tau}. \tag{1.5}$$

The simplest algorithm for finding the position and the velocity of the particle at time $t_{n+1} = (n+1)\tau$ from the corresponding quantities at time $t_n = n\tau$ is obtained after combining Eqs. (1.3), (1.4), and (1.5), and we have

$$x_{n+1} = x_n + \tau v_n, \tag{1.6}$$

$$v_{n+1} = v_n + \frac{\tau}{m} f_n, \tag{1.7}$$

where $f_n = f(x_n)$. If the initial position and velocity of the particle are given and the corresponding quantities at some later time are sought (the initial-value problem), we can obtain them recursively from the algorithm given in Eqs. (1.6) and (1.7). This algorithm is commonly known as the Euler method for the initial-value problem. This simple example illustrates how most algorithms are constructed. First, physical equations are transformed into discrete forms, that is, difference equations. Then the desired physical quantities or solutions of the equations at different variable points are expressed in a recursive manner; that is, the quantities at a later point are expressed in terms of the quantities at earlier points. In the above example, the position and velocity of the particle at t_{n+1} are given by the position and velocity at t_n, provided that the force at any position is explicitly given.

The logical steps in an algorithm can be sequential, parallel, or iterative (implicit). How to utilize the properties of a given problem in constructing a fast and accurate algorithm is a very important issue in computational science. It is my hope that the examples discussed in this book will help students learn how to establish efficient and accurate algorithms as well as how to write clean and structured computer programs for most problems encountered in physics and related fields.

Computer languages

Computer programs are the means through which we communicate with computers. The very first computer programs were written by Ada, the Countess of Lovelace, intended for the analytical engine proposed by Babbage in the mid-1840s. A computer program is a collection of statements, typically written in a specific computer programming language. Computer languages can be divided into two categories: low-level languages that depend on the specific hardware, and high-level languages that are in general free of specific hardware considerations. Simple machine languages and assembly languages were the only ones available before the development of high-level languages. Machine languages are used to program on specific machines. Assembly languages are more advanced than machine languages because they have adopted symbolic addresses, but they are still related to the architecture and wiring of the system. A translating device is needed to translate assembly languages into machine languages before a computer can recognize the instructions. The machine languages and assembly languages do not have portability; that is, a program written for one kind of computers could never be used on others.

The solution to such a problem is very clear. We need high-level languages that are not associated with specific computer hardware and can be run on any computer. Ideal programming languages would be those that are very concise but also close to human languages. Many high-level programming languages are available now, and the choice to use a specific language in scientific computing is more or less a matter of personal taste. Most high-level languages work in a similar manner.

A modern computer program can be viewed as a translation of the algorithm for a specific problem into a set of computer instructions in a specific programming language. The program can be compiled and linked to produce an executable file. Most compilers also have an option to produce an object file that can be linked with other object files and library routines to produce a combined executable file. The compiler is able to detect most errors introduced during programming, that is, the process of writing a program in a high-level language. After running the executable program, the computer will output the results as instructed.

Fortran (For*mula* tran*slation*) was introduced in 1957 as one of the earliest high-level languages and is still used as one of the primary languages in scientific computing. Of course, the Fortran language has evolved from its very early version, known as Fortran 66, to Fortran 77, which has been the most popular language for scientific computing in the last twenty years. The newest version of Fortran, known as Fortran 90, has absorbed many important features for parallel computing. This aspect of Fortran will be discussed in more detail in Chapter 12.

Let us take the algorithm we discussed earlier for a particle moving along the x-axis to illustrate how an algorithm is translated into a program in Fortran. For simplicity, the mass of the particle is taken as 1 and the force is taken to be a harmonic force with a unit elastic constant, $f(x) = -x$. The following Fortran program is an implementation of the algorithm in Eq. (1.7); each statement in the program is almost self-explanatory.

```
        PROGRAM ONE_D_MOTION

C
C Program for the motion of a particle subject to an
C external force f(x) = -x. We have divided the total
C time 2*pi into 10000 intervals with an equal time
C step. The position and velocity of the particle are
C written out at every 500 steps.
C
        PARAMETER (N=10001,IN=500)
        REAL T(N),V(N),X(N)
C
C Assign constants, initial position, and initial
C velocity
C
        PI   = 4.0*ATAN(1.0)
        DT   = 2.0*PI/FLOAT(N-1)
        X(1) = 0.0
        T(1) = 0.0
        V(1) = 1.0
C
C Recursion for position and velocity at later time
C
        DO      100  I = 1, N-1
          T(I+1) = DT*I
          X(I+1) = X(I)+V(I)*DT
          V(I+1) = V(I)-X(I)*DT
    100 CONTINUE
C
C Write the position and velocity every 500 steps
```

```
      C

            WRITE (6,999) (T(I),X(I),V(I),I=1,N,IN)
            STOP
        999 FORMAT (3F16.8)
            END
```

The above program contains some key features of a Fortran program. The arrays are defined at the beginning of the program. Variables beginning with the letters A–H or O–Z are real variables unless defined differently. Variables beginning with I–N are integers unless defined differently. The program has to contain a STOP and an END command. For a modern discussion of the Fortran language and its applications, see Edgar (1992).

After we compile, link, and run this Fortran program, the computer will generate the time-dependent velocity and position of the particle. Figure 1.2 is a plot of the data points from the output of the above program, together with the analytical results. As one can see, the numerical results generated from the program agree well with the analytical results. Since the algorithm we have used here is a very simple one, we have to use a very small time step in order to obtain the results with reasonable accuracy. In Chapter 3, we will introduce and discuss more efficient algorithms for solving differential equations. With these more efficient algorithms, we can usually reach the same accuracy with a very small number of mesh points in one period of the motion, for example, 10 points instead of 10,000.

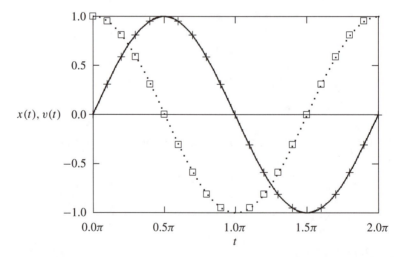

Fig. 1.2. The time-dependent position (+) and velocity (□) of the particle generated from the program ONE_D_MOTION and the corresponding analytical results (solid and dotted lines, respectively).

There are other high-level languages used by scientists and engineers. C is a language most system programmers prefer to use in developing system software because of its high flexibility (Kernighan and Ritchie, 1988). For example, the UNIX operating system (Kernighan and Pike, 1984) now used on almost all workstations and supercomputers is written in C. C++, a newer language based on C, contains valuable extensions in several important aspects (Stroustrup, 1992). Some other high-level languages are also available. Today, Fortran is still the dominant programming language in scientific computing for two very important reasons: Many application packages are available in Fortran, and the structure of Fortran is extremely powerful in dealing with equations.

Since many scientists and engineers use Fortran as their primary programming language, we will adopt it here, too. Even though we will use Fortran in almost all the examples in this book, readers with some knowledge of other high-level languages should have no difficulty in understanding the logical structures and contents of the example programs in the book. The reason is very simple. The logic in all high-level languages is similar to the logic of our own language. All the simple programs written in high-level languages should be self-explanatory, as long as enough commentary lines are provided. Almost all the example programs in this book are written in Fortran 77, with a few extensions that are available on almost all Fortran compilers. Statements such as the GO TO statement and the arithmetic IF statement are avoided intentionally, because they are totally out of date and make it difficult to debug, vectorize, or parallelize the code.

There have been some exciting developments in Fortran recently. Fortran 90, the newest version of Fortran, has most vector and some parallel computing aspects implemented in the statements. Fortran 90 has many extensions over the standard Fortran 77. Most of these extensions are established based on the extensions already adopted by computer manufacturers to enhance their computer performance. Efficient compilers with a full implementation of Fortran 90 are available on all major computer systems now. We will discuss some programming aspects of Fortran 90 in Chapter 12. A more complete discussion can be found in Brainerd, Goldberg, and Adams (1990) and Metcalf and Reid (1990).

High-Performance Fortran (HPF), another new version of the Fortran language with some further extensions beyond Fortran 90, has been developed specifically for parallel computing. There are already some partial implementations of HPF on various parallel computing systems. More implementations are expected in the next few years. The concept of HPF is to transfer the burden of the parallelization from the user to the compiler. But it is still too early to conclude that HPF is the future of scientific computing before it is fully implemented and tested on various scientific problems. For a detailed discussion on HPF, see Koelbel et al. (1994). There are also new developments in parallel and distributed computing with some new environments created with software packages, which will be discussed in Chapter 12.

Exercises

1.1 The value of π can be estimated from the measurements of the length of the sides and the diameters of regular polygons. In general,

$$\pi_n = \pi_\infty + \frac{a_1}{n} + \frac{a_2}{n^2} + \frac{a_3}{n^3} + \cdots,$$

where π_n is the ratio of the perimeter to the diameter of a regular n-sided polygon. Determine the approximate value of $\pi \simeq \pi_\infty$ from $\pi_8 = 3.061467$, $\pi_{16} = 3.121445$, $\pi_{32} = 3.136548$, and $\pi_{64} = 3.140331$.

1.2 Show that the Euler method for Newton's equation in Section 1.3 is accurate up to a term on the order of $(t_{n+1} - t_n)^2$. Discuss how its accuracy can be improved.

1.3 Translate the Fortran program in Section 1.3 for a particle under an elastic force into another computer programming language.

1.4 Modify the Fortran program given in Section 1.3 to study a particle in a uniform gravitational field and a resistive force $f_v = -kv^2$, where v is the velocity of the particle and k is a parameter. Analyze the height dependence of the velocity of a raindrop with different initial heights and k/m, where m is the mass of the raindrop, taken to be a constant for simplicity.

1.5 The Fortran program in Section 1.3 can be generalized to study the motion of a particle in two dimensions. The dynamics of a comet is governed by the gravitational force between the comet and the Sun, $\mathbf{f} = -GMm\mathbf{r}/r^3$, where $G = 6.67 \times 10^{-11}\,\mathrm{N\,m^2/kg^2}$ is the gravitational constant, $M = 1.99 \times 10^{30}\,\mathrm{kg}$ is the mass of the Sun, m is the mass of the comet, $\mathbf{r}$ is the position vector of the comet measured from the Sun, and $r = |\mathbf{r}|$ is the magnitude of $\mathbf{r}$. Write a program for the motion of Halley's comet that has an aphelion (the farthest point from the Sun) distance $5.28 \times 10^{12}\,\mathrm{m}$ and an aphelion velocity $9.12 \times 10^2\,\mathrm{m/s}$. Discuss the error generated by the program in each period of Halley's comet.

2

Basic numerical methods

In this chapter, we will discuss some elementary methods in computational physics. Specifically, we will consider some basic aspects of interpolation and approximation, differentiation and integration, zeros and extremes of a single-variable function, and random number generators. We are going to give only an introductory description of these subjects here as preparation for the later chapters. Some of the subjects covered here would require much more space if discussed in depth. For example, the discussion on random number generators alone could be expanded into a separate book. Therefore, we will focus only on the basics in this chapter.

2.1 Interpolations and approximations

In numerical analysis, the results obtained from computations are always approximations of the desired quantities within some uncertainties. This is similar to experimental observations in physics. Every single physical quantity measured carries some experimental error. We constantly encounter situations in which we need to interpolate a set of discrete data points or to fit them to an adjustable curve. It is extremely important for a physicist to be able to draw conclusions from the information available and to generalize the knowledge gained in order to predict new phenomena.

Interpolation is needed when we want to infer some local information from a set of incomplete or discrete data. Approximation or fitting is needed when we want to know the overall or global behavior of the data. For example, if the speed of a baseball is measured and recorded every 1/100 of a second, we can then estimate the speed of the baseball at any moment by interpolating the recorded data around that time. If we want to know the overall trajectory, then we need to fit the data to a curve. In this section, we will discuss some very basic interpolation and approximation schemes and illustrate how to use them in physics.

Linear interpolation

Consider a given discrete data set from a discrete function $f_i = f(x_i)$ with $i = 1$, $2, \ldots, n + 1$. The simplest way to obtain the approximation of $f(x)$ for $x \in [x_i, x_{i+1}]$ is to construct a straight line between x_i and x_{i+1}. Then $f(x)$ is given by

$$f(x) = f_i + \frac{x - x_i}{x_{i+1} - x_i}(f_{i+1} - f_i) + \Delta f(x), \qquad (2.1)$$

which of course is not accurate enough in most cases but serves as a good start for understanding interpolation schemes. In fact, any value of $f(x)$ in the region $[x_i, x_{i+1}]$ is equal to the sum of the linear interpolation part in the above equation and a quadratic contribution that has a specific curvature and is equal to zero at x_i and x_{i+1}. This means that the error $\Delta f(x)$ in the linear interpolation is given by

$$\Delta f(x) = \frac{\gamma}{2}(x - x_i)(x - x_{i+1}), \qquad (2.2)$$

with γ being a parameter determined by the specific form of $f(x)$. If one draws a quadratic curve passing through $f(x)$, $f(x_i)$, and $f(x_{i+1})$, one can show that

$$\gamma = f''(a), \qquad (2.3)$$

with $a \in [x_i, x_{i+1}]$, as long as $f(x)$ is a smooth function in the region $[x_i, x_{i+1}]$; that is, $f^{(k)}(x)$ exists for any k. The maximum error in the linear interpolation of Eq. (2.1) is then bounded by

$$|\Delta f(x)| \leq \frac{\gamma_1}{8}(x_{i+1} - x_i)^2, \qquad (2.4)$$

where $\gamma_1 = \max[|f''(x)|]$ with $x \in [x_i, x_{i+1}]$. The upper bound of the error in Eq. (2.4) is obtained from Eq. (2.2) with γ replaced by γ_1 and x solved from $d\Delta f(x)/dx = 0$. The accuracy of the linear interpolation can be improved by decreasing the interval $h = x_{i+1} - x_i$.

Let us take $f(x) = \sin x$ as an illustrative example here. Assume that $x_i = \pi/4$ and $x_{i+1} = \pi/2$, so the corresponding $f_i = 0.07071$ and $f_{i+1} = 1.0000$. Let us use the linear interpolation scheme to find the approximate value of $f(x)$ with $x = 3\pi/8$. From Eq. (2.1), we have the interpolated value $f(3\pi/8) \simeq 0.8536$. We know that, of course, $f(3\pi/8) = \sin(3\pi/8) = 0.9239$. The actual difference is $|\Delta f(x)| = 0.070$, which is smaller than the maximum error estimated with Eq. (2.4), 0.077.

The above example is a very simple one showing how most interpolation schemes work. One constructs a continuous curve (a straight line in the above example) from the given discrete set of data, and then one reads off the interpolated value from the curve. The more points one has, the higher the order of the curve can be. For example, one can construct a quadratic curve from three data points and a cubic curve from four data points. One way to achieve higher-order interpolation is through the Lagrange interpolation scheme, which is a generalization of the linear interpolation we have discussed.

The Lagrange interpolation

Let us first make an observation about the linear interpolation discussed in the preceding section. The interpolated function actually passes through the two points used for the interpolation. Now if we use three points for the interpolation, we can always construct a quadratic function that passes through all three points. The error is now given by a term on the order of h^3, where h is the interval between any two nearest points, because an x^3 term can be added to modify the curve to pass through the actual value of the function if it is known. In order to obtain the generalized interpolation formula to pass through the $n + 1$ data points, we rewrite the linear interpolation of Eq. (2.1) in a symmetric form with

$$f(x) = \frac{x - x_{i+1}}{x_i - x_{i+1}} f_i + \frac{x - x_i}{x_{i+1} - x_i} f_{i+1} + \Delta f(x)$$

$$= \sum_{k=i}^{i+1} f_k P_k^{(1)}(x) + \Delta f(x), \tag{2.5}$$

where

$$P_k^{(1)}(x) = \frac{x - x_j}{x_k - x_j}, \tag{2.6}$$

with $j \neq k$. Now we can easily generalize the expression to an nth-order curve that passes through all the $n + 1$ data points,

$$f(x) = \sum_{k=1}^{n+1} f_k P_k^{(n)}(x) + \Delta f(x), \tag{2.7}$$

where $P_k^{(n)}(x)$ is given by

$$P_k^{(n)}(x) = \prod_{j \neq k}^{n+1} \frac{x - x_j}{x_k - x_j}, \tag{2.8}$$

with $k, j = 1, 2, \ldots, n + 1$. In other words, $\Delta f(x_k) = 0$ at all the data points. Following a similar argument for the linear interpolation, one can show that the error in the nth-order Lagrange interpolation is given by

$$\Delta f(x) = \frac{\gamma}{(n + 1)!}(x - x_1)(x - x_2) \cdots (x - x_{n+1}), \tag{2.9}$$

where

$$\gamma = f^{(n+1)}(a), \tag{2.10}$$

with $a \in [x_1, x_{n+1}]$. Therefore, the maximum error is bounded by

$$|\Delta f(x)| \le \frac{\gamma_n}{4(n+1)} h^{n+1}, \tag{2.11}$$

with $\gamma_n = \max[|f^{(n+1)}(x)|]$ with $x \in [x_1, x_{n+1}]$. One can obtain the above upper bound by replacing γ with γ_n and then maximizing the pairs $(x - x_1)(x - x_{n+1})$, $(x - x_2)(x - x_n)$, ..., and $(x - x_{(n+1)/2})(x - x_{(n+3)/2})$ individually for an even $n + 1$. For an odd $n + 1$, one can choose the maximum value nh for the $x - x_i$ that is not paired. One can rewrite Eq. (2.7) in a power series

$$f(x) = \sum_{l=0}^{n} a_l x^l + \Delta f(x), \tag{2.12}$$

with a_l given by a coefficient transform from Eq. (2.7) to Eq. (2.12). Note that the generalized form reduces to the linear case if $n = 1$.

The Aitken method

One way to achieve the Lagrange interpolation efficiently is by performing a sequence of linear interpolations. This scheme was first developed by Aitken (1932). We can first work out n linear interpolations with each constructed from a neighboring pair of the $n + 1$ data points. Then we can use these n interpolated data points to achieve another level of $n - 1$ linear interpolations. We repeat this until we obtain the final result after n levels of consecutive linear interpolations. We can summarize the scheme in the following equation:

$$f_{i...j} = \frac{x - x_j}{x_i - x_j} f_{i...j-1} + \frac{x - x_i}{x_j - x_i} f_{i+1...j}, \tag{2.13}$$

with $f_i = f(x_i)$. Let us show some detail of the above scheme by taking the case of five data points as an example. If one wants to obtain $f(x)$ from a given set $f_i = f(x_i)$ for $i = 1, 2, \ldots, n + 1$, one can carry out n levels of consecutive linear interpolations as shown in Fig. 2.1, in which every column is constructed from the previous column by a linear interpolation of each two adjacent rows. For example,

$$f_{123} = \frac{x - x_3}{x_1 - x_3} f_{12} + \frac{x - x_1}{x_3 - x_1} f_{23}, \tag{2.14}$$

and

$$f_{12345} = \frac{x - x_5}{x_1 - x_5} f_{1234} + \frac{x - x_1}{x_5 - x_1} f_{2345}. \tag{2.15}$$

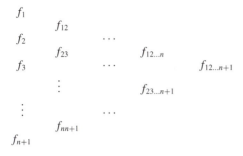

Fig. 2.1. The hierarchy structure in the Aitken scheme for $n+1$ data points.

Table 2.1. *Result of the example with the Aitken method.*

x_i	f_i	f_{ij}	f_{ijk}	f_{ijkl}	f_{ijklm}
0.0	1.000000				
		0.889246			
0.5	0.938470		0.808792		
		0.799852		0.807272	
1.0	0.765198		0.806260		0.807473
		0.815872		0.807717	
1.5	0.511828		0.811725		
		0.857352			
2.0	0.223891				

One can show that the consecutive linear interpolation outlined in Fig. 2.1 recovers the standard Lagrange interpolation, but it is easier to implement. The Aitken method also provides a way of estimating the error of the Lagrange interpolation. If we use the five-point case, that is, $n+1 = 5$, as an example, the error in the Lagrange interpolation scheme is roughly given by

$$\Delta f(x) \approx \frac{|f_{12345} - f_{1234}| + |f_{12345} - f_{2345}|}{2}, \tag{2.16}$$

where the differences are taken from the last two columns on the hierarchy.

Let us take as an actual numerical example evaluating $f(0.9)$ from the given set $f(0.0) = 1.000000$, $f(0.5) = 0.938470$, $f(1.0) = 0.765198$, $f(1.5) = 0.511828$, and $f(2.0) = 0.223891$. These are the values of the Bessel function of the first kind of order zero. The consecutive linear interpolations of the data with the Aitken method are shown in Table 2.1.

The error estimated from the differences of the last two rows of the data in the table is

$$\Delta f(x) \approx \frac{|0.807473 - 0.807273| + |0.807473 - 0.807717|}{2} \simeq 2 \times 10^{-4}.$$

The exact result of $f(0.9)$ is 0.807524. The error of the interpolated value is $|0.807473 - 0.807524| \simeq 5 \times 10^{-5}$, which is a little smaller than the estimate from the differences of the last two rows of the data. The following program is an implementation of the Aitken method for the Lagrange interpolation for the given example of the Bessel function.

```
      PROGRAM INTERPOLATION
C
C Main program for the Lagrange interpolation with the
C Aitken method.
C
      PARAMETER (N=5)
      DIMENSION XI(N),FI(N)
      DATA XI/0.0,0.5,1.0,1.5,2.0/,
     *     FI/1.0,0.938470,0.765198,0.511828,0.223891/
C
      X = 0.9
      CALL AITKEN (N,XI,FI,X,F,DF)
      WRITE (6,999) X,F,DF
      STOP
  999 FORMAT (3F16.8)
      END
C
      SUBROUTINE AITKEN (N,XI,FI,X,F,DF)
C
C Subroutine performing the Lagrange interpolation with
C the Aitken method. F: interpolated value. DF: error
C estimated.
C
      PARAMETER (NMAX=21)
      DIMENSION XI(N),FI(N),FT(NMAX)
C
      IF (N.GT.NMAX) STOP 'Dimension too large'
```

```
DO      100 I = 1, N
FT(I) = FI(I)
100 CONTINUE
C
DO      200 I = 1, N-1
  DO    150 J = 1, N-I
     X1 = XI(J)
     X2 = XI(J+I)
     F1 = FT(J)
     F2 = FT(J+1)
     FT(J) = (X-X1)/(X2-X1)*F2+(X-X2)/(X1-X2)*F1
150   CONTINUE
200 CONTINUE
    F = FT(1)
    DF = (ABS(F-F1)+ABS(F-F2))/2.0
    RETURN
    END
```

The consecutive linear interpolations done above are not very efficient. In practice, the best scheme should reach the last column in Table 2.1 through an optimal path directly instead of calculating all numbers in the table. For example, we may need to calculate only f_{23}, then f_{234}, f_{1234}, and f_{12345} in order to obtain the interpolated value of $f(0.9)$. Moreover, the change in the interpolated value is quite small compared to the actual value of the function during each step of the consecutive linear interpolations. The result will be influenced significantly by the rounding error if the procedure is carried out directly.

We can, however, construct an indirect scheme that improves the interpolated value at every step by updating the difference of the interpolated values from the adjacent rows, that is, by improving the correction of the interpolated value over the last column rather than the interpolated value itself. The effect of the rounding error is then minimized. This procedure is accomplished with the upward and downward correction method (Wong, 1992, pp. 74–6, for example). One can define two corrections at each step,

$$\Delta_{ij}^+ = f_{i-1...j} - f_{i...j}, \qquad (2.17)$$

$$\Delta_{ij}^- = f_{i...j+1} - f_{i...j}, \qquad (2.18)$$

from the differences between two adjacent columns. Δ_{ij}^- is the downward (going down along the triangle in Fig. 2.1) correction and Δ_{ij}^+ is the upward (going up along the triangle in Fig. 2.1) correction. One can show from the definitions of Δ_{ij}^+

and Δ_{ij}^- that they satisfy the following recursion relations,

$$\Delta_{ij}^+ = \frac{x_j - x}{x_j - x_{i-1}}(\Delta_{ij-1}^+ - \Delta_{ij-1}^-), \tag{2.19}$$

$$\Delta_{ij}^- = \frac{x_i - x}{x_{j+1} - x_i}(\Delta_{i+1j}^+ - \Delta_{i+1j}^-), \tag{2.20}$$

with the starting $\Delta_{ii}^\pm = f_i$. x_i is chosen as the data point closest to x. In general, we use the upward correction as the first correction if $x < x_i$. Otherwise, the downward correction is used. Then we alternate the downward and upward corrections in the steps followed until the final result is reached. If the upper (lower) boundary of the triangle in Fig. 2.1 is reached during the process, only downward (upward) corrections can be used afterward.

We can use the numerical values of the Bessel function in Table 2.1 to illustrate the method. Assume that we are still calculating $f(x)$ with $x = 0.9$. One can easily see that the starting point should be $x_i = 1.0$, because it is closest to $x = 0.9$. So the zeroth-order approximation of the interpolated data is $f(x) \approx f(1.0)$. Δ_{33}^+ is then the first correction to $f(x)$. In the next step, we alternate the direction of the correction and use the downward correction, Δ_{23}^- in this example, to further improve $f(x)$. We can continue the procedure with another upward correction and another downward correction to reach the final result. One can write a simple program to accomplish what we have just described. A good practice is to write a subroutine with n, x, x_i, and f_i, for $i = 1, 2, \ldots, n + 1$, being the input and $f(x)$ and $|\Delta f(x)|$ being the output. The subroutine can then be used for any other interpolation task. For the numerical example outlined above, one can write a simple main program with a few lines to set up the problem and use the subroutine as designed. The numerical result with the input data from Table 2.1, using the upward and downward correction method, is exactly the same as the earlier result, $f(0.9) = 0.807473$, as one would expect.

Least-squares approximation

As we have pointed out, interpolation is used mainly to find the local approximation of a given discrete set of data. In many situations in physics one needs to know the global behavior of a set of data in order to understand the trend in a specific measurement or observation. A typical example is a polynomial fit to a set of experimental data with error bars.

The most common approximate scheme is based on the least squares of the differences between the approximation $p_m(x)$ and the data $f(x)$. If $f(x)$ is the data function to be approximated in the region $[a, b]$ and the approximation is an *m*th-order polynomial

$$p_m(x) = \sum_{k=0}^{m} a_k x^k, \tag{2.21}$$

we can construct a function of a_k for $k = 0, 1, \ldots, m$ as

$$\Delta[a_k] = \int_a^b [p_m(x) - f(x)]^2 \, dx \tag{2.22}$$

for the continuous data function $f(x)$ and

$$\Delta[a_k] = \sum_{i=1}^{n+1} [p_m(x_i) - f(x_i)]^2, \tag{2.23}$$

for the discrete data function $f(x_i)$ for $i = 1, 2, \ldots, n+1$. The least-squares approximation is obtained with $\Delta[a_k]$ minimized with respect to all the $m+1$ coefficients. This general minimization problem will be discussed in more detail in Chapter 4. Here we have used the notation $\Delta[a_k]$ for a quantity that is a function of a set of independent variables $a_0, a_1, \ldots, a_m$. This notation will be used throughout this book. Here we would like to approach the problem with orthogonal polynomials. In principle, we can express the polynomial $p_m(x)$ in terms of a set of real orthogonal polynomials with

$$p_m(x) = \sum_{k=0}^{m} \alpha_k u_k(x), \tag{2.24}$$

where $u_k(x)$ is a set of real orthogonal polynomials which satisfy

$$\int_a^b u_k(x) w(x) u_l(x) \, dx = \langle u_k \mid u_l \rangle = \delta_{kl} \mathcal{N}_k, \tag{2.25}$$

with $w(x)$ being the weight whose form depends on the specific set of orthogonal polynomials. δ_{kl} is the Kronecker δ-function, which equals one for $k = l$ and zero for $k \neq l$. $\mathcal{N}_k$ is the normalization constant. α_k with $k = 0, 1, \ldots, m$ is another set of coefficients that can be formally related to a_j with $j = 0, 1, \ldots, m$ by a matrix transformation and is determined with $\Delta[\alpha_k]$ minimized. $\Delta[\alpha_k]$ is the same quantity defined in Eq. (2.22) or Eq. (2.23) with $p_m(x)$ from Eq. (2.24). If we want the polynomials to be orthonormal, we can simply divide $u_k(x)$ by $\sqrt{\mathcal{N}_k}$. We will also use the notation in the above equation for the discrete case with

$$\langle u_k \mid u_l \rangle = \sum_{i=1}^{n+1} u_k(x_i) w(x_i) u_l(x_i) = \delta_{kl} \mathcal{N}_k. \tag{2.26}$$

The orthogonal polynomials can be generated with the following recursion:

$$u_{k+1}(x) = (x - g_k) u_k(x) - h_k u_{k-1}(x), \tag{2.27}$$

where the coefficients g_k and h_k are given by

$$g_k = \frac{\langle xu_k \mid u_k \rangle}{\langle u_k \mid u_k \rangle}, \tag{2.28}$$

$$h_k = \frac{\langle u_k \mid u_k \rangle}{\langle u_{k-1} \mid u_{k-1} \rangle}, \tag{2.29}$$

with the starting $u_0(x) = 1$ and $h_0 = 0$. To convince oneself, one can take the above polynomials and show that they are orthogonal regardless of whether they are continuous or discrete; that is, the polynomials generated from the above equations satisfy $\langle u_k \mid u_l \rangle = \delta_{kl} \mathcal{N}_k$. For simplicity, we will just consider the case with $w(x) = 1$. The formalism developed here can easily be generalized to the cases with $w(x) \neq 1$. We will have more discussions on orthogonal polynomials in Chapter 5 when we introduce the special functions and Gaussian quadratures. The least-squares approximation is obtained if we find all the coefficients α_k that minimize the function $\Delta[\alpha_k]$. In other words, we would like to have

$$\frac{\partial \Delta[\alpha_k]}{\partial \alpha_j} = 0 \tag{2.30}$$

and

$$\frac{\partial^2 \Delta[\alpha_k]}{\partial \alpha_j^2} > 0 \tag{2.31}$$

for $j = 0, 1, \ldots, m$. The first-order derivative of $\Delta[\alpha_k]$ can easily be carried out. After exchanging the summation and the integration in $\partial \Delta[\alpha_k]/\partial \alpha_j = 0$, we have

$$\alpha_j = \frac{\langle u_j \mid f \rangle}{\langle u_j \mid u_j \rangle} \tag{2.32}$$

for $j = 0, 1, \ldots, m$, which ensures a minimum value of $\Delta[\alpha_k]$, because $\partial^2 \Delta[\alpha_k]/\partial \alpha_j^2 = 2\langle u_j \mid u_j \rangle$ is always greater than 0. One can always construct a set of discrete orthogonal polynomials numerically in the region $[a,b]$. The following subroutine is a simple example for obtaining a set of orthogonal polynomials $u_k(x_i)$ and the coefficients α_k for a given set of discrete data $f(x_i)$ with x_i evenly spaced.

```
         SUBROUTINE PFIT(N,M,X,F,A,U)
   C
   C Subroutine generating orthonormal polynomials U(M,N)
   C up to (M-1)th order and coefficients A(M), for the
   C least-squares approximation of the function F(N) at
```

```
C X(N). Other variables used: G(K) for g_k, H(K) for
C h_k, S(K) for <u_k|u_k>.
C
      PARAMETER (NMAX=101,MMAX=101)
      DIMENSION G(MMAX),H(MMAX),S(MMAX),X(N),F(N),
     *          U(M,N),A(M)
C
      IF(N.GT.NMAX) STOP 'Too many points'
      IF(M.GT.MMAX) STOP 'Order too high'
C
C Set up the zeroth-order polynomial u_0
C
      DO     100  I = 1, N
        U(1,I) = 1.0
  100 CONTINUE
      DO    200   I = 1, N
        TMP  = U(1,I)*U(1,I)
        S(1) = S(1)+TMP
        G(1) = G(1)+X(I)*TMP
        A(1) = A(1)+U(1,I)*F(I)
  200 CONTINUE
      G(1) = G(1)/S(1)
      H(1) = 0.0
      A(1) = A(1)/S(1)
C
C Set up the first-order polynomial u_1
C
      DO     300  I = 1, N
        U(2,I) = X(I)*U(1,I)-G(1)*U(1,I)
        S(2)   = S(2)+U(2,I)**2
        G(2)   = G(2)+X(I)*U(2,I)**2
        A(2)   = A(2)+U(2,I)*F(I)
  300 CONTINUE
      G(2) = G(2)/S(2)
      H(2) = S(2)/S(1)
      A(2) = A(2)/S(2)
C
      IF(M.LT.3) RETURN
C
```

```
C Higher-order polynomials u_k from the recursion
C relation
C
      DO      700  I = 2, M-1
        DO    500  J = 1, N
          U(I+1,J) = X(J)*U(I,J)-G(I)*U(I,J)-H(I)
     *                *U(I-1,J)
          U(I+1,J) = X(J)*U(I,J)-G(I)*U(I,J)-H(I)
     *                *U(I-1,J)
          S(I+1)   = S(I+1)+U(I+1,J)**2
          G(I+1)   = G(I+1)+X(J)*U(I+1,J)**2
          A(I+1)   = A(I+1)+U(I+1,J)*F(J)
 500    CONTINUE
        G(I+1) = G(I+1)/S(I+1)
        H(I+1) = S(I+1)/S(I)
        A(I+1) = A(I+1)/S(I+1)
 700  CONTINUE
      RETURN
      END
```

Note that this subroutine is only for discrete functions. For continuous functions, one needs to generate the orthogonal polynomials with a continuous variable and a quadrature for the integrals. Such polynomials will be generated in Chapter 5, and integration quadratures will be discussed in Section 2.2 and in Section 5.8.

The Millikan experiment

Here we would like to use a set of data from the famous oil drop experiment of Millikan as an example to illustrate how one can actually use the least-squares approximation discussed above. In 1910, Millikan (1910) published the famous work on the oil drop experiment in *Science*. Based on the measurements of the charges carried by all the oil drops, Milliken concluded that the charge carried by any object is a multiple of a fundamental charge, the charge of an electron (for negative charges) or the charge of a proton (for positive charges). In that article Milliken extracted the fundamental charge by taking the average of the measured charges carried by all the oil drops. We would like to take the data of Milliken and make a least-squares fit to a straight line here. Based on the fit, we can estimate the fundamental charge and the accuracy of the Millikan measurement. Millikan did not use the method we are discussing here to reach the conclusion, of course.

Table 2.2. *Data from the Millikan experiment.*

n	4	5	6	7	8	9	10	11
e_n	6.558	8.206	9.880	11.50	13.14	14.82	16.40	18.04

n	12	13	14	15	16	17	18	
e_n	19.68	21.32	22.96	24.60	26.24	27.88	29.52	

Each measured data point from the Millikan experiment is assigned with an integer. The measured charges e_n (in units of 10^{-19} C) and the corresponding integers n are listed in Table 2.2. From the average charges of the oil drops, Millikan concluded that the fundamental charge is 1.65×10^{-19} C, which was very close to the accepted value of the fundamental charge, now known to be 1.60×10^{-19} C. Let us take the data obtained by Millikan from Table 2.2 and apply the least-squares approximation discussed in the previous subsection for a discrete function to find the fundamental charge and the order of the errors in the measurements. One can take the linear equation

$$e_n = ne_0 + \Delta e \tag{2.33}$$

as the approximation of the measured data. Here Δe represents the error associated with the measurement. Even though Δe cannot provide an exact estimate of the experimental error, it does reflect the size of the error. The following program applies the least-squares approximation subroutine and calculates the fundamental charge e_0 and the estimated error bar $|\Delta e|$, based on the data from the Millikan experiment.

```
      PROGRAM MILLIKAN
C
C Main program for a linear fit of the Millikan
C experimental data on the fundamental charge
C e_0 from e_k = k*e_0 + de.
C
      PARAMETER (N=15,M=2)
      DIMENSION X(N),F(N),A(M),U(M,N)
      DATA X /4.0,5.0,6.0,7.0,8.0,9.0,10.0,11.0,
     *         12.0,13.0,14.0,15.0,16.0,17.0,18.0/
      DATA F /6.558,8.206,9.880,11.50,13.14,14.82,16.40,
     *         18.04,19.68,21.32,22.96,24.60,26.24,27.88,
     *         29.52/
```

```
C
          CALL PFIT(N,M,X,F,A,U)
          SUMO = 0.0
          SUMT = 0.0
          DO        100  K = 1, N
            SUMO = SUMO+U(1,K)**2
            SUMT = SUMT+X(K)*U(1,K)**2
  100 CONTINUE
          EO   = A(2)
          DE   = A(1)-A(2)*SUMT/SUMO
          WRITE (6,999) EO,DE
          STOP
  999 FORMAT (2F16.8)
          END
```

After we compile and run the above program, we obtain $e_0 = 1.64 \times 10^{-19}$ C, and the intersect on the y-axis gives us a rough estimate of the error bar $|\Delta e| \approx 0.03 \times 10^{-19}$ C. The Millikan data and the least-squares approximation of Eq. (2.33) with e_0 and $|\Delta e|$ obtained from the above program are plotted in Fig. 2.2. As one can see, the measured data are very accurately represented by the straight line of the least-squares approximation.

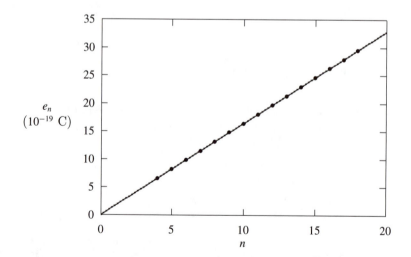

Fig. 2.2. The Millikan measurements of charges in oil drop experiments (dots) and the least-squares approximation of a linear fit (solid line).

2.2 Differentiation and integration

One basic tool that we will often use in this book is the Taylor expansion of a function $f(x)$ around a point x_0,

$$f(x) = f(x_0) + (x - x_0)f'(x_0) + \frac{(x - x_0)^2}{2!} f''(x_0)$$

$$+ \cdots + \frac{(x - x_0)^n}{n!} f^{(n)}(x_0) + \cdots . \qquad (2.34)$$

The above expansion can be generalized to describe a multivariable function $f(x, y, \ldots)$ around the point $(x_0, y_0, \ldots)$,

$$f(x, y, \ldots) = f(x_0, y_0, \ldots) + (x - x_0)f_x(x_0, y_0, \ldots)$$

$$+ (y - y_0)f_y(x_0, y_0, \ldots) + \frac{(x - x_0)^2}{2!} f_{xx}(x_0, y_0, \ldots)$$

$$+ \frac{(y - y_0)^2}{2!} f_{yy}(x_0, y_0, \ldots) + \frac{2(x - x_0)(y - y_0)}{2!}$$

$$\times f_{xy}(x_0, y_0, \ldots) + \cdots , \qquad (2.35)$$

where the indices are for partial derivatives, for example, $f_{xy} = \partial^2 f / \partial x \partial y$.

Numerical differentiation

The first-order derivative of a single-variable function $f(x)$ around a point x_k is defined from the limit

$$f'(x_k) = \lim_{\Delta x \to 0} \frac{f(x_k + \Delta x) - f(x_k)}{\Delta x}, \qquad (2.36)$$

if it exists. Now if we divide the space into discrete points x_k with an evenly spaced interval $h = x_{k+1} - x_k$ and label the function at the lattice points as $f_k = f(x_k)$, we obtain the simplest expression for the first-order derivative

$$f'_k = \frac{f_{k+1} - f_k}{h} + O(h). \qquad (2.37)$$

We have used the notation $O(h)$ for a term on the order of h. Similar notation will be used throughout this book. The above formula is referred to as the *two-point formula* for the first-order derivative; it can be easily derived by taking the Taylor expansion of f_{k+1} around x_k. The accuracy can be improved if we expand f_{k+1} and f_{k-1} around x_k and take the difference

$$f_{k+1} - f_{k-1} = 2hf'_k + \frac{h^3}{6} f_k^{(3)} + \cdots . \qquad (2.38)$$

After a simple rearrangement, we have

$$f_k' = \frac{f_{k+1} - f_{k-1}}{2h} + O(h^2), \tag{2.39}$$

which is commonly known as the *three-point formula* for the first-order derivative. The accuracy of the expression will increase to a higher order in h if more points are used. For example, a *five-point formula* can be derived by including the expansions of f_{k+2} and f_{k-2} around x_k. If we use the combinations

$$f_{k+1} - f_{k-1} = 2hf_k' + \frac{h^3}{3} f_k^{(3)} + O(h^5) \tag{2.40}$$

and

$$f_{k+2} - f_{k-2} = 4hf_k' + \frac{8h^3}{3} f_k^{(3)} + O(h^5) \tag{2.41}$$

to cancel the $f_k^{(3)}$ terms, we have

$$f_k' = \frac{1}{12h}(f_{k-2} - 8 f_{k-1} + 8 f_{k+1} - f_{k+2}) + O(h^4). \tag{2.42}$$

One can, of course, make the accuracy even higher by including more points, but in many cases this is not a good practice. For real problems, the derivatives at points close to the boundaries are important and need to be calculated accurately. The effect of the errors in the derivatives of the boundary points will accumulate into other points when one tries to integrate an equation. The more points involved in the expressions of the derivatives, the more difficulties one encounters in achieving accurate derivatives at the boundaries. Another way to increase the accuracy is by decreasing the interval h. This is very practical on vector computers. The algorithms for first-order or second-order derivatives usually are fully vectorized, so a vector processor can carry out many points in just one computer clock cycle.

The approximate expressions for the second-order derivative can be obtained with different combinations of f_i. The *three-point formula* for the second-order derivative is given by the combination

$$f_{k+1} - 2f_k + f_{k-1} = h^2 f_k'' + O(h^4), \tag{2.43}$$

with the Taylor expansions of $f_{k\pm1}$ around x_k. Note that the third-order term with $f_k^{(3)}$ also vanishes because of the cancellation in the combination. The above equation gives the second-order derivative as

$$f_k'' = \frac{f_{k+1} - 2f_k + f_{k-1}}{h^2} + O(h^2). \tag{2.44}$$

Similarly, we can combine the expansions of $f_{k\pm2}$ and $f_{k\pm1}$ around x_k and f_k to cancel the f_k', $f_k^{(3)}$, $f_k^{(4)}$, and $f_k^{(5)}$ terms; we have

$$f_k'' = \frac{1}{12h^2}(-f_{k-2} + 16f_{k-1} - 30f_k + 16f_{k+1} - f_{k+2}) + O(h^4) \quad (2.45)$$

as the five-point formula for the second-order derivative. The difficulty in dealing with the points around the boundaries still remains. One can use the interpolation formula we developed in the last section to extrapolate the derivatives to the boundary points. For example, one can use the linear interpolation of the derivatives of the two points next to a boundary point to extrapolate the derivative at the boundary. The following program shows an example of calculating the first-order and second-order derivatives with the three-point formulas. We have assumed that a function $f(x)$ is given at discrete points with evenly spaced intervals. We have also used the linear interpolation scheme to extrapolate the derivatives at the boundary points.

```
      PROGRAM DERIVATIVES
C
C Main program for derivatives of f(x) = sin(x).
C F1: f'; F2: f";  D1: error in f'; and D2: error
C in f".
C
      PARAMETER (N=101)
      DIMENSION X(N),F(N),F1(N),D1(N),F2(N),D2(N)
C
      PI = 4.0*ATAN(1.0)
      H  = PI/(2.0*100)
      DO      100  I = 1, N
        X(I) = H*(I-1)
        F(I) = SIN(X(I))
  100 CONTINUE
      CALL THREE(N,H,F,F1,F2)
      DO      300  I = 1, N
        D1(I) = F1(I)-COS(X(I))
        D2(I) = F2(I)+SIN(X(I))
        WRITE (6,999) X(I),F1(I),D1(I),F2(I),D2(I)
  300 CONTINUE
      STOP
  999 FORMAT (5F10.6)
      END
C
```

Table 2.3. *Derivatives obtained in the example.*

x	f'	$\Delta f'$	f''	$\Delta f''$
0	1.000206	0.000206	0.000015	0.000015
$\pi/10$	0.951017	−0.000039	−0.309087	−0.000070
$\pi/5$	0.808985	−0.000032	−0.587736	0.000049
$3\pi/10$	0.587762	−0.000023	−0.809013	0.000004
$2\pi/5$	0.309003	−0.000014	−0.951055	0.000001
$\pi/2$	0.000006	0.000006	−1.000335	−0.000335

```
      SUBROUTINE THREE(N,H,FI,F1,F2)
C
C Subroutine for the first- and second-order derivatives
C with three-point formulas. Extrapolations are made at
C boundaries.
C FI: input f(x); H: interval; F1: f'; and F2: f".
C
      DIMENSION FI(N),F1(N),F2(N)
C
C f' and f" from three-point formulas
C
      DO      100  I = 2, N-1
        F1(I) = (FI(I+1)-FI(I-1))/(2.0*H)
        F2(I) = (FI(I+1)-2.0*FI(I)+FI(I-1))/(H*H)
  100 CONTINUE
C
C Linear extrapolation for the boundary points
C
      F1(1) = 2.0*F1(2)-F1(3)
      F1(N) = 2.0*F1(N-1)-F1(N-2)
      F2(1) = 2.0*F2(2)-F2(3)
      F2(N) = 2.0*F2(N-1)-F2(N-2)
      RETURN
      END
```

We have used $f(x) = \sin x$ in the region of $x \in [0, \pi/2]$ with 101 mesh points in the above example program. The derivatives at the boundaries are extrapolated with the linear interpolation scheme discussed in the last section. The values of the two points next to the boundary point are used. We summarize the numerical results from the above program in Table 2.3, together with their errors. As one can see, the extrapolated data are of the same order of accuracy as other calculated values for f' at $x = \pi/2$ and f'' at $x = 0$ because the error in the linear Lagrange interpolation scheme is of the same order as the error in the three-point formulas for derivatives, that is, $O(h^2)$. The other two boundary values are much less accurate, because the functions now are no longer linear at the boundaries but rather quadratic. A higher-order interpolation scheme would have solved this problem.

Numerical integration

Let us turn to numerical integration. In general, we want to obtain the numerical value of an integral defined in the region $[a, b]$ as

$$S = \int_a^b f(x)\,dx. \tag{2.46}$$

We can divide the region $[a, b]$ into n slices with an evenly spaced interval h. If we take the lattice points as x_k with $k = 0, 1, \ldots, n$, we can write the integral as a summation of integrals over all the slices:

$$\int_a^b f(x)\,dx = \sum_{k=0}^{n-1} \int_{x_k}^{x_{k+1}} f(x)\,dx. \tag{2.47}$$

Of course, if we can develop a numerical scheme that evaluates the integral over several slices accurately, we will have solved the problem. Let us first consider each slice separately. The simplest quadrature is obtained if we approximate $f(x)$ in the region $[x_k, x_{k+1}]$ linearly, that is, $f(x) \simeq f_k + (x - x_k)(f_{k+1} - f_k)/h$. After integrating every slice with this linear function, we have

$$S = \frac{h}{2} \sum_{k=0}^{n-1} (f_k + f_{k+1}) + O(h^2), \tag{2.48}$$

where $O(h^2)$ comes from the error in the linear interpolation of the function. The above quadrature is commonly referred to as the *trapezoid rule*, which has an overall accuracy up to $O(h^2)$.

We can obtain a quadrature with a higher accuracy by working on two slices together. If we apply the Lagrange interpolation to the function $f(x)$ in the region of $[x_{k-1}, x_{k+1}]$, we have

$$f(x) = \frac{(x - x_k)(x - x_{k+1})}{(x_{k-1} - x_k)(x_{k-1} - x_{k+1})} f_{k-1} + \frac{(x - x_{k-1})(x - x_{k+1})}{(x_k - x_{k-1})(x_k - x_{k+1})} f_k$$

$$+ \frac{(x - x_{k-1})(x - x_k)}{(x_{k+1} - x_{k-1})(x_{k+1} - x_k)} f_{k+1} + O(h^3). \tag{2.49}$$

If we carry out the integral for every pair of slices together with the integrand given from the above equation, we have

$$S = \frac{h}{3} \sum_{l=0}^{n/2-1} (f_{2l} + 4f_{2l+1} + f_{2l+2}) + O(h^4), \qquad (2.50)$$

which is known as the *Simpson rule*. The third-order term vanishes because of cancellation. In order to pair up all the slices, we have to have an even number of slices. What happens if we have an odd number of slices, or an even number of points between $[a, b]$? One solution is to separate the last slice with

$$\int_{b-h}^{b} f(x)\, dx = \frac{h}{12}(-f_{n-2} + 8f_{n-1} + 5f_n). \qquad (2.51)$$

The expression for $f(x)$ in Eq. (2.49) has been used in the last interval in order to obtain the above result. The following program is an implementation of the Simpson rule for calculating an integral over $\sin x$.

```
        PROGRAM INTEGRAL
C
C Main program for evaluation of an integral with
C integrand sin(x) in the region of [0,pi/2].
C
        PARAMETER (N=9)
        DIMENSION X(N),F(N)
        PI = 4.0*ATAN(1.0)
        H  = PI/2.0/(N-1)
        DO      100 I = 1, N
          X(I) = H*(I-1)
          F(I) = SIN(X(I))
    100 CONTINUE
        CALL SIMP(N,H,F,S)
        WRITE (6,999) S
        STOP
    999 FORMAT (F16.8)
        END
C
        SUBROUTINE SIMP (N,H,FI,S)
C
C Subroutine for integration over f(x) with the
```

```
C Simpson rule. FI: integrand f(x); H: interval;
C S: integral.
C
      DIMENSION FI(N)
C
      S  = 0.0
      S0 = 0.0
      S1 = 0.0
      S2 = 0.0
      DO     100 I = 2, N-1, 2
         S0 = S0+FI(I-1)
         S1 = S1+FI(I)
         S2 = S2+FI(I+1)
  100 CONTINUE
      S = H*(S0+4.0*S1+S2)/3.0
C
C If N is even, add the last slice separately
C
      IF(MOD(N,2).EQ.0) S = S
    *   + H*(5.0*FI(N)+8.0*FI(N-1)-FI(N-2))/12.0
      RETURN
      END
```

We have used $f(x) = \sin x$ as the integrand in the above example program and $[0, \pi/2]$ as the integration region. The output of the above program is 1.000008, which has six digits of accuracy compared with the exact result, 1, in this case. Note that we have used only nine mesh points to reach such a high accuracy.

2.3 Zeros and extremes of a single-variable function

In physics, we often encounter situations in which we need to find the possible value of x that ensures the equation $f(x) = 0$, where $f(x)$ can be either an explicit or an implicit function of x. If such a value exists, we call it a *root* of the equation. An associated problem is to find the maxima and/or minima of a function $g(x)$. Relevant situations in physics include the equilibrium position of an object, the potential surface of a field, and the quantized energy levels of confined structures. In this section, we will discuss only single-variable problems and leave multivariable cases to be discussed in Chapter 4, after we gain some basic knowledge of matrix operations.

Bisection method

If we know that there is a root in the region $x \in [a, b]$ for $f(x) = 0$, we can use the *bisection method* to find it within a required accuracy. The bisection method is the most intuitive method, and the idea is very simple. Because there is a root in the region, $f(a)f(b) < 0$. We can divide the region into two equal parts with $x_1 = (a + b)/2$. Then we have either $f(a)f(x_1) < 0$ or $f(x_1)f(b) < 0$. If $f(a)f(x_1) < 0$, the next trial value is $x_2 = (a + x_1)/2$; otherwise, $x_2 = (x_1 + b)/2$. This procedure is repeated until the improvement on x_i is less than the given tolerance Δ.

Let us take $f(x) = e^x \ln x - x^2$ as an example to illustrate how the bisection method works. We know that when $x = 1$, $f(x) = -1$ and when $x = 2$, $f(x) = e^2 \ln 2 - 4 \approx 1$. So there is at least one value $x_0 \in [1, 2]$ that would make $f(x_0) = 0$. In the neighborhood of x_0, one would have $f(x_0 + \Delta) > 0$ and $f(x_0 - \Delta) < 0$. The following program is an implementation of the bisection method described for $f(x) = e^x \ln x - x^2$.

```
        PROGRAM BISECTION
C
C This program uses the bisection method to find the
C root of f(x) = exp(x)*ln(x) - x*x = 0.
C
        DL = 1.0E-06
        A  = 1.0
        B  = 2.0
        DX = B-A
        ISTEP = 0
        DO    100   WHILE (ABS(DX).GT.DL)
          X0 = (A+B)/2.0
          IF ((F(A)*F(X0)).LT.0) THEN
            B  = X0
            DX = B-A
          ELSE
            A  = X0
            DX = B-A
          END IF
          ISTEP = ISTEP+1
    100 END DO
        WRITE (6,999) ISTEP,X0,DX
        STOP
```

```
  999 FORMAT (I4,2F16.8)
      END
C
      FUNCTION F(X)
        F = EXP(X)*ALOG(X)-X*X
      RETURN
      END
```

The root obtained from the above program is $x_0 = 1.694600 \pm 0.000001$ in only twenty iterations. The error comes from the final improvement in the search. This error can be reduced to a lower value by making the tolerance smaller.

The Newton method

This method is based on linear approximation of a smooth function around its root. We can formally expand the function $f(x_0) = 0$ in the neighborhood of the root x_0 through the Taylor expansion introduced in Section 2.2,

$$f(x_0) \simeq f(x) + (x_0 - x)f'(x) + \cdots = 0, \tag{2.52}$$

where x can be viewed as a trial value for the root of x_0 at the nth step and the approximate value of the next step x_{n+1} can be derived from

$$f(x_{n+1}) = f(x_n) + (x_{n+1} - x_n)f'(x_n) \simeq 0, \tag{2.53}$$

that is,

$$x_{n+1} = x_n - f_n/f_n'. \tag{2.54}$$

Here we have used the notation $f_n = f(x_n)$. The above iterative scheme is known as the *Newton method*. It is also referred to as the *Newton–Raphson method* in literature. To see how this method works, we can still apply it to the example used in the preceding subsection in the bisection method with $f(x) = e^x \ln x - x^2$. The following program is an implementation of the Newton method.

```
      PROGRAM NEWTON
C
C This program applies the Newton method to find the
C root of f(x) = exp(x)*ln(x) - x*x = 0.
C
      DL = 1.0E-06
      A  = 1.0
```

```
      B   = 2.0
      DX  = B-A
      X0  = (A+B)/2.0
      ISTEP = 0
      DO    100   WHILE (ABS(DX).GT.DL)
        X1 = X0-F(X0)/DF(X0)
        DX = X1-X0
        X0 = X1
        ISTEP = ISTEP+1
  100 END DO
      WRITE (6,999) ISTEP,X0,DX
      STOP
  999 FORMAT (I4,2F16.8)
      END
C

      FUNCTION F(X)
        F = EXP(X)*ALOG(X)-X*X
      RETURN
      END
C

      FUNCTION DF(X)
        DF = EXP(X)*(ALOG(X)+1.0/X)-2.0*X
      RETURN
      END
```

The root obtained from the above program is $x_0 = 1.69460094$ after only five iterations. The error here is close to the limit of the single precision floating-point number on a 32-bit computer. One can clearly see that the Newton method is more efficient, because the error is now much smaller even though we have started the search exactly at the same point and have gone through fewer steps. The reason is very simple. Each new step size ($|x_{n+1} - x_n|$) in the Newton method is determined according to Eq. (2.54) by the ratio of the value and the slope of the function $f(x)$ at x_n. For the same function value, a large step is created for the small slope case, whereas a small step is made for the large slope case. This ensures an efficient update and also avoids possible overshooting. There is no such mechanism in the bisection method.

Secant method

In many cases, especially in the case of $f(x)$ with an implicit dependence on x, an analytic expression for the first-order derivative needed in the Newton method may not exist or may be very difficult to obtain. One has to find an alternative scheme to achieve a similar algorithm. One way to do it is to replace f'_n in Eq. (2.54) with the two-point formula for the first-order derivative, which gives

$$x_{n+1} = x_n - (x_n - x_{n-1})f_n/(f_n - f_{n-1}). \tag{2.55}$$

This iterative scheme is commonly known as the *secant method*, or the *discrete Newton method*. The disadvantage of the method is that one needs two points in order to start the search process. The advantage of the method is that $f(x)$ can now be implicitly given without the need for the first-order derivative. We can still use the function $f(x) = e^x \ln x - x^2$ as an example, in order to make a comparison. The following program is an implementation of the secant method just discussed.

```
      PROGRAM ROOT
C
C Main program to use the secant method to find the
C root of f(x) = exp(x)*ln(x) - x*x = 0.
C
      DL = 1.0E-06
      A  = 1.0
      B  = 2.0
      DX = (B-A)/10.0
      X0 = (A+B)/2.0
      CALL SECANT (DL,X0,DX,ISTEP)
      WRITE (6,999) ISTEP,X0,DX
      STOP
  999 FORMAT (I4,2F16.8)
      END
C
      SUBROUTINE SECANT (DL,X0,DX,ISTEP)
C
C Subroutine for the root of f(x) = 0 with the
C secant method.
C
      ISTEP = 0
      X1 = X0+DX
      DO    100  WHILE (ABS(DX).GT.DL)
        D  = F(X1)-F(X0)
        X2 = X1-(X1-X0)*F(X1)/D
        X0 = X1
        X1 = X2
```

```
          DX = X1-X0
          ISTEP = ISTEP+1
  100 END DO
      RETURN
      END
C

      FUNCTION F(X)
        F = EXP(X)*ALOG(X)-X*X
      RETURN
      END
```

The root obtained from the above program is $x_0 = 1.6946010 \pm 0.0000001$ after five iterations. As expected, the secant method is more efficient than the bisection method but less efficient than the Newton method, because of the two-point approximation of the first-order derivative. However, if the expression for the first-order derivative cannot easily be obtained, the secant method becomes very useful and powerful.

Extremes of a single-variable function

With the knowledge of the solution of a nonlinear equation $f(x) = 0$, we can develop numerical schemes to obtain minima or maxima of a function $g(x)$. We know that an extreme of $g(x)$ happens at the point with

$$f(x) = \frac{dg(x)}{dx} = 0, \tag{2.56}$$

which is a minimum (maximum) if $f'(x) = g''(x)$ is greater (less) than zero. So all the root-search schemes discussed so far can be generalized here to search for the extremes of a single-variable function.

However, in each step of updating the value of x, we need to make a judgment whether $g(x_{n+1})$ is increasing (decreasing) if we are searching for the maxima (minima) of the function. If it is, we accept the update. If it is not, we reverse the update; that is, instead of $x_{n+1} = x_n + \Delta x_n$, $x_{n+1} = x_n - \Delta x_n$ is used. With the Newton method, the increment is $\Delta x_n = -f_n/f'_n$, and with the secant method, the increment is $\Delta x_n = -(x_n - x_{n-1})f_n/(f_n - f_{n-1})$.

Let us illustrate these ideas with a simple example of finding the bond length of a diatomic molecule NaCl from the interaction potential between the two ions (Na^+ and Cl^- in this case). Assuming that the interaction potential is $V(r)$ when the two ions are separated with a distance r, the bond length r_0 is the equilibrium distance when $V(r)$ is at its minimum. We can model the interaction potential between Na^+ and Cl^- with

$$V(r) = -\frac{e^2}{r} + \alpha e^{-r/\rho}, \tag{2.57}$$

where e is the charge of a proton and α and ρ are parameters of this effective interaction. The first term in Eq. (2.57) comes from the Coulomb interaction between the two ions but the second term is the result of the electron distribution in the system. We will use $\alpha = 1.09 \times 10^3$ eV, which is taken from the experimental value for solid NaCl (Kittel, 1986, pp. 86–92), and $\rho = 0.330$ Å, which is a little larger than the corresponding parameter for solid NaCl ($\rho = 0.321$ Å), because there is less charge screening in an isolated molecule. At equilibrium, the force between the two ions,

$$f(r) = -\frac{dV(r)}{dr} = -\frac{e^2}{r^2} + \frac{\alpha}{\rho}e^{-r/\rho}, \tag{2.58}$$

is zero. Therefore, we search for the root of $f(x) = dg(x)/dx = 0$, with $g(x) = -V(x)$ and $x = r$. We will force the algorithm to move toward the maximum of $g(x)$ [minimum of $V(x)$]. The following program is an implementation of the algorithm with the secant method to find the bond length of NaCl.

```
      PROGRAM BOND
C
C Main program to calculate the bond length of NaCl.
C
      COMMON /CSTNT/ E2,A0,R0
      DL = 1.0E-06
      A0 = 1090.0
      R0 = 0.33
      E2 = 14.4
      X0 = 2.0
      DX = 0.1
      CALL M_SECANT (DL,X0,DX,ISTEP)
      WRITE (6,999) ISTEP,X0,DX
      STOP
  999 FORMAT (I4,2F16.8)
      END
C
      SUBROUTINE M_SECANT (DL,X0,DX,ISTEP)
C
```

```
C Subroutine for the root of f(x) = dg(x)/dx = 0 with
C the secant method with the search toward the
C maximum of g(x).
C
      ISTEP = 0
      G0 = G(X0)
      X1 = X0+DX
      G1 = G(X1)
      IF(G1.LT.G0) X1 = X0-DX
      DO    100  WHILE (ABS(DX).GT.DL)
        D   = F(X1)-F(X0)
        DX  = -(X1-X0)*F(X1)/D
        X2  = X1+DX
        G2  = G(X2)
        IF(G2.LT.G1) X2 = X1-DX
        X0  = X1
        X1  = X2
        G1  = G2
        ISTEP = ISTEP+1
  100 END DO
      X0 = X2
      RETURN
      END
C
      FUNCTION G(X)
      COMMON /CSTNT/ E2,A0,R0
        G   = E2/X-A0*EXP(-X/R0)
      RETURN
      END
C
      FUNCTION F(X)
      COMMON /CSTNT/ E2,A0,R0
        F   = E2/(X*X)-A0*EXP(-X/R0)/R0
      RETURN
      END
```

The bond length obtained from the program BOND is $r_0 = 2.36\,\text{Å}$. We have used $e^2 = 14.4\,\text{Å eV}$ for convenience. The subroutine M_SECANT is modified slightly from the subroutine SECANT used earlier in this section in order to force the search toward the maximum of $g(x)$. We will still obtain the same result with M_SECANT replaced by SECANT for this simple example, since there is only one minimum of $V(x)$ around the point where we start the search. The other minimum of $V(x)$ is at $x = 0$, which is also a singularity. For a more general $g(x)$, modifications introduced in M_SECANT are necessary.

Another relevant issue is that in many cases one does not have the explicit function $f(x) = g'(x)$ if $g(x)$ is not an explicit function of x. However, one can construct the first-order and second-order derivatives numerically from the three-point formulas, for example. We will discuss this point in more detail in Section 4.4 when we introduce the schemes for multivariable problems. The single-variable problem discussed here is just a special case.

2.4 Classical scattering

Scattering processes are very important in physics. From the microscopic scale, such as protons and neutrons in nuclei, to the astronomical scale, such as galaxies and stars, scattering processes play an important role in the physical phenomena observed in such systems. Generally, a many-body process can be viewed as a sum of many two-body scattering processes if coherent scattering does not play a significant role.

In this section, we will apply the computational methods we have developed in this chapter to study the classical scattering process of two point particles interacting with each other through a two-body potential. Most scattering processes with realistic interaction potentials cannot be solved analytically. Therefore, numerical solutions to a scattering problem become extremely valuable if one wants to understand the physical process of particle–particle interaction. We will assume that the interaction potential between the two particles is spherically symmetric and therefore that the total angular momentum and energy of the system are conserved during the scattering process.

The Lagrangian of a two-particle system

The Lagrangian for a general two-body system can be written as

$$\mathcal{L} = \frac{m_1}{2} v_1^2 + \frac{m_2}{2} v_2^2 - U(\mathbf{r}_1, \mathbf{r}_2), \tag{2.59}$$

where m_i, $\mathbf{r}_i$, $v_i = |d\mathbf{r}_i/dt|$ with $i = 1, 2$ are, respectively, the mass, position vector, and speed of the ith particle and U is the interaction potential between the two particles, which we take to be spherically symmetric, that is, $U(\mathbf{r}_1, \mathbf{r}_2) = U(r_{12})$ with $r_{12} = |\mathbf{r}_2 - \mathbf{r}_1|$ being the distance between the two particles.

We can always make a coordinate transformation from $\mathbf{r}_1$ and $\mathbf{r}_2$ to the relative coordinate $\mathbf{r}$ and the center-of-mass coordinate $\mathbf{r}_c$ with

$$\mathbf{r} = \mathbf{r}_2 - \mathbf{r}_1, \tag{2.60}$$

$$\mathbf{r}_c = \frac{m_1\mathbf{r}_1 + m_2\mathbf{r}_2}{m_1 + m_2}. \tag{2.61}$$

Then we can write the Lagrangian of the system in terms of the new coordinates and their corresponding speeds as

$$\mathcal{L} = \frac{M}{2}v_c^2 + \frac{m}{2}v^2 - U(r), \tag{2.62}$$

where $r = r_{12}$ and $v = |d\mathbf{r}/dt|$ are the distance and relative speed between the two particles, $M = m_1 + m_2$ is the total mass of the system, $m = m_1m_2/(m_1 + m_2)$ is the reduced mass of the two particles, and $v_c = |d\mathbf{r}_c/dt|$ is the speed of the center of mass. If we study the scattering in the center-of-mass coordinate with $d\mathbf{r}_c/dt = \mathbf{0}$, the system is then represented by a system of a single particle with a mass of m in a central potential $U(r)$. In general, a two-particle system with a spherically symmetric interaction potential can be viewed as a single particle with a reduced mass moving in a central potential that is identical to the interaction potential.

Cross section of scattering

Now we need to study only the scattering process of a particle with a mass m in a central potential $U(r)$. Assume that the particle is coming in from the left with an impact parameter b, the shortest distance between the particle and the potential center if $U(r) = 0$. The total cross section of a scattering process is given by

$$\sigma = \int \sigma(\theta)\,d\Omega, \tag{2.63}$$

where $\sigma(\theta)$ is the differential cross section, or the probability of a particle's being found in the solid angle element $d\Omega = 2\pi \sin\theta\,d\theta$ at the deflection angle θ.

If the particles are coming in with a flux density I, then the number of particles per unit time within the range db of the impact parameter b is $2\pi Ib\,db$. Because all the incoming particles in this area will go out in the solid angle element $d\Omega$ with the probability $\sigma(\theta)$, we have

$$2\pi Ib\,db = I\sigma(\theta)\,d\Omega, \tag{2.64}$$

which gives the differential cross section as

$$\sigma(\theta) = \frac{b}{\sin\theta}\left|\frac{db}{d\theta}\right|. \tag{2.65}$$

The reason for taking the absolute value of $db/d\theta$ in the above equation is that $db/d\theta$ can be positive or negative depending on the form of the potential and the impact parameter. However, $\sigma(\theta)$ has to be positive because it is a probability. One can relate this center-of-mass cross section to the cross section measured in the laboratory through an inverse coordinate transformation of Eq. (2.60) and Eq. (2.61), which relates $\mathbf{r}$ and $\mathbf{r}_c$ back to $\mathbf{r}_1$ and $\mathbf{r}_2$. We will not discuss this transformation here; interested readers can find it in any standard advanced mechanics textbook.

Numerical evaluation of the cross section

Since the interaction between two particles is described by a spherically symmetric potential, the angular momentum and the total energy of the system are conserved during the scattering. Formally, we have

$$l = mbv_0 = mr^2\dot\phi \tag{2.66}$$

and

$$E = \frac{m}{2}v_0^2 = \frac{m}{2}(\dot{r}^2 + r^2\dot\phi^2) \tag{2.67}$$

as the total momentum and total energy, which are constant. Here r is the radial coordinate, ϕ is the polar angle, and v_0 is the initial impact velocity. A dot over a symbol here means the time derivative of that quantity; for example, $\dot{r} = dr/dt$. Combining Eq. (2.66) and Eq. (2.67) with

$$\frac{d\phi}{dr} = \frac{d\phi}{dt}\frac{dt}{dr}, \tag{2.68}$$

we have

$$\frac{d\phi}{dr} = \pm\frac{b}{r^2}\frac{1}{\sqrt{1 - b^2/r^2 - U(r)/E}}, \tag{2.69}$$

which provides a relation between ϕ and r for given E, b, and $U(r)$. Here the $+$ and $-$ signs correspond to two different but symmetric parts of the trajectory. The equation above can be used to calculate the deflection angle θ through

$$\theta = \pi - 2\Delta\phi, \tag{2.70}$$

where $\Delta\phi$ is the change in the polar angle when r changes from infinity to its minimum value. From Eq. (2.69), one has

$$\begin{aligned}
\Delta\phi &= \int_{r_m}^{\infty} \frac{b}{r^2}\frac{dr}{\sqrt{1 - b^2/r^2 - U(r)/E}} \\
&= -\int_{\infty}^{r_m} \frac{b}{r^2}\frac{dr}{\sqrt{1 - b^2/r^2 - U(r)/E}},
\end{aligned} \tag{2.71}$$

where r_m is the minimum distance between the trajectory of the particle and the potential center. If one uses the energy conservation and angular momentum conservation discussed earlier in Eq. (2.67) and Eq. (2.66), one can show that r_m is given from

$$1 - \frac{b^2}{r_m^2} - \frac{U(r_m)}{E} = 0, \tag{2.72}$$

which is the result of zero r-component velocity, that is, $\dot{r} = 0$. Because the change in the polar angle $\Delta\phi = \pi/2$ for $U(r) = 0$, one can rewrite Eq. (2.70) as

$$\theta = 2b \left[\int_b^\infty \frac{b}{r^2} \frac{dr}{\sqrt{1 - b^2/r^2}} - \int_{r_m}^\infty \frac{b}{r^2} \frac{dr}{\sqrt{1 - b^2/r^2 - U(r)/E}} \right]. \tag{2.73}$$

The real reason for rewriting the constant π into an integral in the expression for θ is a numerical strategy to reduce possible errors coming from the truncation of the integration region at both ends of the second term. The integrand in the first integral diverges as $r \to b$ in much the same way as the integrand in the second integral does as $r \to r_m$. The errors from the first and second terms cancel each other because they are of opposite signs.

Here we would like to demonstrate how to calculate the differential cross section for a given potential. Let us take the Yukawa potential

$$U(r) = \frac{\kappa}{r} e^{-r/a} \tag{2.74}$$

as an illustrative example. Here κ and a are positive parameters that reflect respectively the range and the strength of the potential and can be adjusted. We can use the secant method to solve Eq. (2.72) to obtain r_m for the given b and E. Then we can use the Simpson rule to carry out the integrals in Eq. (2.73). After that, we can apply the three-point formula for the first-order derivative to obtain $d\theta/db$. Finally, we put all these together to obtain the differential cross section of Eq. (2.65).

For simplicity, we choose $E = m = \kappa = 1$. The following program is an implementation of the scheme outlined above.

```
      PROGRAM SCATTERING
C
C This is the main program for the scattering problem.
C
      PARAMETER(N=10001,M=21)
      DIMENSION FI(N),THETA(M),SIG(M),SIG1(M)
      COMMON /CCNST/ B,E,A
      B0 = 0.01
```

```
          DB = 0.5
          DX = 0.01
          E  = 1.0
          A  = 100.0
          DL = 1.0E-06
          DO      100  I = 1, M
            B  = B0+(I-1)*DB
C
C Calculate the first term of theta
C
          DO      80   J = 1, N
            X     = B+DX*J
            FI(J) = 1.0/(X*X*SQRT(FB(X)))
   80     CONTINUE
          CALL SIMP(N,DX,FI,G1)
C
C Find r_m from 1 - b*b/(r*r) - U/E = 0
C
          X0  = B
          DX0 = DX
          CALL SECANT(DL,X0,DX0,ISTEP)
C
C Calculate the second term of theta
C
          DO      90   J = 1, N
            X = X0+DX*J
            FI(J) = 1.0/(X*X*SQRT(F(X)))
   90     CONTINUE
          CALL SIMP(N,DX,FI,G2)
          THETA(I) = 2.0*B*(G1-G2)
  100 CONTINUE
C
C Calculate d_theta/d_b
C
      CALL THREE(M,DB,THETA,SIG,SIG1)
C
C Put the cross section in log form with the exact
C result of the Coulomb scattering (RUTH)
C
```

```
        DO      120  I = M, 1, -1
        B       = B0+(I-1)*DB
        SIG(I) = B/ABS(SIG(I))/SIN(THETA(I))
        RUTH    = 1.0/SIN(THETA(I)/2.0)**4/16.0
        SI      = ALOG(SIG(I))
        RU      = ALOG(RUTH)
        WRITE (6,999) THETA(I),SI,RU
  120 CONTINUE
        STOP
  999 FORMAT (3F16.8)
        END
C

        FUNCTION F(X)
        COMMON /CCNST/ B,E,A
        F = 1.0-B*B/(X*X)-U(X)/E
        RETURN
        END
C

        FUNCTION FB(X)
        COMMON /CCNST/ B,E,A
        FB = 1.0-B*B/(X*X)
        RETURN
        END
C

        FUNCTION U(X)
        COMMON /CCNST/ B,E,A
        U = 1.0/X*EXP(-X/A)
        RETURN
        END
```

The subroutines called in the program are exactly those given in Sections 2.2 and 2.3. We have also included the analytical result for the Coulomb scattering, which is a special case of the Yukawa potential with $a \to \infty$. The differential cross section for the Coulomb scattering is

$$\sigma(\theta) = \left(\frac{\kappa}{4E}\right)^2 \frac{1}{\sin^4(\theta/2)}, \qquad (2.75)$$

which is commonly referred to as the Rutherford formula. The results for different values of a from the above program are shown in Fig. 2.3. As one can see, when

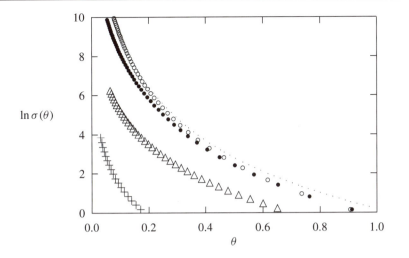

Fig. 2.3. This figure shows the differential cross section for scattering from the Yukawa potential with $a = 0.1$ (crosses), $a = 1$ (triangles), $a = 10$ (solid circles), and $a = 100$ (open circles), together with the analytical result for the Coulomb scattering (dots). Other parameters $E = m = \kappa = 1$ are used.

a is increased, the differential cross section becomes closer to that of the Coulomb scattering, as one would expect.

2.5 Random number generators

In practice, many numerical simulations need to use random number generators either for setting up initial configurations or for performing simulations. There is no such thing as *random* in a computer program. A computer will execute the program exactly the same way if the program is started in exactly the same way. A random number generator here really means a pseudo-random number generator that can generate a sequence of numbers that has a very long period and mimics a given distribution. In this section we will discuss some basic random number generators used in computational physics and computer simulations.

Uniform random number generators

The most useful random number generators are those with a uniform distribution in the region [0, 1]. The three most important criteria for a good uniform random number generator are the following (Park and Miller, 1988).

First, a good generator should have a long period, which should be close to the range of the integers on the computer. For example, on 32-bit computers, a good generator should have a period close to $2^{31} - 1 = 2,147,483,647$. The range of the integers on a 32-bit computer is $[-2^{31}, 2^{31} - 1]$. Note that one bit is used for the sign of the integers.

Second, a good generator should have good *randomness*. There should be only a very small correlation among all the numbers generated in sequence. If $\langle A \rangle$ represents the statistical average of the variable A, the n-point correlation function $\langle x_{n_1} x_{n_2} \ldots x_{n_l} \rangle$ for $l = 2, 3, 4, \ldots$, should be very small. One way to illustrate the behavior of the correlation function $\langle x_n x_{n+l} \rangle$ is to plot x_n and x_{n+l} in an xy-plane. A good random number generator will have a very uniform distribution of the points for any $l \neq 0$. A poor generator may show strips, lattices, or other inhomogeneous distributions.

Finally, a good generator has to be very fast. In practice, one will need a lot of random numbers in order to have good statistical results. The speed of the generator can become a very important factor.

The simplest uniform random number generator is built from the so-called linear congruent scheme. The random numbers are generated in sequence from the linear relation

$$x_{n+1} = (ax_n + b) \bmod c, \tag{2.76}$$

where a, b, and c are *magic* numbers: The choice of these numbers determines the quality of the generator. One option, $a = 7^7 = 16,807$, $b = 0$, and $c = 2^{31} - 1 = 2,147,483,647$, has been tested and found to be an excellent choice for 32-bit computers. It has the full period of $2^{31} - 1$ and is very fast. The correlation function $\langle x_{n_1} x_{n_2} \ldots x_{n_l} \rangle$ is very small. In Fig. 2.4, we plot x_n and x_{n+10} generated using the linear congruent method with the above selection of magic numbers. As

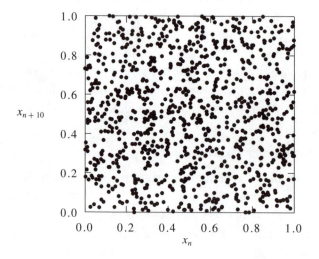

Fig. 2.4. The plot of (x_n, x_{n+10}) for $x_i \in [0, 1]$ with 1,000 points shown.

one can see, the plot is very homogeneous and random. There are no strips or lattice structures in the plot.

Implementation of this random number generator on a computer is not always trivial, because of the different numerical range of each computer. For example, most 32-bit computers would have integers in $[-2^{31}, 2^{31} - 1]$. If the number runs out of this range, the computer will change that integer to zero. If a computer could modulate the integers by $2^{31} - 1$ automatically, one could easily have a random number generator with the above magic numbers a, b, and c simply by taking consecutive numbers from $x_{n+1} = 16,807 x_n$ with any initial choice in the range $1 < x_0 < 2^{31} - 1$. The range of this generator would be $[0, 2^{31} - 1]$. However, this automatic modulation would cause some serious problems in other situations. For example, when a quantity is out of range due to a bug in the program, the computer would still wrap it back without producing any error messages. This is why, in practice, computers do not modulate numbers automatically, so we have to devise a scheme to modulate the numbers generated in sequence. The following routine is an implementation of the uniform random number generator discussed above.

```
FUNCTION RANF()
DATA IA/16807/,IC/2147483647/,IQ/127773/,IR/2836/
COMMON /CSEED/ ISEED
   IH = ISEED/IQ
   IL = MOD(ISEED,IQ)
   IT = IA*IL-IR*IH
   IF(IT.GT.0) THEN
      ISEED = IT
   ELSE
      ISEED = IC+IT
   END IF
   RANF() = ISEED/FLOAT(IC)
RETURN
END
```

In this program, $IQ = IC/IA$, $IR = MOD(IC, IA)$, and $ISEED \in [1, 2^{31} - 1]$. A program in Pascal similar to the above function routine was given by Park and Miller (1988). One can easily show that the above function routine would modulate the numbers with $c = 2^{31} - 1$ on any computer with 32 bits or more. To use this function routine, one needs to have the same common block in the calling routine, so the new value of ISEED will be returned. To convince oneself that the function given above does implement the algorithm correctly, one can set ISEED = 1 first, and then after 10,000 steps, one should have ISEED = 1,043,618,065 (Park and Miller, 1988).

In order to start the random number generator differently every time, one needs to have a systematic way of obtaining a different initial seed (the initial value of ISEED). Otherwise, one would end up with exactly the same result if one starts the program with exactly the same initial seed. Almost every computer has intrinsic function routines to report the current time in an integer form, and we can use this integer to construct an initial seed (Anderson, 1990). For example, most computers would have

$$
\begin{array}{ll}
\text{year:} & 0 \le i_y \le 99, \\
\text{month:} & 1 \le i_m \le 12, \\
\text{day:} & 1 \le i_d \le 31, \\
\text{hour:} & 0 \le i_h \le 23, \\
\text{minute:} & 0 \le i_n \le 59, \\
\text{second:} & 0 \le i_s \le 59.
\end{array}
$$

Then one can choose

$$
\text{ISEED} = i_y + 70\big(i_m + 12\{i_d + 31[i_h + 23(i_n + 59 i_s)]\}\big), \tag{2.77}
$$

as the initial seed, which is roughly in the region of $[0, 2^{31} - 1]$. The results should never be the same as long as the jobs are started at least a second apart but within 100 years. For example, the following segment from a Fortran program on a UNIX machine is an implementation of the above initial seed.

```
INTEGER*4 time,STIME,T(9)
STIME = time(%REF())
CALL gmtime_(STIME,T)
ISEED = T(6)+70*(T(5)+12*(T(4)
*              +31*(T(3)+23*(T(2)+59*T(1))))) 
IF (MOD(ISEED,2).EQ.0) ISEED = ISEED-1
```

The intrinsic function TIME and subroutine GMTIME_ are used to obtain the system time and to transform it into the needed integer array. An even initial seed is usually avoided in order to have the full period of the generator. This initial seed will be used in Chapter 9.

Uniform random number generators are very important in scientific computing, and good ones are extremely difficult to find. The function routine given here is considered one of the best uniform random number generators.

Other distributions

As soon as we obtain good uniform random number generators, we can use them to create other types of random number generators. For example, we can use a uniform random number generator to create an exponential distribution or a Gaussian distribution.

All the exponential distributions can be cast into their simplest form

$$p(x) = e^{-x} \tag{2.78}$$

after proper choice of units and coordinates. For example, if a system has energy levels of E_0, E_1, ..., E_N, the probability of the system's being at energy level E_i at temperature T is given by

$$p(E_i, T) \propto e^{-(E_i - E_0)/k_B T}, \tag{2.79}$$

where k_B is the Boltzmann constant. If we choose $k_B T$ as the energy unit and E_0 as the zero point, the above equation reduces to Eq. (2.78).

One way to generate the exponential distribution is to relate it to a uniform distribution. For example, if we have a uniform distribution $f(y) = 1$ for $y \in [0, 1]$, we can relate it to an exponential distribution by

$$f(y)\,dy = dy = p(x)\,dx = e^{-x}\,dx, \tag{2.80}$$

which gives

$$y(x) - y(0) = 1 - e^{-x}. \tag{2.81}$$

We can set $y(0) = 0$ and invert the above equation to have

$$x = -\ln(1 - y), \tag{2.82}$$

which relates the exponential distribution of x to the uniform distribution of $y \in [0, 1]$. The following function routine is an implementation of the exponential random number generator given in the above equation and constructed from a uniform random number generator.

```
FUNCTION ERNF()
COMMON /CSEED/ ISEED
  ERNF = -ALOG(1.0-RANF())
RETURN
END
```

The uniform random number generator obtained earlier is used by this function routine.

As we have pointed out, another useful distribution used in physics is the Gaussian distribution

$$g(x) = \frac{1}{\sqrt{2\pi}\sigma} e^{-x^2/2\sigma^2}, \tag{2.83}$$

where σ is the variance of the distribution, which we can take as 1 for the moment. The distribution with $\sigma \neq 1$ can be obtained via the rescaling of x by σ in the generator with $\sigma = 1$. We can use a uniform distribution $f(\phi) = 1$ for $\phi \in [0, 2\pi]$ and an exponential distribution $p(t) = \exp(-t)$ for $t \in [0, \infty]$ to obtain two Gaussian distributions $g(x)$ and $g(y)$. We can relate the product of a uniform distribution and an exponential distribution to a product of two Gaussian distributions by

$$\frac{1}{2\pi} f(\phi) \, d\phi \, p(t) \, dt = g(x) \, dx \, g(y) \, dy, \tag{2.84}$$

which gives

$$e^{-t} dt \, d\phi = e^{-(x^2+y^2)/2} \, dx \, dy. \tag{2.85}$$

The above equation can be viewed as the coordinate transform from the polar system (ρ, ϕ) with $\rho = \sqrt{2t}$ into the rectangular system (x, y), that is,

$$x = \sqrt{2t} \cos \phi, \tag{2.86}$$

$$y = \sqrt{2t} \sin \phi, \tag{2.87}$$

which are two Gaussian distributions if t is taken from an exponential distribution and ϕ is taken from a uniform distribution in the region $[0, 2\pi]$. With the availability of the exponential random number generator and uniform random number generator, we can construct two Gaussian random numbers immediately from Eqs. (2.86) and (2.87). The exponential random number generator itself can be obtained from a uniform random number generator as discussed in this subsection. The following subroutine is an implementation of the algorithm given in Eqs. (2.86) and (2.87), which generates two Gaussian random numbers from two uniform random numbers in [0,1].

```
SUBROUTINE GRNF(X,Y)
COMMON /CSEED/ ISEED
   PI = 4.0*ATAN(1.0)
   R1 = -ALOG(1.0-RANF())
   R2 = 2.0*PI*RANF()
   R1 = SQRT(2.0*R1)
   X  = R1*COS(R2)
   Y  = R1*SIN(R2)
RETURN
END
```

In principle, one can generate any given distribution function numerically. For the cases of Gaussian distribution and exponential distribution, we construct the new generators with the integral transformations in order to relate the distributions sought to the distributions known. A general numerical procedure can be devised by dealing with the integral transformation numerically. Interested readers can find a discussion on this numerical procedure in Koonin (1986, pp. 192–4).

Percolation in two dimensions

Let us use two-dimensional percolation as an example to illustrate how a random number generator is utilized in computer simulations. When atoms are added to a solid surface, they will first occupy the sites with the lowest potential energy to form small two-dimensional clusters. If the probability of occupying each empty site is still high, the clusters will grow on the surface and eventually form a single layer, or a percolated two-dimensional network. If the probability of occupying each empty site is low, the clusters will grow into islandlike three-dimensional clusters. The general problem of the formation of a two-dimensional network can be cast into a simple model with each site carrying a fixed occupancy probability.

Assume that we have a two-dimensional square lattice with $N \times M$ lattice points. Then we generate $N \times M$ random numbers $x_{ij} \in [0, 1]$ for $i = 1, 2, \ldots, N$ and $j = 1, 2, \ldots, M$. We can compare the random number x_{ij} with the assigned occupancy probability $p \in [0, 1]$. The site is occupied if $p > x_{ij}$; otherwise the site remains empty. Clusters are formed by the occupied sites. A site in each cluster is, at least, a nearest neighbor of another site in the same cluster. We can gradually change p from 0 to 1. As p increases, the sizes of the clusters will increase and some clusters will also merge into larger clusters. When p reaches a critical probability p_c, there will be at least one cluster of the occupied sites, which will reach all the boundaries of the lattice. We call p_c the percolation threshold. The following subroutine is the core part of the simulation of two-dimensional percolation, which assigns a false value to an empty site and a true value to an occupied site.

```
SUBROUTINE PERCOLATION (L,N,M,P)
LOGICAL L(N,M)
COMMON /CSEED/ ISEED
DO     200  I = 1, N
  DO   100  J = 1, M
    X = RANF()
    IF(P.GT.X) THEN
      L(I,J) = .TRUE.
```

```
          ELSE
              L(I,J) = .FALSE.
          END IF
      100    CONTINUE
      200 CONTINUE
     RETURN
     END
```

Here L(I, J) is a logical array that contains true values at all the occupied sites and false values at all the empty sites. One can use the above subroutine with a program that has p increased from 0 to 1 and can sort out the sizes of all the clusters formed by the occupied lattice sites. In order to obtain good statistical averages, the procedure should be carried out many times. For more discussions on percolation, see Stauffer and Aharony (1992).

The above example is an extremely simple application of the uniform random number generator. In Chapter 7 and Chapter 9, we will discuss more on the application of random number generators in other simulations. Interested readers can find more on different random number generators in Knuth (1981), Park and Miller (1988), and Anderson (1990).

Exercises

2.1 Show that the error in the nth-order Lagrange interpolation scheme is bounded by

$$|\Delta f(x)| \le \frac{\gamma_n}{4(n+1)} h^{n+1},$$

with $\gamma_n = \max[|f^{(n+1)}(x)|]$ for $x \in [x_1, x_{n+1}]$ for both the even and odd $n+1$.

2.2 Write a program that implements the Lagrange interpolation scheme directly. Test it by evaluating $f(0.3)$ and $f(0.5)$ from the given data taken from the table of the error function with $f(0.0) = 0$, $f(0.4) = 0.428392$, $f(0.8) = 0.742101$, $f(1.2) = 0.910314$, and $f(1.6) = 0.970348$. Examine the accuracy of the interpolation by comparing the results obtained from the interpolation with the exact values $f(0.3) = 0.328627$ and $f(0.5) = 0.520500$.

2.3 Implement the upward and downward update scheme for the Aitken method in a subroutine and test it with the numerical values in Exercise 2.2.

2.4 Use the subroutine given in the subsection on the least-squares approximation to the data set $f(0.0) = 1.000000$, $f(0.2) = 0.912005$, $f(0.4) = 0.671133$, $f(0.6) = 0.339986$, $f(0.8) = 0.002508$, $f(0.9) = -0.142449$, and $f(1.0) = -0.260052$ and the corresponding points of $f(-x) = f(x)$ as the input. The data are taken from the Bessel function table with $f(x) = J_0(3x)$. Compare the results with the well-known approximate formula for the Bessel function,

$$f(x) = 1 - 2.2499997x^2 + 1.2656208x^4 - 0.3163866x^6$$
$$+ 0.0444479x^8 - 0.039444x^{10} + 0.0002100x^{12}.$$

2.5 Modify the program for the linear fitting of the Millikan experiments in Section 2.1 to analyze the accuracy of the fitting. Assume that the error bars in the experimental measurements are $\Delta e_n = 0.05e_n$. The accuracy of a fit is determined from

$$\Delta^2 = \frac{1}{N} \sum_{i=1}^{N} \frac{|f(x_i) - p_n(x_i)|^2}{\sigma_i^2},$$

where σ_i is the error bar of $f(x_i)$. If $\Delta \ll 1$, the fit is considered very good.

2.6 Write a program that calculates the first-order and second-order derivatives from the five-point formulas. Check its accuracy with the function $f(x) = \sin(x)$ with 101 uniform points in $[0, \pi/2]$. Discuss the procedure in dealing with the points at boundaries.

2.7 Derive the Simpson rule with a pair of slices with an equal interval by using the Taylor expansion of $f(x)$ around x_k in the region $[x_{k-1}, x_{k+1}]$ and the three-point formulas for the first- and second-order derivatives. Show that the contribution of the last slice is also correctly given by the formula in Section 2.2 under such an approach.

2.8 Develop a program that can calculate the integral with a given integrand $f(x)$ in the region $[a, b]$ by the Simpson rule with a nonuniform mesh. Check its accuracy with the integral $\int_0^\infty e^{-x} dx$ with $x_n = nhe^{n\alpha}$, where h and α are small constants.

2.9 Apply the subroutine developed in Section 2.3 with the secant method to solve $f(x) = e^{x^2} \ln x^2 - x = 0$. Discuss the procedure for cases with more than one root.

2.10 Write a subroutine that returns the minimum of a single-variable function $g(x)$ in a given region $[a, b]$. Assume that the first-order and second-order derivatives of $g(x)$ are not explicitly given. Test the subroutine with some well-known functions, for example, $g(x) = x^2$.

2.11 Modify the program in Section 2.4 to evaluate the differential cross section of the Lennard–Jones potential

$$U(r) = 4\epsilon_0 \left[\left(\frac{\sigma}{r} \right)^{12} - \left(\frac{\sigma}{r} \right)^{6} \right].$$

Choose ϵ_0 as the energy unit and σ as the length unit.

2.12 Write a program that can generate clusters of occupied sites in a two-dimensional square lattice with $n \times n$ sites. Determine $p_c(n)$, the threshold probability with a cluster across the whole lattice. Then determine p_c for an infinite lattice from

$$p_c(n) = p_c + \frac{a_1}{n} + \frac{a_2}{n^2} + \frac{a_3}{n^3} + \cdots,$$

where a_i and p_c can be solved from the $p_c(n)$ obtained.

2.13 Calculate the time needed for a meterstick to fall to a horizontal position on a smooth surface if it is released from a position with an angle $\theta \in [0, \pi/2]$ to the horizontal. What happens if $\theta = \pi/2$?

3

Ordinary differential equations

Many problems in computational physics appear in the form of differential equations. For example, the problem of a particle moving in the one-dimensional space discussed in Chapter 1 appears in the form of Newton's equation, which is a second-order differential equation.

In this chapter, we will introduce some basic methods for solving ordinary differential equations. We will discuss numerical methods for solving partial differential equations in Chapter 6 and some special techniques to solve Newton's equation and the Schrödinger equation for many-body systems in Chapters 7, 8, and 9.

In general, we can classify ordinary differential equations into three major categories:

(1) Initial-value problems, which involve time-dependent equations with given initial conditions;

(2) Boundary-value problems, which involve differential equations with specified boundary values;

(3) Eigenvalue problems, which involve solutions for selected parameters in the equations.

In reality, a problem may involve more than just one of the categories listed above. A common situation is that one has to separate several variables by introducing multipliers so that the initial-value problem is isolated from the boundary-value or eigenvalue problem. One can then solve the boundary-value or eigenvalue problem first to determine the multipliers, which in turn are used to solve the related initial-value problem. We will discuss separating variables in Chapter 6. In this chapter, we will discuss some basic numerical methods for all three categories listed above and illustrate how one can use these basic methods to deal with problems encountered in physics and other fields.

3.1 Initial-value problems

Typically, initial-value problems involve dynamical systems, for example, the motion of the Moon, the Earth, and the Sun; the dynamics of a rocket; or the propagation of oceanic waves. The behavior of a dynamical system can be described by a set of first-order differential equations

$$\frac{d\mathbf{y}}{dt} = \mathbf{g}(\mathbf{y}, t), \tag{3.1}$$

with

$$\mathbf{y} = (y_1, y_2, \dots, y_n) \tag{3.2}$$

being the dynamical variable vector and

$$\mathbf{g}(\mathbf{y}, t) = [g_1(\mathbf{y}, t), g_2(\mathbf{y}, t), \dots, g_n(\mathbf{y}, t)] \tag{3.3}$$

being the generalized velocity vector. Here n is the total number of dynamical variables. In principle, one can obtain the solution of the above equation numerically if the initial values $\mathbf{y}(t = 0) = \mathbf{y}_0$ are given and a solution exists. For the case of a particle moving in a one-dimensional space under a harmonic force discussed in Chapter 1, the dynamical behavior of the particle is governed by Newton's equation

$$f = ma, \tag{3.4}$$

where a is the acceleration of the particle, m is the mass of the particle and f is the force exerted on the particle. This equation can be viewed as a special case of Eq. (3.1) with the vector dimensions $n = 2$; that is, $y_1 = x$ and $y_2 = v = dx/dt$, and $g_1 = v = y_2$ and $g_2 = f/m = -kx/m = -ky_1/m$. Then we can rewrite Newton's equation into the form of Eq. (3.1),

$$\frac{dy_1}{dt} = y_2, \tag{3.5}$$

$$\frac{dy_2}{dt} = -\frac{k}{m}y_1. \tag{3.6}$$

If the initial position $y_1(0)$ and the initial velocity $y_2(0)$ are given, we can solve the problem numerically as demonstrated in Chapter 1.

In fact, most higher-order differential equations can be transformed into a set of coupled first-order differential equations in the form of Eq. (3.1). The higher-order derivatives are usually redefined into dynamical variables during the transformation. The velocity in Newton's equation discussed above is such an example.

3.2 The Euler and Picard methods

For convenience of notation, we will work out our numerical schemes for all the cases with only one dynamical variable, that is, treating $\mathbf{y}$ and $\mathbf{g}$ in Eq. (3.1) as one-dimensional vectors or scalars. Extending the formalism developed here to multivariable cases is straightforward. We will illustrate such an extension in Section 3.5.

Intuitively, one can solve Eq. (3.1) numerically as was done in Chapter 1 for the problem of a particle moving in a one-dimensional space by taking the average field as the time derivative; that is,

$$\frac{dy}{dt} \simeq \frac{y_{n+1} - y_n}{t_{n+1} - t_n} \simeq g(y_n, t_n). \tag{3.7}$$

Note that the index here is not the component index of the vector used in Eqs. (3.5) and (3.6) but is an indication of time steps. We will also use $g_n = g(y_n, t_n)$ to simplify the notation. The approximation of the first-order derivative in Eq. (3.7) is equivalent to the two-point formula, which has a maximum accuracy of $O(|t_{n+1} - t_n|)$. If we take an even-spaced mesh with a time step τ and rearrange the terms in Eq. (3.7), we obtain the simplest algorithm for initial-value problems,

$$y_{n+1} = y_n + \tau g_n + O(\tau^2), \tag{3.8}$$

which is commonly known as the *Euler method* and was used in the example of a particle moving in a one-dimensional space in Chapter 1. The accuracy of this algorithm is relatively low. At the end of the calculation, that is, after a total of N steps, the error accumulated in the calculation is on the order of $N O(\tau^2) \simeq O(\tau)$. To illustrate the relative error in this algorithm, let us use the same program for the one-dimensional motion given in Chapter 1. Now if we use a time step of 0.02π instead of 0.0002π, the program can accumulate a sizable error. The results for the position and velocity of the particle in the first period are given in Fig. 3.1. The results from a better algorithm with a corrector discussed in Section 3.3 and the exact results are also shown for comparison. As one can see, the accuracy of the Euler algorithm is very low. We have 100 points in one period of the motion, which is a typical number of points chosen in most numerical calculations. If we go on to the second period, the error will increase even further. For problems without periodic motion, the results at a later time would be even worse. We can conclude that this algorithm cannot be adopted in actual numerical solutions of most physical problems. How can we improve the algorithm so it will become practical?

We can formally rewrite Eq. (3.1) into an integral

$$y_{n+m} = y_n + \int_{t_n}^{t_{n+m}} g(y, t) \, dt, \tag{3.9}$$

which is the exact solution of Eq. (3.1) if the integral in the equation can be carried out exactly for any given integers n and m. Since we cannot carry out the integral in Eq. (3.9) exactly in general, we have to approximate it. The accuracy in the

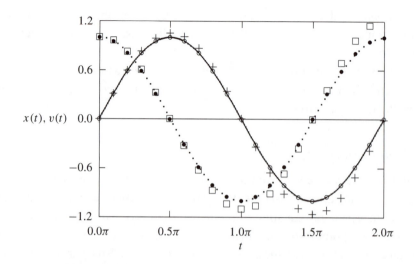

Fig. 3.1. The position (+) and velocity (□) of the particle moving in a one-dimensional space under a harmonic force calculated with the Euler method with a time step of 0.02π compared with the position (○) and velocity (•) calculated with the predictor–corrector method and the exact results (solid and dotted lines).

approximation of the integral determines the accuracy of the solution. If we take the simplest case of $m = 1$ and approximate $g(y, t) \simeq g_n$ in the integral, we recover the Euler algorithm of Eq. (3.8). The *Picard method* is an iterative scheme that uses the result from the last iterative step as the integrand on the right side of Eq. (3.9). One can use the initial value as the starting point, that is, $y^{(0)}(t) = y(0)$, and then iterate the solution through Eq. (3.9). In most cases, one also needs to use a numerical quadrature for the integral. For example, if one chooses $m = 1$ and uses the trapezoid rule, one obtains an improved Euler algorithm

$$y_{n+1} = y_n + \frac{\tau}{2}(g_n + g_{n+1}) + O(\tau^3). \tag{3.10}$$

Note that $g_{n+1} = g(y_{n+1}, t_{n+1})$ contains y_{n+1}, which has to be obtained iteratively. However, the Picard method will require a lot of computing time if the initial guess is not very close to the actual solution. The Picard scheme may not even converge if certain conditions are not satisfied. Can we avoid such tedious iteration by an intelligent guess of the solution?

3.3 Predictor–corrector methods

One way to get out of the situation is to use the so-called predictor–corrector method. One can apply a less accurate algorithm to predict the next value y_{n+1} first, for example, using the Euler algorithm of Eq. (3.8), and then apply a better algorithm to improve the new value, for example, using the improved Euler algorithm of Eq. (3.10). If we apply this method to the particle moving in a one-dimensional space with the same time step of 0.02π, we obtain much better results, as shown in Fig. 3.1. The following program is the application of this simplest predictor–corrector method to a particle moving in a one-dimensional space under a harmonic force.

```
      PROGRAM ONE_D_MOTION2
C
C Simplest predictor-corrector algorithm applied to
C a particle in one dimension under an elastic force.
C
      PARAMETER (N=101,IN=5)
      REAL T(N),V(N),X(N)
      PI  = 4.0*ATAN(1.0)
      DT  = 2.0*PI/100
      X(1)= 0.0
      T(1)= 0.0
      V(1)= 1.0
      DO      100  I = 1, N-1
        T(I+1) = I*DT
C
C Predictor for position and velocity
C
        X(I+1) = X(I)+V(I)*DT
        V(I+1) = V(I)-X(I)*DT
C
C Corrector for position and velocity
C
        X(I+1) = X(I)+(V(I)+V(I+1))*DT/2.0
        V(I+1) = V(I)-(X(I)+X(I+1))*DT/2.0
  100 CONTINUE
      WRITE (6,999) (T(I),X(I),V(I),I=1,N,IN)
      STOP
  999 FORMAT (3F16.8)
      END
```

Another way to improve an algorithm is by increasing the number of mesh points m in Eq. (3.9). Thus we can apply a better quadrature to the integral. For example, if we take $m = 2$ in Eq. (3.9) and then use the linear interpolation scheme to approximate $g(y, t)$ in the integral from g_n and g_{n+1}, we have

$$g(y, t) = \frac{(t - t_n)}{\tau} g_{n+1} - \frac{(t - t_{n+1})}{\tau} g_n + O(\tau^2). \tag{3.11}$$

Now if we carry out the integral with $g(y, t)$ given from this equation, we obtain a new algorithm

$$y_{n+2} = y_n + 2\tau g_{n+1} + O(\tau^3), \tag{3.12}$$

which has an accuracy one order higher than that of the Euler algorithm. However, one needs the values of the first two points in order to start this algorithm, because $g_{n+1} = g(y_{n+1}, t_{n+1})$. Usually, the second point can be obtained by expanding the differential equation around the initial point. Of course, one can always include more points in the integral of Eq. (3.9) to obtain algorithms with apparently higher accuracy. But one will need the values of more points in order to start the algorithm. Usually it is not very practical if one needs more than two points in order to start the algorithm, because the errors accumulated from the approximations of the first few points will eliminate the apparently high accuracy of the algorithm.

We can make the accuracy even higher by using a better quadrature. For example, we can take $m = 2$ in Eq. (3.9) and apply the Simpson rule, discussed in Section 2.2, to the integral. Then we have

$$y_{n+2} = y_n + \frac{\tau}{3}(g_{n+2} + 4g_{n+1} + g_n) + O(\tau^5). \tag{3.13}$$

This implicit algorithm can be used as the corrector if the algorithm of Eq. (3.12) is used as the predictor. One can go further by including more points in the quadrature of Eq. (3.9), and interested readers can find many multiple-point formulas in Davis and Polonsky (1965).

3.4 The Runge–Kutta method

The accuracy of the methods we have discussed so far can be improved only by including more starting points, which is not always practical, because a problem associated with a dynamical system usually has only the first point, namely, the initial condition given. A very practical method that requires only the first point in order to start or improve the algorithm is the *Runge–Kutta method*, which is derived from the two different Taylor expansions of the dynamical variables and their derivatives in Eq. (3.1).

Formally, one can expand $y(t + \tau)$ in terms of the quantities at t with the Taylor expansion

$$y(t + \tau) = y + \tau y' + \frac{\tau^2}{2} y'' + \frac{\tau^3}{3!} y^{(3)} + \cdots$$

$$= y + \tau g + \frac{\tau^2}{2} (g_t + g g_y)$$

$$+ \frac{\tau^3}{6} (g_{tt} + 2 g g_{ty} + g^2 g_{yy} + g g_y^2 + g_t g_y) + \cdots, \tag{3.14}$$

where the indices are for partial derivatives. For example, $g_{yt} = \partial^2 g / \partial y \partial t$. One can also formally write the solution at $t + \tau$ as

$$y(t + \tau) = y(t) + \alpha_1 k_1 + \alpha_2 k_2 + \cdots + \alpha_n k_n, \tag{3.15}$$

with

$$k_1 = \tau g(y, t), \tag{3.16}$$

$$k_2 = \tau g(y + v_{21} k_1, t + v_{21} \tau), \tag{3.17}$$

$$k_3 = \tau g(y + v_{31} k_1 + v_{32} k_2, t + v_{31} \tau + v_{32} \tau), \tag{3.18}$$

$$\vdots$$

$$k_n = \tau g \left(y + \sum_{l=1}^{n-1} v_{nl} k_l, t + \tau \sum_{l=1}^{n-1} v_{nl} \right), \tag{3.19}$$

where α_i (with $i = 1, 2, \ldots, n$) and v_{ij} (with $i = 2, 3, \ldots, n$ and $j < i$) are parameters to be determined. One can expand Eq. (3.15) into a power series of τ by carrying out the Taylor expansions for all k_i with $i = 1, 2, \ldots, n$. Then one can compare the resulting expression of $y(t + \tau)$ from Eq. (3.15) with the expansion in Eq. (3.14) term by term. A set of equations for α_i and v_{ij} is obtained by keeping the coefficients for the terms with the same power of τ on both sides equal. By truncating the expansion to the term of $O(\tau^n)$, one obtains n equations but with $n + n(n-1)/2$ parameters (α_i and v_{ij}) to be determined. Thus, there are still options in choosing the parameters.

Let us illustrate this scheme by working out the case for $n = 2$ in detail. If only the terms up to $O(\tau^2)$ are kept in Eq. (3.14), we have

$$y(t + \tau) = y + \tau g + \frac{\tau^2}{2} (g_t + g g_y). \tag{3.20}$$

We can obtain an expansion up to the same order by truncating Eq. (3.15) at $n = 2$,

$$y(t + \tau) = y(t) + \alpha_1 k_1 + \alpha_2 k_2, \tag{3.21}$$

with

$$k_1 = \tau g(y, t), \tag{3.22}$$

$$k_2 = \tau g(y + v_{21}k_1, t + v_{21}\tau). \tag{3.23}$$

Now if we perform a Taylor expansion for k_2 up to the term of $O(\tau^2)$, we have

$$k_2 = \tau g + v_{21}\tau^2(g_t + gg_y). \tag{3.24}$$

Substituting k_1 and k_2 into Eq. (3.21) yields

$$y(t + \tau) = y(t) + (\alpha_1 + \alpha_2)\tau g + \alpha_2\tau^2 v_{21}(g_t + gg_y). \tag{3.25}$$

If we compare this expression with that of Eq. (3.20) term by term, we have

$$\alpha_1 + \alpha_2 = 1, \tag{3.26}$$

$$\alpha_2 v_{21} = \frac{1}{2}. \tag{3.27}$$

As pointed out earlier, there are only n (2 in this case) equations available but $n + n(n - 1)/2$ (3 in this case) parameters to be determined. We do not have a unique solution for all the parameters; thus we have flexibility in assigning their values as long as they satisfy the n equations. One could choose $\alpha_1 = \alpha_2 = 1/2$ and $v_{21} = 1$, or $\alpha_1 = 1/3, \alpha_2 = 2/3$, and $v_{21} = 3/4$. The flexibility in choosing the parameters provides one more way to increase the numerical accuracy in practice. One can adjust the parameters according to the specific problems involved.

The most commonly used Runge–Kutta method is the one with Eqs. (3.14) and (3.15) truncated at the terms of $O(\tau^4)$. We will give the result here and leave the derivation as an exercise to the readers. This well-known fourth-order Runge–Kutta algorithm is given by

$$y(t + \tau) = y(t) + \frac{1}{6}(k_1 + k_2 + k_3 + k_4), \tag{3.28}$$

where

$$k_1 = \tau g(y, t), \tag{3.29}$$

$$k_2 = \tau g\left(y + \frac{k_1}{2}, t + \frac{\tau}{2}\right), \tag{3.30}$$

$$k_3 = \tau g\left(y + \frac{k_2}{2}, t + \frac{\tau}{2}\right), \tag{3.31}$$

$$k_4 = \tau g(y + k_3, t + \tau). \tag{3.32}$$

One can easily show that the above selection of parameters does satisfy the required equations. As pointed out earlier in this section, this selection is not unique and can be modified according to the problem under study.

3.5 Chaotic dynamics of a driven pendulum

Before discussing numerical methods for solving boundary-value and eigenvalue problems, let us apply the numerical methods we have been discussing so far for solving initial-value problems to a simple example of nonlinear dynamical systems.

Consider a pendulum consisting of a light rod of length l and a point mass m attached to the end. Assume that the pendulum is confined to a vertical plane, acted upon by a driving force f_d and a resistive force f_r. The motion of the pendulum is described by Newton's equation along the tangential direction of the circular motion of the point mass,

$$ma = f_g + f_d + f_r, \tag{3.33}$$

where $f_g = -mg \sin \theta$ is the contribution of gravity along the direction of motion, with θ as the angle made by the rod with respect to the vertical line and $a = l d^2\theta/dt^2$ is the acceleration along the tangential direction. Assume that the time dependency of the driving force is periodic with

$$f_d(t) = f_d^0 \cos \omega_0 t, \tag{3.34}$$

where f_d^0 is the amplitude and ω_0 is the angular frequency of the driving force. The resistive force is taken as a linear function of the velocity, $f_r = -kv$, with $v = l d\theta/dt$ being the velocity and k being a parameter. This is a reasonable assumption for a particle moving in a dense medium. If we rewrite Eq. (3.33) into a dimensionless form with $\sqrt{l/g}$ being the time unit, we have

$$\frac{d^2\theta}{dt^2} + q\frac{d\theta}{dt} + \sin\theta = b\cos\omega_0 t, \tag{3.35}$$

where $q = k/m$ and $b = f_d^0/ml$ are redefined parameters. As discussed at the beginning of this chapter, we can rewrite the derivatives into variables. We thus can transform higher-order differential equations into a set of first-order differential equations. If we choose $y_1 = \theta$ and $y_2 = \omega = d\theta/dt$, we have

$$\frac{dy_1}{dt} = y_2, \tag{3.36}$$

$$\frac{dy_2}{dt} = -qy_2 - \sin y_1 + b\cos\omega_0 t, \tag{3.37}$$

which are in the form of Eq. (3.1). In principle, we can use any method discussed so far to solve this equation. However, considering the accuracy required for long-time behavior, we will use the fourth-order Runge–Kutta method in this case.

As we will show later from the numerical solutions of Eqs. (3.36) and (3.37), in different regions of the parameter space (q, b, ω_0), the system has quite different dynamics. Specifically, in some parameter regions the motion of the pendulum is totally chaotic.

If we generalize the fourth-order Runge–Kutta method discussed in the preceding section to multivariable cases, we have

$$\mathbf{y}_{n+1} = \mathbf{y}_n + \frac{1}{6}(\mathbf{k}_1 + 2\mathbf{k}_2 + 2\mathbf{k}_3 + \mathbf{k}_4), \tag{3.38}$$

with

$$\mathbf{k}_1 = \tau \mathbf{g}(\mathbf{y}_n, t_n), \tag{3.39}$$

$$\mathbf{k}_2 = \tau \mathbf{g}\left(\mathbf{y}_n + \frac{\mathbf{k}_1}{2}, t_n + \frac{\tau}{2}\right), \tag{3.40}$$

$$\mathbf{k}_3 = \tau \mathbf{g}\left(\mathbf{y}_n + \frac{\mathbf{k}_2}{2}, t_n + \frac{\tau}{2}\right), \tag{3.41}$$

$$\mathbf{k}_4 = \tau \mathbf{g}(\mathbf{y}_n + \mathbf{k}_3, t_n + \tau), \tag{3.42}$$

where $\mathbf{y}_n$ for any n and $\mathbf{k}_l$ for $l = 1, 2, 3, 4$ are multidimensional vectors. As one can see, generalizing an algorithm from the single-variable case to the multivariable case is straightforward. Other algorithms we have discussed can be generalized in exactly the same fashion. In the pendulum problem we have two dynamical variables, the angle between the rod and the vertical line, θ, and its first-order derivative $\omega = d\theta/dt$. Any physical quantities that are functions of θ are periodic. For example, $\omega(\theta) = \omega(\theta \pm 2n\pi)$, where n is an integer. Therefore, we can confine θ in the region $[-\pi, \pi]$. If θ is out of this region, it can be transformed back with $\theta' = \theta \pm 2n\pi$ without losing any generality. The following program is an implementation of the fourth-order Runge–Kutta algorithm as applied to the driven pendulum under damping.

```
            PROGRAM PENDULUM
      C
      C Main program for a driven pendulum under damping
      C solved with the fourth-order Runge-Kutta algorithm.
      C Parameters: Q, B, and W (omega_0).
      C
            PARAMETER (N=1000,L=100,M=1)
            DIMENSION Y(2,N)
            COMMON /CONST/ Q,B,W
      C
            PI = 4.0*ATAN(1.0)
            H  = 3.0*PI/L
            Q  = 0.5
```

```
      B  = 0.9
      W  = 2.0/3.0
      Y(1,1) = 0.0
      Y(2,1) = 2.0
C
C Using the Runge-Kutta algorithm to integrate the
C equation
C
      DO      100  I = 1, N-1
        T  = H*I
        Y1 = Y(1,I)
        Y2 = Y(2,I)
        DK11 = H*G1(Y1,Y2,T)
        DK21 = H*G2(Y1,Y2,T)
        DK12 = H*G1((Y1+DK11/2.0),(Y2+DK21/2.0),
     *            (T+H/2.0))
        DK22 = H*G2((Y1+DK11/2.0),(Y2+DK21/2.0),
     *            (T+H/2.0))
        DK13 = H*G1((Y1+DK12/2.0),(Y2+DK22/2.0),
     *            (T+H/2.0))
        DK23 = H*G2((Y1+DK12/2.0),(Y2+DK22/2.0),
     *            (T+H/2.0))
        DK14 = H*G1((Y1+DK13),(Y2+DK23),(T+H))
        DK24 = H*G2((Y1+DK13),(Y2+DK23),(T+H))
        Y(1,I+1) = Y(1,I)+(DK11+2.0*(DK12+DK13)
     *                  +DK14)/6.0
        Y(2,I+1) = Y(2,I)+(DK21+2.0*(DK22+DK23)
     *                  +DK24)/6.0
C
C Bring theta back to the region [-pi,pi]
C
        IF(ABS(Y(1,I+1)).GT.PI) THEN
          Y(1,I+1) = Y(1,I+1)-2.0*PI*ABS(Y(1,I+1))/
     *            Y(1,I+1)
        END IF
  100 CONTINUE
C
      WRITE (6,999) (Y(1,I),Y(2,I),I=1,N,M)
      STOP
  999 FORMAT (2F16.8)
```

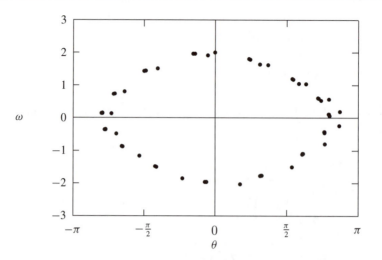

Fig. 3.2. The angular velocity ω versus the angle θ. The points are generated from the program PENDULUM with the parameters $\omega_0 = 2/3$, $q = 0.5$, and $b = 0.9$. The system under the given condition is apparently periodic. 1,000 points from 10,000 time steps are shown here.

```
        END
C

        FUNCTION G1(Y1,Y2,T)
        COMMON /CONST/ Q,B,W
          G1 = Y2
        RETURN
        END
C

        FUNCTION G2(Y1,Y2,T)
        COMMON /CONST/ Q,B,W
          G2 = -Q*Y2-SIN(Y1)+B*COS(W*T)
        RETURN
        END
```

In Figs. 3.2 and 3.3, we show two typical numerical results. The dynamical behavior of the pendulum shown in Fig. 3.2 is periodic in the selected parameter region, and the dynamical behavior shown in Fig. 3.3 is chaotic in another parameter region. One can modify the program developed here to explore the dynamics of the pendulum through the whole parameter space and many important aspects of chaos. Interested readers can find discussions on these aspects in Baker and Gollub (1990).

Several interesting features have appeared in the results shown in Figs. 3.2 and 3.3. In Fig. 3.2, the motion of the system is periodic, with a period $T = 2T_0$,

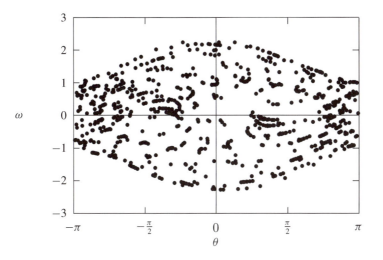

Fig. 3.3. The same plot as in Fig. 3.2, with $\omega_0 = 2/3$, $q = 0.5$, and $b = 1.15$. The system at this point of the parameter space is apparently chaotic. 1,000 points from 10,000 time steps are shown here.

where T_0 is the period of the driving force. If one explored other parameter regions, one would find other periodic motions with $T = nT_0$, where n is an even integer. The reason that n is an even integer is actually due to the bifurcation behavior of the system. The dynamical behavior shown in Fig. 3.3 appears to be totally irregular; however, detailed analysis shows that the phase-space diagram (ω–θ plot) has self-similarity at all length scales, as indicated by fractal structures.

3.6 Boundary-value and eigenvalue problems

Another class of problems in physics requires knowledge of solutions to differential equations with values of physical quantities or their derivatives given at the boundaries of a specified region. This applies to the solution of the Poisson equation with a given charge distribution and known boundary values of the electrostatic potential or of the stationary Schrödinger equation with a given potential and boundary conditions.

A typical boundary-value problem in physics is usually given in a second-order differential equation

$$u'' = f(u, u'; x), \tag{3.43}$$

where u is a function of x, u' and u'' are the first-order and second-order derivatives of u with respect to x, and $f(u, u'; x)$ is a function of u, u', and x. Either u or u' is given at each boundary. Note that one can always choose a coordinate system so that the boundaries of the system are at $x = 0$ and $x = 1$ without losing any generality.

For example, if the actual boundaries are at $x = x_1$ and $x = x_2$ for a given problem, one can always bring them back to $x' = 0$ and $x' = 1$ with a transformation

$$x' = (x - x_1)/(x_2 - x_1). \tag{3.44}$$

For problems in one dimension, one can have a total of four possible types of boundary conditions:

(1) $u(0) = u_0$ and $u(1) = u_1$;
(2) $u(0) = u_0$ and $u'(1) = v_1$;
(3) $u'(0) = v_0$ and $u(1) = u_1$;
(4) $u'(0) = v_0$ and $u'(1) = v_1$.

The boundary-value problem is more difficult to solve than the similar initial-value problem with the same differential equation. For example, if we want to solve

$$u'' = f(u, u'; x), \tag{3.45}$$

with the initial conditions $u(0) = u_0$ and $u'(0) = v_0$, we can first transform the differential equation into a set of first-order differential equations with a redefinition of the first-order derivative into a new variable. The solution will follow if we adopt one of the algorithms discussed earlier in this chapter. However, for the boundary-value problem, we know only $u(0)$ or $u'(0)$, which is not sufficient to start any algorithms for the initial-value problems without any further work.

Typical eigenvalue problems are even more complicated, because at least one more parameter, that is, the eigenvalue, is involved in the equation. For example,

$$u'' = f(u, u'; x; \lambda), \tag{3.46}$$

with a set of given boundary conditions, defines an eigenvalue problem. Here the eigenvalue λ can have only some selected values in order to yield acceptable solutions of the equation under the given boundary conditions.

Let us take the longitudinal vibrations along an elastic rod as an example here. The equation describing the stationary solution of elastic waves is given by

$$u'' = -k^2 u, \tag{3.47}$$

where $u(x)$ is the displacement from the equilibrium at x and the allowed values of the wavevector k are the eigenvalues of the problem. The wavevector in the equation is related to the sound speed c along the rod and the allowed angular frequency ω by the dispersion relation

$$\omega = ck. \tag{3.48}$$

If both ends ($x = 0$, $x = 1$) of the rod are fixed, the boundary conditions are then $u(0) = u(1) = 0$. If one end ($x = 0$) is fixed and the other end ($x = 1$) is free, the boundary conditions are then $u(0) = 0$ and $u'(1) = 0$. For this problem, one can obtain an analytical solution. For example, if both ends of the rod are fixed, one will have the eigenfunctions

$$u_n(x) = \sqrt{2} \sin k_n x \tag{3.49}$$

as possible solutions of the differential equation. Here the eigenvalues are given by

$$k_n^2 = (n\pi)^2, \tag{3.50}$$

where $n = \pm 1, \pm 2, \ldots, \pm \infty$. The complete solution of the longitudinal waves along the elastic rod is given by a linear combination of all the eigenfunctions with their associated initial solutions,

$$u(x, t) = \sum_{n=-\infty}^{\infty} c_n u_n(x) e^{in\pi ct}, \tag{3.51}$$

where c_n are coefficients to be determined by the initial conditions. We will come back to this problem in Chapter 6 when we discuss the solutions of a partial differential equation.

3.7 The shooting method

A simple method for solving the boundary-value problem of Eq. (3.43) and the eigenvalue problem of Eq. (3.46) with a set of given boundary conditions is the so-called *shooting method*. We will first discuss how it works for the boundary-value problem and then generalize it to the eigenvalue problem.

We first convert the second-order differential equation into two first-order differential equations by defining $y_1 = u$ and $y_2 = u'$; that is,

$$\frac{dy_1}{dx} = y_2, \tag{3.52}$$

$$\frac{dy_2}{dx} = f(y_1, y_2; x), \tag{3.53}$$

with a given set of boundary conditions. To illustrate the method, let us assume that the boundary conditions are $u(0) = u_0$ and $u(1) = u_1$. Any other types of boundary conditions can be solved in a similar manner.

The key here is to make the problem look like an initial-value problem by introducing an adjustable parameter; the solution is then obtained by varying the parameter. Since $u(0)$ is given already, we can make a guess for the first-order derivative at $x = 0$, for example, $u'(0) = \delta$. Here δ is the parameter to be adjusted. For a specific δ, we can integrate the equation to $x = 1$ with any of the algorithms discussed earlier for the initial-value problem. Since the initial choice of δ can hardly be the actual derivative at $x = 0$, the value of the function $u_\delta(1)$, resulting from the integration with $u'(0) = \delta$ to $x = 1$, would not be the same as u_1. The idea of the shooting method is to use one of the root search algorithms to find the appropriate δ that ensures $F(\delta) = u_\delta(1) - u_1 = 0$ within a given tolerance Δ.

Let us take an actual numerical example to illustrate the scheme. Assume that we want to solve the following differential equation:

$$u'' = -\frac{\pi^2}{4}(u + 1), \tag{3.54}$$

with the given boundary conditions $u(0) = 0$ and $u(1) = 1$. We can define the new variables $y_1 = u$ and $y_2 = u'$; then we have

$$\frac{dy_1}{dx} = y_2, \tag{3.55}$$

$$\frac{dy_2}{dx} = -\frac{\pi^2}{4}(y_1 + 1). \tag{3.56}$$

Now assume that this equation set has the initial values $y_1(0) = 0$ and $y_2(0) = \delta$. Here δ is a parameter to be adjusted in order to have $|F(\delta)| = |u_\delta(1) - 1| < \Delta$, with Δ being the tolerance of the solution. We can combine the secant method for root search and the fourth-order Runge–Kutta method for the initial-value problem to solve the above equation. The following program is an implementation of such a combined approach to the boundary-value problem defined in Eq. (3.54) or Eqs. (3.55) and (3.56) with the given boundary conditions.

```
      PROGRAM SHOOTING
C
C Program for the boundary-value problem with the
C shooting method. The Runge-Kutta and secant methods
C are used.
C
      PARAMETER (N=101,KMAX=10)
      DIMENSION Y(2,N)
C
C Initialization of the problem
C
      XL      = 0.0
      XU      = 1.0
      DX      = 0.01
      TOL     = 1.0E-06
```

```
        H       = (XU-XL)/(N-1)
        YL      = 0.0
        YU      = 1.0
        X0      = (YU-YL)/(XU-XL)
        X1      = X0+DX
        Y(1,1) = YL
C
C The secant search for the root
C
        DO 500 K = 1, KMAX
C
C The Runge-Kutta calculation of the first
C trial solution
C
          Y(2,1) = X0
          DO      100  I = 1, N-1
            X  = XL+H*I
            Y1 = Y(1,I)
            Y2 = Y(2,I)
            DK11 = H*G1(Y1,Y2,X)
            DK21 = H*G2(Y1,Y2,X)
            DK12 = H*G1((Y1+DK11/2.0),(Y2+DK21/2.0),
     *               (X+H/2.0))
            DK22 = H*G2((Y1+DK11/2.0),(Y2+DK21/2.0),
                     (X+H/2.0))
            DK13 = H*G1((Y1+DK12/2.0),(Y2+DK22/2.0),
     *               (X+H/2.0))
            DK23 = H*G2((Y1+DK12/2.0),(Y2+DK22/2.0),
     *               (X+H/2.0))
            DK14 = H*G1((Y1+DK13),(Y2+DK23),(X+H))
            DK24 = H*G2((Y1+DK13),(Y2+DK23),(X+H))
            Y(1,I+1) = Y(1,I)+(DK11+2.0*(DK12+DK13)
     *                     +DK14)/6.0
            Y(2,I+1) = Y(2,I)+(DK21+2.0*(DK22+DK23)
     *                     +DK24)/6.0
   100    CONTINUE

          F0 = Y(1,N)-1.0
C
C The Runge-Kutta calculation of the second
```

```
C trial solution
C
        Y(2,1) = X1
        DO      200  I = 1, N-1
          X  = XL+H*I
          Y1 = Y(1,I)
          Y2 = Y(2,I)
          DK11 = H*G1(Y1,Y2,X)
          DK21 = H*G2(Y1,Y2,X)
          DK12 = H*G1((Y1+DK11/2.0),(Y2+DK21/2.0),
     *                (X+H/2.0))
          DK22 = H*G2((Y1+DK11/2.0),(Y2+DK21/2.0),
                      (X+H/2.0))
          DK13 = H*G1((Y1+DK12/2.0),(Y2+DK22/2.0),
     *                (X+H/2.0))
          DK23 = H*G2((Y1+DK12/2.0),(Y2+DK22/2.0),
     *                (X+H/2.0))
          DK14 = H*G1((Y1+DK13),(Y2+DK23),(X+H))
          DK24 = H*G2((Y1+DK13),(Y2+DK23),(X+H))
          Y(1,I+1) = Y(1,I)+(DK11+2.0*(DK12+DK13)
     *                       +DK14)/6.0
          Y(2,I+1) = Y(2,I)+(DK21+2.0*(DK22+DK23)
     *                       +DK24)/6.0
   200    CONTINUE
        F1 = Y(1,N)-1.0
C
        D = F1 - F0
        IF (ABS(D).LT.TOL) THEN
          WRITE (6,999) (H*(I-1),Y(1,I),I=1,N,5)
          STOP
C
C Update by the secant method
C
        ELSE
          X2 = X1-F1*(X1-X0)/D
          X0 = X1
          X1 = X2
        END IF
   500 CONTINUE
C
```

```
  999 FORMAT (2F16.8)
      END
C
      FUNCTION G1(Y1,Y2,T)
        G1 = Y2
      RETURN
      END
C
      FUNCTION G2(Y1,Y2,T)
        PI =   4.0*ATAN(1.0)
        G2 = -PI*PI(Y1+1.0)/4.0
      RETURN
      END
```

The boundary-value problem solved from the above program can also be solved exactly with an analytical solution

$$u(x) = \cos\frac{\pi x}{2} + 2\sin\frac{\pi x}{2} - 1. \tag{3.57}$$

One can easily check that the above expression does satisfy the equation and the boundary conditions. Here we plot both the numerical result obtained from the shooting method and the analytical solution in Fig. 3.4. As one can see, the shooting

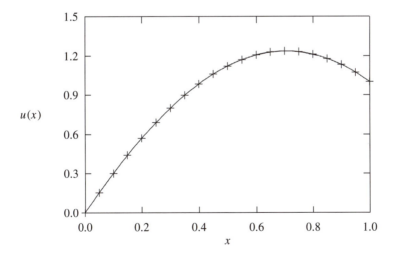

Fig. 3.4. The numerical solution of the boundary-value problem of Eq. (3.54) by the shooting method (+) compared with the analytical solution (solid line) of the same problem.

method provides a very accurate solution of the boundary-value problem. It is also a very general method for the boundary-value problem.

Boundary-value problems with other types of boundary conditions can be solved in a similar manner. For example, if $u'(0) = v_0$ and $u(1) = u_1$ are given, we can make a guess on $u(0) = \delta$ and then integrate the equation set of y_1 and y_2 to $x = 1$. The root to be sought is from $F(\delta) = u_\delta(1) - u_1 = 0$. Here $u_\delta(1)$ is the numerical result of the equation with $u(0) = \delta$. If $u'(1) = v_1$ is given, the equation $F(\delta) = u'_\delta(1) - v_1 = 0$ is solved instead.

One can apply the shooting method to solve the eigenvalue problem, too. The parameter to be adjusted in the eigenvalue problem is no longer a parameter introduced but the eigenvalue of the problem. For example, if $u(0) = u_0$ and $u(1) = u_1$ are given, we can integrate the equation with $u'(0) = \delta$ with δ as a small quantity. Then we search for the root of $F(\lambda) = u_\lambda(1) - u_1 = 0$. When $|F(\lambda)| < \Delta$, we obtain an approximate eigenvalue λ and the corresponding eigenvector from the normalized solution of $u_\lambda(x)$. The introduced parameter δ is not relevant here, because it will be automatically modified to be the first-order derivative when the solution is normalized. In other words, one can choose the first-order derivative at the boundary arbitrarily and it will not affect the results as long as the solutions are made orthonormal.

3.8 Linear equations and the Sturm–Liouville problem

Many eigenvalue or boundary-value problems are in the form of linear equations,

$$u'' + d(x)u' + q(x)u = s(x), \tag{3.58}$$

where $d(x), q(x)$, and $s(x)$ are functions of x. Assume that the boundary conditions are $u(0) = u_0$ and $u(1) = u_1$. If all $d(x), q(x)$, and $s(x)$ are smooth, one can solve the equation with the shooting method developed in the preceding section. In fact, one can show that an extensive search for the parameter δ from $F(\delta) = u_\delta(1) - u_1 = 0$ is not necessary in this case, because of the superposition principle of the linear equation: Any linear combination of the solutions is still a solution of the equation. One needs only two trial solutions $u_{\delta_0}(x)$ and $u_{\delta_1}(x)$, with δ_0 and δ_1 being two different parameters. The correct solution of the equation is given by

$$u(x) = au_{\delta_0}(x) + bu_{\delta_1}(x), \tag{3.59}$$

with a and b determined from $u(0) = u_0$ and $u(1) = u_1$. Note that $u_{\delta_0}(0) = u_{\delta_1}(0) = u(0) = u_0$. So we have

$$a + b = 1, \tag{3.60}$$

$$u_{\delta_0}(1)a + u_{\delta_1}(1)b = u_1, \tag{3.61}$$

which can easily be solved to give

$$a = \frac{u_{\delta_1}(1) - u_1}{u_{\delta_1}(1) - u_{\delta_0}(1)},$$ (3.62)

$$b = \frac{u_1 - u_{\delta_0}(1)}{u_{\delta_1}(1) - u_{\delta_0}(1)}.$$ (3.63)

With a and b given from the above equation, we have the solution of the differential equation from Eq. (3.59).

An important class of linear equations in physics is referred to as the Sturm–Liouville problem, defined by

$$[p(x)u'(x)]' + q(x)u(x) = s(x),$$ (3.64)

which has the first-order derivative term combined with the second-order derivative term. $p(x)$, $q(x)$, and $s(x)$ are the coefficient functions of x. For most actual problems, $s(x) = 0$ and $q(x) = -r(x) + \lambda w(x)$, with λ being the eigenvalue of the equation. $r(x)$ and $w(x)$ are the redefined coefficient functions. The Legendre equation, the Bessel equation, and the related equations in physics are examples of the Sturm–Liouville problem.

Our goal here is to construct an accurate algorithm that can integrate the Sturm–Liouville equation, that is, Eq. (3.64). In Chapter 2, we obtained the three-point formulas for the first-order and second-order derivatives from the combinations

$$\Delta_1 = \frac{u_{n+1} - u_{n-1}}{2h} = u'_n + \frac{h^2 u_n^{(3)}}{6} + O(h^4)$$ (3.65)

and

$$\Delta_2 = \frac{u_{n+1} + u_{n-1} - 2u_n}{h^2} = u''_n + \frac{h^2 u_n^{(4)}}{12} + O(h^4).$$ (3.66)

Now if we multiply Eq. (3.65) by p'_n and Eq. (3.66) by p_n and add them together, we have

$$p'_n \Delta_1 + p_n \Delta_2 = (p_n u'_n)' + \frac{h^2}{12}(p_n u_n^{(4)} + 2p'_n u_n^{(3)}) + O(h^4).$$ (3.67)

If we replace the first term on the right-hand side with $s_n - q_n u_n$ and drop the second term, we obtain the simplest numerical algorithm for the Sturm–Liouville equation,

$$(2p_n + hp'_n)u_{n+1} + (2p_n - hp'_n)u_{n-1} = 4p_n u_n + 2h^2(s_n - q_n u_n),$$ (3.68)

which is accurate up to $O(h^4)$. One has to be very careful with this apparent local accuracy in the algorithm. In reality, because of the repeated use of the three-point formula, the local accuracy delivered is usually lower than $O(h^4)$. Before we discuss how to improve the accuracy of this algorithm, let us illustrate it with an example. The Legendre equation is given by

$$\frac{d}{dx}\left[(1-x^2)\frac{du}{dx}\right] + l(l+1)u = 0, \tag{3.69}$$

with $l = 0, 1, \ldots, \infty$ and $x \in [-1, 1]$. The solutions of the Legendre equation are the Legendre polynomials $P_l(x)$. Let us assume that we do not know the value of l but know the first two points of $P_1(x) = x$; then we can treat the problem as an eigenvalue problem.

The following program is an implementation of the simplest algorithm for the Sturm–Liouville equation, in combination with the bisection method for the root search, to solve for the eigenvalue $l = 1$ of the Legendre equation.

```
        PROGRAM S_L_LEGENDRE
C
C Main program for solving the Legendre equation with
C the simplest algorithm for the Sturm-Liouville
C equation and the bisection method for the root search.
C
        PARAMETER (N=501)
        REAL U(N)
C
C Initialization of the problem
C
        H     =  2.0/(N-1)
        AK    =  0.5
        BK    =  1.5
        DK    =  0.5
        DL    =  1.0E-06
        ISTEP =  0
        U(1)  = -1.0
        U(2)  = -1.0+H
        EK    =  AK
        CALL SMPL(N,H,EK,U)
        FO = U(N)-1.0
C
C Bisection method for the root
C
        DO      200  WHILE (ABS(DK).GT.DL)
          EK = (AK+BK)/2.0
```

```
      CALL SMPL (N,H,EK,U)
      F1 = U(N)-1.0
      IF ((F0*F1).LT.0) THEN
         BK = EK
         DK = BK-AK
      ELSE
         AK = EK
         DK = BK-AK
         F0 = F1
      END IF
      ISTEP = ISTEP+1
  200 END DO
C
      WRITE (6,999) ISTEP,EK,DK,F1
  999 FORMAT (I4,3F16.8)
      STOP
      END
C
      SUBROUTINE SMPL(N,H,EK,U)
C
C The simplest algorithm for the Sturm-Liouville
C equation
C
      REAL U(N)
C
      H2 = 2.0*H*H
      Q = EK*(1.0+EK)
      DO       100  I = 2, N-1
        X   =  (I-1)*H-1.0
        P   =  2.0*(1.0-X*X)
        P1 = -2.0*X*H
        U(I+1) = ((2.0*P-H2*Q)*U(I)+(P1-P)*U(I-1))/
     *            (P1+P)
  100 CONTINUE
      RETURN
      END
```

The eigenvalue obtained from the above program is 1.000091, which contains an error of 9×10^{-5} in comparison with the exact result of $l = 1$. The error comes from both the rounding error and the inaccuracy of the algorithm.

If we want to have higher accuracy in the algorithm for the Sturm–Liouville equation, we can differentiate the equation twice. Then we have

$$pu^{(4)} + 2p'u^{(3)} = s'' - 3p''u'' - p^{(3)}u' - p'u^{(3)} - q''u - 2q'u' - qu'',$$

(3.70)

where $u^{(3)}$ on the right-hand side can be replaced with

$$u^{(3)} = \frac{1}{p}(s' - p''u' - 2p'u'' - q'u - qu'),$$

(3.71)

which is the result of differentiating the Sturm–Liouville equation once. If we combine Eqs. (3.67), (3.70), and (3.71), we obtain a better algorithm

$$c_{n+1}u_{n+1} + c_{n-1}u_{n-1} = c_n u_n + d_n + O(h^6),$$

(3.72)

where c_{n+1}, c_n, c_{n-1}, and d_n are given by

$$c_{n+1} = 24p_n + 12hp'_n + 2h^2 q_n + 6h^2 p''_n - 4h^2 (p'_n)^2/p_n + h^3 p_n^{(3)}$$
$$+ 2h^3 q'_n - h^3 p'_n q_n/p_n - h^3 p'_n p''_n/p_n,$$

(3.73)

$$c_n = 48p_n - 20h^2 q_n - 8h^2 (p'_n)^2/p_n + 12h^2 p''$$
$$+ 2h^4 p'_n q'_n/p_n - 2h^4 q''_n,$$

(3.74)

$$c_{n-1} = 24p_n - 12hp'_n + 2h^2 q_n + 6h^2 p''_n - 4h^2 (p'_n)^2/p_n - h^3 p_n^{(3)}$$
$$- 2h^3 q'_n + h^3 p'_n q_n/p_n + h^3 p'_n p''_n/p_n,$$

(3.75)

$$d_n = 24h^2 s_n + 2h^4 s''_n - 2h^4 p'_n s'_n/p_n,$$

(3.76)

which can be evaluated easily if $p(x)$, $q(x)$, and $s(x)$ are explicitly given. In the case where some of the derivatives that are needed are not easily obtained analytically, one can evaluate them numerically. In order to maintain the high accuracy of the algorithm, one needs to use comparable numerical formulas.

For the special case with $p(x) = 1$, the above coefficients reduce to much simpler forms. Without sacrificing the high accuracy of the algorithm, we can apply the three-point formulas to the first-order and second-order derivatives of $q(x)$ and $s(x)$. Then we have

$$c_{n+1} = 1 + \frac{h^2}{12}\left(\frac{1}{2}q_{n+1} + q_n - \frac{1}{2}q_{n-1}\right),$$

(3.77)

$$c_n = 2 - \frac{5h^2}{6}\left(\frac{1}{10}q_{n+1} + \frac{4}{5}q_n + \frac{1}{10}q_{n-1}\right),$$

(3.78)

$$c_{n-1} = 1 + \frac{h^2}{12}\left(-\frac{1}{2}q_{n+1} + q_n + \frac{1}{2}q_{n-1}\right),$$

(3.79)

$$d_n = \frac{h^2}{12}(s_{n+1} + 10s_n + s_{n-1}),$$

(3.80)

which are slightly different from the Numerov algorithm, which is an extremely accurate scheme for linear differential equations without the first-order derivative term, that is, Eq. (3.58) with $d(x) = 0$ or the Sturm–Liouville equation with $p(x) = 1$. Many equations in physics have this form, for example, the Poisson equation with spherical symmetry or the one-dimensional Schrödinger equation.

The Numerov algorithm is derived from the three-point formula for the second-order derivative from Eq. (3.66) with the fourth-order derivative of $u(x)$ from taking the second-order derivative of the differential equation,

$$u^{(4)}(x) = \frac{d^2}{dx^2}[-q(x)u(x) + s(x)].$$ (3.81)

If we apply the three-point formula to the above equation by keeping all the terms on the right-hand side together, then we obtain the Numerov algorithm (Koonin, 1986, pp. 50–1) with the recursion of Eq. (3.72) and the coefficients given by

$$c_{n+1} = 1 + \frac{h^2}{12}q_{n+1},$$ (3.82)

$$c_n = 2 - \frac{5h^2}{6}q_n,$$ (3.83)

$$c_{n-1} = 1 + \frac{h^2}{12}q_{n-1},$$ (3.84)

$$d_n = \frac{h^2}{12}(s_{n+1} + 10s_n + s_{n-1}).$$ (3.85)

Note that even though the apparent local accuracy of the Numerov algorithm and the algorithm we derived earlier in this section for the Sturm–Liouville equation is $O(h^6)$, the actual global accuracy of the algorithm is only $O(h^4)$ because of the repeated use of the three-point formulas. For more discussion on this issue, see Simos (1993). The algorithms discussed here can be applied to initial-value problems as well as to boundary-value or eigenvalue problems. The Numerov algorithm and the algorithm for the Sturm–Liouville equation usually have lower accuracy than the fourth-order Runge–Kutta algorithm when applied to the same problem, and this is different from what the apparent local accuracies imply. For more numerical examples of the Sturm–Liouville equation and the Numerov algorithm in the related problems, see Pryce (1993) and Onodera (1994).

3.9 The one-dimensional Schrödinger equation

The solutions associated with the one-dimensional Schrödinger equation are of importance in understanding quantum mechanics and quantum processes. For example, the energy levels of electrons in a superlattice of GaAs/Ga$_{1-x}$Al$_x$As and the electron transport properties of semiconductor devices are essential for microlaser device design. In this section, we will apply the numerical methods we have developed so far to solve both the eigenvalue and transport problems defined through the one-dimensional Schrödinger equation

$$-\frac{\hbar^2}{2m}\frac{d^2\phi(x)}{dx^2} + V(x)\phi(x) = E\phi(x), \tag{3.86}$$

where m is the mass of the particle, $\hbar$ is the Planck constant, E is the energy, $\phi(x)$ is the wavefunction, and $V(x)$ is the external potential. We can rewrite the Schrödinger equation into

$$\phi''(x) + \frac{2m}{\hbar^2}[E - V(x)]\phi(x) = 0, \tag{3.87}$$

which is in the same form as the Sturm–Liouville equation with $p(x) = 1$, $q(x) = 2m[E - V(x)]/\hbar^2$, and $s(x) = 0$.

The eigenvalue problem

For the eigenvalue problem, the particle is confined by the potential well $V(x)$, so that $\phi(x) \to 0$ with $|x| \to \infty$. A sketch of a typical $V(x)$ is shown in Fig. 3.5. In order to solve this eigenvalue problem, one can integrate the equation with the

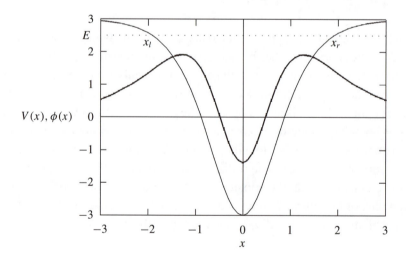

Fig. 3.5. The eigenvalue problem of the one-dimensional Schrödinger equation. Here a potential well $V(x)$ (solid thin line) and the corresponding eigenvalue E (dotted line) and eigenfunction (solid thick line) are illustrated. Turning points x_l and x_r are indicated, too.

Numerov algorithm from left to right or from right to left of the potential region. However, since the wavefunction goes to zero as $|x| \to \infty$, the integration will require integrating from an exponentially increasing region to an oscillatory region and then into an exponentially decreasing region. This is illustrated in Fig. 3.5. The error accumulated will become significant if one integrates the solution from the oscillatory region into the exponentially decreasing region. The reason is that an exponentially increasing solution is also a possible solution of the equation and can easily enter the numerical integration to destroy the accuracy of the algorithm. The rule of thumb is to avoid integrating into the exponential regions, that is, to carry out the solutions from both sides and then match them in the well region. Usually the matching is done at one of the turning points, where the energy is equal to the potential energy, such as x_l and x_r in Fig. 3.5. The so-called matching here is to adjust the trial eigenvalue until the solution integrated from the right, $\phi_r(x)$, and the solution integrated from the left, $\phi_l(x)$, satisfy the continuity conditions at one of the turning points. If we choose the right turning point as the matching point, the continuity conditions are

$$\phi_l(x_r) = \phi_r(x_r), \tag{3.88}$$

$$\phi'_l(x_r) = \phi'_r(x_r). \tag{3.89}$$

If we combine these two conditions, we have

$$\frac{\phi'_l(x_r)}{\phi_l(x_r)} = \frac{\phi'_r(x_r)}{\phi_r(x_r)}. \tag{3.90}$$

If we use the three-point formula for the first-order derivatives in the above equation, we have

$$F(E) = \frac{[\phi_l(x_r + h) - \phi_l(x_r - h)] - [\phi_r(x_r + h) - \phi_r(x_r - h)]}{2h\phi(x_r)} = 0, \tag{3.91}$$

which can be ensured by a root search scheme. Note that $F(E)$ is a function of E only because $\phi_l(x_r) = \phi_r(x_r)$ can be used to rescale the wavefunctions.

Here we outline the numerical procedure for solving the eigenvalue problem of the one-dimensional Schrödinger equation and leave the programming details to the readers as an exercise.

(1) Choose the region of the numerical solution. This region should be large enough compared to the effective region of the potential to have negligible effect on the solution.

(2) Provide a reasonable guess for the lowest eigenvalue E_0.

(3) Integrate the equation for $\phi_l(x)$ from the left to the point $x_r + h$ and for $\phi_r(x)$ from the right to $x_r - h$. One can choose zero as the value of the

first point $\phi_l(x)$ and $\phi_r(x)$ and small quantity as the value of the second point $\phi_l(x)$ and $\phi_r(x)$ to start the integration, for example, with the Numerov algorithm. Before matching the solutions, rescale one of them to ensure that $\phi_l(x_r) = \phi_r(x_r)$. For example, one can multiply $\phi_l(x)$ with $\phi_r(x_r)/\phi_l(x_r)$ up to $x = x_r + h$. This rescaling will also ensure that the solutions have the correct node structure, that is, changing the sign of $\phi_l(x)$ if it is incorrect.

(4) Evaluate $F(E_0) = [\phi_r(x_r - h) - \phi_r(x_r + h) - \phi_l(x_r - h) + \phi_l(x_r + h)]/ 2h\phi_r(x_r)$.

(5) Apply a root search method to obtain E_0 from $F(E_0) = 0$ within a given tolerance.

(6) Carry out the above steps for the next eigenvalue. One can start the search with a slightly higher value than the last eigenvalue. One needs to make sure that no eigenstate is missed. This can easily be done after one finds the second eigenstate and then performs a few more calculations between.

Now let us see an actual example with a particle bound in a potential well

$$V(x) = \frac{\hbar^2}{2m}\alpha^2\lambda(\lambda - 1)\left[\frac{1}{2} - \frac{1}{\cosh^2(\alpha x)}\right], \tag{3.92}$$

where both α and λ are given parameters. The Schrödinger equation with this potential can be solved exactly with the eigenvalues

$$E_n = \frac{\hbar^2}{2m}\alpha^2[\lambda(\lambda - 1)/2 - (\lambda - 1 - n)^2], \tag{3.93}$$

with $n = 0, 1, 2, \ldots$. We have solved this problem numerically in the region $[-5, 5]$ with 501 points uniformly spaced. The potential well, eigenvalue, and eigenfunction shown in Fig. 3.5 are from this problem with $\alpha = 1$, $\lambda = 4$, and $n = 3$. We have also used $\hbar = m = 1$ in the numerical solution for convenience.

Quantum scattering

Now let us turn to the problem of unbound states, that is, the scattering problem. Assume that the potential is not zero in the region $[0, a]$ and that the incident particle comes from the left. We can write the general solution of the Schrödinger equation outside the potential region as

$$\phi(x) = \begin{cases} \phi_1(x) = e^{ikx} + Ae^{-ikx} & \text{for } x < 0, \\ \phi_3(x) = Be^{ik(x-a)} & \text{for } x > a, \end{cases} \tag{3.94}$$

where A and B are the parameters to be determined and k can be found from

$$E = \frac{\hbar^2 k^2}{2m}, \tag{3.95}$$

with E being the energy of the incident particle. The solution in the region $[0, a]$, $\phi_2(x)$, can be obtained numerically. During the process of solving $\phi_2(x)$, one will also obtain A and B, which are necessary for calculating the reflectivity $|A|^2$ and transmissivity $|B|^2$. The boundary conditions at $x = 0$ and $x = a$ are

$$\phi_1(0) = \phi_2(0), \tag{3.96}$$

$$\phi_2(a) = \phi_3(a), \tag{3.97}$$

$$\phi_1'(0) = \phi_2'(0), \tag{3.98}$$

$$\phi_2'(a) = \phi_3'(a), \tag{3.99}$$

which give

$$\phi_2(0) = 1 + A, \tag{3.100}$$

$$\phi_2(a) = B, \tag{3.101}$$

$$\phi_2'(0) = ik(1 - A), \tag{3.102}$$

$$\phi_2'(a) = ikB. \tag{3.103}$$

One should note that the wavefunction is now a complex function, as are the parameters A and B. We would like to outline here a combined numerical scheme that utilizes either the Numerov or the Runge–Kutta method to integrate the equation and a minimization scheme to adjust the solution to the desired accuracy. The minimization schemes for multivariable functions will be discussed in Chapter 4. Here we first outline the scheme through the Numerov algorithm.

(1) For a given particle energy $E = \hbar^2 k^2/2m$, guess a complex parameter $A = A_r + i A_i$. One can use the analytical results for a square potential that has the same range and average strength of the given potential as the initial guess. Since the convergence is very fast, the initial guess is not important.

(2) Perform the Numerov integration of the Schrödinger equation

$$\phi_2''(x) + \left[k^2 - \frac{2m}{\hbar^2} V(x)\right]\phi_2(x) = 0 \tag{3.104}$$

from $x = 0$ to $x = a$ with the second point given by the Taylor expansion of $\phi_2(x)$ around $x = 0$ to the second order,

$$\phi_2(h) = 1 + A + ik(1 - A)h - k^2(1 + A)h^2/2, \tag{3.105}$$

where h is the discrete interval size. Truncation at the first order would also work, but with a little less accuracy.

(3) One can then obtain the approximation for B after the first integration to the point $x = a$,

$$B = \phi_2(a). \tag{3.106}$$

(4) Using this B value, one can integrate the equation from $x = a$ to $x = 0$ with the same Numerov algorithm with the second point given by the Taylor expansion of $\phi_2(x)$ around $x = a$,

$$\phi_2(a - h) = \phi_2(a) - ik\phi_2(a)h - k^2\phi_2(a)h^2/2. \tag{3.107}$$

(5) From the backward integration, we can obtain a new $A^{\text{new}} = A_r^{\text{new}} + i A_i^{\text{new}}$ from $\phi_2^{\text{new}}(0) = 1 + A^{\text{new}}$, where the indices r and i are used for the real and imaginary parts of A. One can then construct a real function $f(A_r, A_i) = (A_r - A_r^{\text{new}})^2 + (A_i - A_i^{\text{new}})^2$. Note that A_r^{new} and A_i^{new} are implicit functions of (A_r, A_i).

(6) Now the problem becomes an optimization problem of minimizing $f(A_r, A_i)$ with A_r and A_i varied. We will discuss such optimization problems in Chapter 4, because they are related to matrix operations. Here we just want to point out that as soon as one has an optimization scheme, the problem becomes very simple.

Now let us illustrate it with an actual example. Assume that we are interested in the quantum scattering of a double barrier potential

$$V(x) = \begin{cases} V_0 & \text{for } 0 \le x \le x_1; \ x_2 \le x \le a, \\ 0 & \text{otherwise.} \end{cases} \tag{3.108}$$

This is a very interesting problem, because it is one of the basic elements conceived for the next generation of electronic devices. A program can easily be completed with a combination of the Numerov algorithm and any optimization scheme. We have performed numerical studies of the above model with the electron effective mass taken as the value for the GaAs/Ga$_{1-x}$Al$_x$As structures. In Fig. 3.6, we plot the transmissivity of the electron obtained for $V_0 = 0.3\,\text{eV}$, $x_1 = 25\,\text{Å}$, $x_2 = 75\,\text{Å}$, and $a = 125\,\text{Å}$. One can see that there is a significant increase in the transmissivity around a resonance energy $E \approx 0.09\,\text{eV}$, which is a virtual energy level. The second peak appears at an energy slightly above the barriers. The study of the symmetric barrier case (Pang, 1995) shows that the transmissivity can reach at least 0.999997 at the resonance energy $E \simeq 0.089535\,\text{eV}$ for $x_1 = 50\,\text{Å}$, $x_2 = 100\,\text{Å}$, and $a = 150\,\text{Å}$. One can also use the Runge–Kutta algorithm to integrate the equation if we choose $y_1(x) = \phi_2(x)$ and $y_2(x) = \phi_2'(x)$. Using the initial condition for $\phi_2(x)$ and $\phi_2'(x)$ at $x = 0$, one can integrate the equation to $x = a$. From $y_1(a)$ and

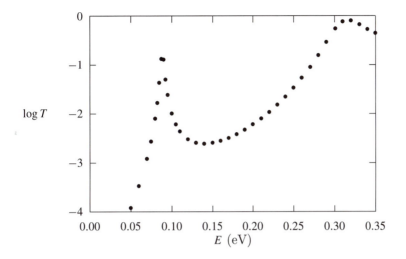

Fig. 3.6. The energy dependence of the transmissivity for a double barrier potential with the barrier height 0.3 eV, barrier widths 25 and 50 Å, and the width of the well between the barriers 50 Å. The transmissivity is plotted on a logarithmic scale.

$y_2(a)$, one can obtain two different values of B,

$$B_1 = y_1(a), \tag{3.109}$$

$$B_2 = -\frac{i}{k}y_2(a), \tag{3.110}$$

which are implicit functions of the initial guess of $A = A_r + iA_i$. This means one can construct an implicit function $f(A_r, A_i) = |B_1 - B_2|^2$ and then optimize it. Note that both B_1 and B_2 are complex, as are the functions $y_1(x)$ and $y_2(x)$. The Runge–Kutta algorithm in this case is much more accurate, because no approximation for the second point is needed. For more details on the application of the Runge–Kutta algorithm and the optimization method in the scattering problem, see Pang (1995).

The procedure can be simplified in the simple potential case, as suggested by R. Zimmermann (personal communication), if one realizes that the Schrödinger equation in question is a linear equation. One can take $B = 1$ and then integrate the equation from $x = a$ back to $x = -h$ with either the Numerov or the Runge–Kutta scheme. The solution at $x = 0$ and $x = -h$ satisfies

$$\phi(x) = A_1 e^{ikx} + A_2 e^{-ikx}, \tag{3.111}$$

and we can solve for A_1 and A_2 with the numerical results of $\phi(0)$ and $\phi(-h)$. The reflectivity and transmissivity are then given by $R = |A_2/A_1|^2$ and $T = 1/|A_1|^2$, respectively. Note that it is not necessary now to have the minimization in the scheme. However, for a more general potential, for example, a nonlinear potential, a combination of the Runge–Kutta integration scheme and an optimization scheme is necessary.

Exercises

3.1 Derive the fourth-order Runge–Kutta algorithm for solving a differential equation

$$\frac{dy}{dt} = f(y, t)$$

with given initial conditions. Discuss the options in the selection of the coefficients.

3.2 Construct a subroutine that solves a differential equation

$$\frac{d\mathbf{y}}{dt} = \mathbf{f}(\mathbf{y}, t)$$

with the fourth-order Runge–Kutta method. Here the dimensions of the variable $\mathbf{y}$ is treated as an input parameter and the initial conditions $\mathbf{y}(0) = \mathbf{y}_0$ are also the input of the subroutine.

3.3 Modify the program given in Section 3.5 for the driven pendulum to obtain a bifurcation diagram (ω–b plot at a fixed θ) with $q = 1/2$ and $\omega_0 = 2/3$ and in the region of $b \in [1, 1.5]$.

3.4 The Duffing model is given by

$$\frac{d^2x}{dt^2} + g\frac{dx}{dt} + x^3 = b\cos t,$$

which is clearly nonlinear. Write a program to solve the Duffing model in different parameter region of g and b. Discuss the behavior of the system from the phase diagram of (x, v) with $v = dx/dt$. Is there any parameter region in which the system is chaotic?

3.5 Apply the Numerov algorithm to solve

$$y'' = -4\pi^2 y,$$

with $y(0) = 1$ and $y'(0) = 0$. Discuss the accuracy of the result by comparing it with the solution obtained via the fourth-order Runge–Kutta algorithm and with the exact result.

3.6 Develop a program that applies the Numerov algorithm and the secant method to solve the one-dimensional Schrödinger equation. Find the two lowest eigenvalues and eigenfunctions for an electron in the potential well

$$V(x) = \begin{cases} V_0 x/x_0 & \text{for } 0 < x < x_0, \\ V_0 & \text{otherwise.} \end{cases}$$

Atomic units (a.u.), that is, $m = e = \hbar = c = 1$, $x_0 = 5$ a.u., and $V_0 = 50$ a.u. can be used.

3.7 Following the procedure discussed in Section 3.9, apply the Numerov algorithm to solve the quantum scattering problem in one dimension. The optimization can be achieved by the bisection method through varying A_r and A_i, respectively, within the region $[-1, 1]$. Test the program with the potential given in the text.

3.8 Find the angle dependence of the angular velocity and the center-of-mass velocity of a meterstick falling to a horizontal position if it is released from an initial position with an angle $\theta_0 \in [0, \pi/2]$ to the horizontal on a surface having a kinetic friction coefficient $0 \leq \mu \leq 0.9$.

3.9 Implement the algorithm for the Sturm–Liouville equation derived in Section 3.8 in a subroutine. Test the subroutine by applying it to the spherical Bessel equation

$$(x^2 y')' + [x^2 - l(l + 1)]y = 0,$$

where l is a positive integer. Use the known analytic solutions to set up the first two points. Is the anticipated global accuracy $O(h^5)$ [from the local accuracy $O(h^6)$ in the algorithm] delivered?

4

Numerical methods for matrices

Matrix operations are necessary in many numerical and analytical problems. Schemes developed to solve matrix problems can be applied to related problems in ordinary and partial differential equations. For example, an eigenvalue problem given in the form of a partial differential equation can be rewritten as a matrix problem. A boundary-value problem after discretization is essentially a linear algebra problem.

4.1 Matrices in physics

There are a lot of problems in physics that can be formulated in a matrix form. Here we would like to give a few examples to illustrate the importance of matrix operations and related numerical schemes.

If one wants to study the vibrational spectrum of a molecule with n degrees of vibrational freedom, the first approach is to investigate the harmonic oscillations of the system, that is, to expand the potential energy up to the second order of the generalized coordinates around the equilibrium structure,

$$U(q_1, q_2, \ldots, q_n) \simeq \frac{1}{2} \sum_{j,k}^{n} A_{jk} q_j q_k, \tag{4.1}$$

where q_j or q_k are the generalized coordinates and A_{jk} are the potential parameters that can be obtained through, for example, a quantum chemistry calculation. We have taken the equilibrium potential as the zero point of the potential energy. Usually the kinetic energy of the system is in the quadratic form of the generalized velocities,

$$T(\dot{q}_1, \dot{q}_2, \ldots, \dot{q}_n) \simeq \frac{1}{2} \sum_{j,k}^{n} M_{jk} \dot{q}_j \dot{q}_k, \tag{4.2}$$

where M_{jk} are the elements of the generalized mass matrix, which depend upon the specifics of the molecule. Now if we apply the Lagrange equation

$$\frac{\partial \mathcal{L}}{\partial q_j} - \frac{d}{dt}\frac{\partial \mathcal{L}}{\partial \dot{q}_j} = 0, \tag{4.3}$$

with $\mathcal{L} = T - U$ being the Lagrangian of the system, we have

$$\sum_{j=1}^{n}(A_{jk}q_j + M_{jk}\ddot{q}_j) = 0, \tag{4.4}$$

with $k = 1, 2, \ldots, n$. If we assume that the time dependence of the generalized coordinates is oscillatory with an angular frequency ω,

$$q_j = x_j e^{i\omega t}, \tag{4.5}$$

we have

$$\sum_{j=1}^{n}\left[A_{jk} - M_{jk}\omega^2\right]x_j = 0, \tag{4.6}$$

which is a homogeneous linear equation set. In order to have a nontrivial (not all zero) solution of the equation set, the determinant of the coefficient matrix has to vanish; that is,

$$|\mathbf{A} - \mathbf{M}\omega^2| = 0, \tag{4.7}$$

with $\mathbf{A}$ and $\mathbf{M}$ being the potential parameter matrix and the generalized mass matrix, respectively. The vibrational frequencies ω_k with $k = 1, 2, \ldots, n$ are then obtained by solving the above secular equation.

Another example we would like to illustrate here is that of electrical circuit network problems. We can apply the Kirchhoff rules to obtain a set of equations for the voltages and currents, and then we can solve the equation set to find the unknowns. Let us take the unbalanced Wheatstone bridge, as shown in Fig. 4.1, as an example. There is a total of three independent loops. We can choose the first as going through the source and two upper bridges and the second and third as the loops on the left and right of the ammeter. Each loop results in one of three independent equations,

$$r_v i_1 + r_1 i_2 + r_2 i_3 = v_0, \tag{4.8}$$

$$-r_x i_1 + (r_1 + r_x + r_a)i_2 - r_a i_3 = 0, \tag{4.9}$$

$$-r_3 i_1 - r_a i_2 + (r_2 + r_3 - r_a)i_3 = 0, \tag{4.10}$$

where r_1, r_2, r_3, r_x, r_a, and r_v are the resistances of the upper left bridge, upper right bridge, lower right bridge, lower left bridge, ammeter, and external source; i_1, i_2, and i_3 are the currents through the source, upper left bridge, and upper right

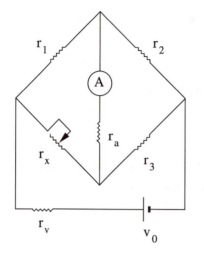

Fig. 4.1. The unbalanced Wheatstone bridge with all the resistors indicated.

bridge; and v_0 is the voltage of the external source. The above equation set can be rewritten in a matrix form

$$\mathbf{Ri} = \mathbf{v}, \tag{4.11}$$

where $\mathbf{R}$ is the resistance coefficient matrix and $\mathbf{i} = (i_1, i_2, i_3)$ and $\mathbf{v} = (v_0, 0, 0)$ are the current and voltage arrays. If we multiply both sides of the equation by the inverse matrix $\mathbf{R}^{-1}$, we have

$$\mathbf{i} = \mathbf{R}^{-1}\mathbf{v}, \tag{4.12}$$

which is solved if we know $\mathbf{R}^{-1}$. We will see later in this chapter that, in general, it is not necessary to know the inverse of the coefficient matrix in order to solve a linear equation set. But the scheme for obtaining the inverse of a matrix is almost the same as the scheme for solving a linear equation set.

A third example is in the calculation of the electronic structure of a many-body electron system. Let us examine a very simple but still meaningful system H_3^+. Three protons are arranged in a regular triangle. Two electrons in the system are shared by all three protons. Assuming that we can describe the system by a simple Hamiltonian containing one term for the hopping of an electron from one site to another and another for the doubly occupied site, we have

$$\mathcal{H} = -t \sum_{i \neq j} a_{i\sigma}^+ a_{j\sigma} + U \sum n_{i\uparrow} n_{i\downarrow}, \tag{4.13}$$

where $a_{i\sigma}^+$ and $a_{j\sigma}$ are creation and annihilation operators of an electron with either spin up ($\sigma = \uparrow$) or down ($\sigma = \downarrow$) and $n_{i\sigma} = a_{i\sigma}^+ a_{i\sigma}$ is the corresponding occupancy at the ith site. This Hamiltonian is called the Hubbard model when it is used to describe highly correlated electronic systems. The parameters t and U in the Hamiltonian can be obtained from either a quantum chemistry calculation or an experimental measurement. Since each site has only one orbital, we have a total of fifteen possible states for the two electrons under the single particle representation,

$$
\begin{aligned}
&|\phi_1\rangle = a_{1\uparrow}^+ a_{1\downarrow}^+ |0\rangle, \quad |\phi_2\rangle = a_{2\uparrow}^+ a_{2\downarrow}^+ |0\rangle, \quad |\phi_3\rangle = a_{3\uparrow}^+ a_{3\downarrow}^+ |0\rangle, \\
&|\phi_4\rangle = a_{1\uparrow}^+ a_{2\downarrow}^+ |0\rangle, \quad |\phi_5\rangle = a_{1\uparrow}^+ a_{3\downarrow}^+ |0\rangle, \quad |\phi_6\rangle = a_{2\uparrow}^+ a_{3\downarrow}^+ |0\rangle, \\
&|\phi_7\rangle = a_{1\downarrow}^+ a_{2\uparrow}^+ |0\rangle, \quad |\phi_8\rangle = a_{1\downarrow}^+ a_{3\uparrow}^+ |0\rangle, \quad |\phi_9\rangle = a_{2\downarrow}^+ a_{3\uparrow}^+ |0\rangle, \\
&|\phi_{10}\rangle = a_{1\uparrow}^+ a_{2\uparrow}^+ |0\rangle, \quad |\phi_{11}\rangle = a_{1\uparrow}^+ a_{3\uparrow}^+ |0\rangle, \quad |\phi_{12}\rangle = a_{2\uparrow}^+ a_{3\uparrow}^+ |0\rangle, \\
&|\phi_{13}\rangle = a_{1\downarrow}^+ a_{2\downarrow}^+ |0\rangle, \quad |\phi_{14}\rangle = a_{1\downarrow}^+ a_{3\downarrow}^+ |0\rangle, \quad |\phi_{15}\rangle = a_{2\downarrow}^+ a_{3\downarrow}^+ |0\rangle,
\end{aligned}
\tag{4.14}
$$

with $|0\rangle$ being the vacuum. The Hamiltonian is now a 15×15 matrix, $\mathcal{H}_{ij} = \langle \phi_i | \mathcal{H} | \phi_j \rangle$. We leave it as an exercise to the reader to figure out the Hamiltonian elements. The problem then becomes a matrix eigenvalue problem

$$
\mathcal{H} |\psi_i\rangle = \epsilon_i |\psi_i\rangle,
\tag{4.15}
$$

with $|\psi_i\rangle$ being an eigenvector of $\mathcal{H}$ with the eigenvalue ϵ_i. We could have simplified the problem by exploiting the symmetry of the system. The total spin and the z-component of the total spin commute with the Hamiltonian, so they are good quantum numbers. One can reduce the Hamiltonian matrix into block diagonal form with a maximum block size of 2×2. After one obtains all the eigenvalues and eigenvectors of the system, one can analyze the electronic, optical, and magnetic properties of H_3^+ easily.

4.2 Basic matrix operations

A matrix $\mathbf{A}$ is defined through its elements A_{ij}, with $i = 1, 2, \ldots, n$ and $j = 1, 2, \ldots, m$, and is called an $n \times m$ matrix. $\mathbf{A}$ is called a square matrix if $n = m$. We will consider mainly square matrices in this chapter.

A variable array $\mathbf{x} = (x_1, x_2, \ldots, x_n)$ can be viewed as an $n \times 1$ matrix. A typical set of linear algebraic equations is given by

$$
A_{i1}x_1 + A_{i2}x_2 + \cdots + A_{in}x_n = b_i,
\tag{4.16}
$$

for $i = 1, 2, \ldots, n$, where x_i are the unknowns to be solved, A_{ij} are given coefficients, and b_i are given constants. Eq. (4.16) can be expressed in a matrix form

$$
\mathbf{A}\mathbf{x} = \mathbf{b},
\tag{4.17}
$$

with $\mathbf{A}\mathbf{x}$ defined from the standard matrix multiplication

$$C_{ij} = \sum_k A_{ik} B_{kj}, \tag{4.18}$$

for $\mathbf{C} = \mathbf{A}\mathbf{B}$. The summation over k requires the number of columns of the first matrix to be the same as the number of rows of the second matrix. Otherwise, the product does not exist. Basic matrix operations are extremely important. For example, the inverse of a matrix $\mathbf{A}$ (written as $\mathbf{A}^{-1}$) is defined by

$$\mathbf{A}^{-1}\mathbf{A} = \mathbf{A}\mathbf{A}^{-1} = \mathbf{I}, \tag{4.19}$$

where $\mathbf{I}$ is a unit matrix with the elements $I_{ij} = \delta_{ij}$. The determinant of an $n \times n$ matrix $\mathbf{A}$ is defined as

$$|\mathbf{A}| = \sum_{i=1}^{n} (-1)^{i+j} A_{ij} |\mathbf{R}_{ij}|, \tag{4.20}$$

for $j = 1, 2, \ldots, n$, where $|\mathbf{R}_{ij}|$ is the determinant of the residual matrix of $\mathbf{A}$ with its ith row and jth column removed. The determinant of a 1×1 matrix is the element itself. The determinant of a (lower or upper) triangular matrix is the product of the diagonal elements. The inverse of $\mathbf{A}$ can be obtained through

$$A_{ij}^{-1} = (-1)^{i+j} \frac{|\mathbf{R}_{ij}|}{|\mathbf{A}|}. \tag{4.21}$$

If a matrix has an inverse or nonzero determinant, it is called a nonsingular matrix. Otherwise, it is a singular matrix.

The trace of a matrix $\mathbf{A}$ (written as Tr $\mathbf{A}$) is the summation of all the diagonal elements of $\mathbf{A}$,

$$\text{Tr } \mathbf{A} = \sum_{i=1}^{n} A_{ii}, \tag{4.22}$$

which is extremely useful in quantum statistics. The transpose of a matrix $\mathbf{A}$ (written as $\mathbf{A}^T$) has elements with the row and column indices of $\mathbf{A}$ interchanged, that is,

$$A_{ij}^T = A_{ji}. \tag{4.23}$$

Another useful matrix operation is to add one row (or column) to another row (or column) multiplied by a factor λ,

$$A_{ij}' = A_{ij} + \lambda A_{kj}, \tag{4.24}$$

for $j = 1, 2, \ldots, n$, where i and k are row indices, which can be the same. This operation preserves the determinant, that is, $|\mathbf{A}'| = |\mathbf{A}|$, and can be represented by a matrix multiplication

$$\mathbf{A}' = \mathbf{M}\mathbf{A}, \tag{4.25}$$

with $M_{jj} = 1$, $M_{ik} = \lambda$, and other $M_{jl} = 0$. If we interchange any two rows or columns of a matrix, its determinant changes only its sign.

The matrix eigenvalue problem is defined by the equation

$$\mathbf{A}\mathbf{x} = \lambda\mathbf{x}, \tag{4.26}$$

with $\mathbf{x}$ and λ as an eigenvector and its corresponding eigenvalue of the matrix. The matrix eigenvalue problem can also be viewed as a linear equation set problem solved iteratively. For example, if we want to solve the eigenvalues and eigenvectors of the above equation, we can solve the recursion relation

$$\mathbf{A}\mathbf{x}^{(n+1)} = \lambda^{(n)}\mathbf{x}^{(n)}, \tag{4.27}$$

where $\lambda^{(n)}$ and $\mathbf{x}^{(n)}$ are the approximate eigenvalue and eigenvector at the nth iteration. Later in this chapter we will discuss this iterative scheme in more detail, but here we just want to point out that at each iteration the problem is equivalent to the solution of the linear equation set in Eq. (4.17). One can also show that the eigenvalues of a matrix under a similarity transformation

$$\mathbf{B} = \mathbf{S}^{-1}\mathbf{A}\mathbf{S} \tag{4.28}$$

are the same as the ones of the original matrix $\mathbf{A}$, because

$$\mathbf{B}\mathbf{y} = \lambda\mathbf{y} \tag{4.29}$$

is equivalent to

$$\mathbf{A}\mathbf{x} = \lambda\mathbf{x}, \tag{4.30}$$

if we choose $\mathbf{x} = \mathbf{S}\mathbf{y}$. This, of course, assumes that the matrix $\mathbf{S}$ is not singular. This also means that

$$|\mathbf{B}| = |\mathbf{A}| = \prod_{i=1}^{n} \lambda_i. \tag{4.31}$$

These basic aspects of matrix operations are quite important and will be used throughout this chapter. Readers who are not familiar with these aspects should consult one of the standard textbooks, for example, Lewis (1991).

4.3 Linear equation systems

A matrix is called an upper (lower) triangular matrix if the elements below (above) the diagonal are all zero. The simplest scheme in solving matrix problems is the

Gaussian elimination scheme, which uses very basic matrix operations to transform the coefficient matrix into an upper (lower) triangular matrix first and then finds the solution of the equation set by means of backward (forward) substitution. The inverse and the determinant of a matrix can be obtained in a similar scheme.

Let us here take the solution of the linear equation set

$$\mathbf{A}\mathbf{x} = \mathbf{b} \tag{4.32}$$

as an example. If we assume that $|\mathbf{A}| \neq 0$ and $\mathbf{b} \neq \mathbf{0}$, then the system will have a unique solution. In order to show the steps of the Gaussian elimination scheme, we will label the original matrix $\mathbf{A}^{(1)} = \mathbf{A}$ with $\mathbf{A}^{(r)}$ as the resultant matrix after $r - 1$ matrix operations. Similar notation is used for the transformed $\mathbf{b}$ as well.

The basic idea of the Gaussian elimination scheme is to transform the original linear equation set to one that has an upper triangular coefficient matrix but the same set of solutions. As we will discuss in this section, in each step of transformation we eliminate the elements of a column that are not part of the upper triangle. The final linear equation set is then given by

$$\mathbf{A}^{(n)}\mathbf{x} = \mathbf{b}^{(n)}, \tag{4.33}$$

with $\mathbf{A}^{(n)}$ being an upper triangular matrix, that is, $A_{ij}^{(n)} = 0$ for $i < j$. The procedure is quite simple. We first multiply the first equation by $-A_{i1}^{(1)}/A_{11}^{(1)}$, that is, all the elements of the first row in the coefficient matrix and b_1, and then add it to the ith equation for $i > 1$. The first element of every row except the first row in the coefficient matrix is then eliminated. Now we denote the new matrix $\mathbf{A}^{(2)}$ and then multiply the second equation by $-A_{i2}^{(2)}/A_{22}^{(2)}$. After we add the modified second equation to all the other equations with $i > 2$, the second element of every row except the first row and the second row of the coefficient matrix is eliminated. This procedure can be continued with the third, fourth, ..., and $(n - 1)$th equation, and then the coefficient matrix becomes an upper triangular matrix $\mathbf{A}^{(n)}$. A linear equation set with an upper triangular coefficient matrix can easily be solved with backward substitutions.

Since all the diagonal elements are used in the denominators, the scheme would have failed if any of them had happened to be zero or a very small quantity. The problem can be circumvented in most cases by interchanging the rows and/or columns to have the elements used for rescaling as the ones with largest magnitudes possible. This is the so-called pivoting procedure. This procedure will not change the solutions of the linear equation set but will put them into a different order. Here we will consider only the partial pivoting scheme, which searches the pivoting element only from the given column. A full pivoting scheme would have searched the pivoting element from all the remaining elements. Bearing in mind the consideration of computing speed and accuracy, the partial pivoting scheme seems to be a good compromise. In most cases, the coefficients from different equations are of the same order of magnitude. We can then compare the magnitudes of the coefficients directly. In some other cases, however, the coefficients can be significantly different and their rescaled magnitudes, that is, the ones divided by the largest magnitude in the same equation, are used for comparison.

We first search for the largest (rescaled) element $|A_{i1}^{(1)}|$ for $i = 1, 2, \ldots, n$. Assuming that the element obtained is $A_{k_1 1}^{(1)}$, we then interchange the first row and the k_1th row and eliminate the first element of each row except the first row. Similarly, we can search for the second pivoting element from the largest (rescaled) element $|A_{i2}^{(2)}|$ for $i = 2, 3, \ldots, n$. Assuming that the element obtained is $A_{k_2 2}^{(2)}$ with $k_2 > 1$, we can then interchange the second row and the k_2th row and eliminate the second element of each row except the first and second rows. This procedure is continued to complete the elimination and transform the original matrix into an upper triangular matrix.

Physically, we do not need to interchange the rows of the matrix when searching for pivoting elements. An index can be used to record the positions of the pivoting rows. The following subroutine is a demonstration of the partial pivoting Gaussian elimination scheme.

```
      SUBROUTINE ELGS(A,N,IR,INDX)
C
C Subroutine for the partial pivoting Gaussian
C elimination. A is the coefficient matrix in the
C input and the transformed matrix in the output.
C INDX records the pivoting order.
C IR is the indicator for the rescaling.
C
      PARAMETER (NMAX=100)
      DIMENSION A(N,N),INDX(N),C(NMAX)
      LOGICAL IR
C
      IF (N.GT.NMAX) STOP 'The dimension is too large.'
C
C Initialize the index
C
      DO      50    I = 1, N
         INDX(I) = I
   50 CONTINUE
```

```
C
C Rescale the coefficients
C
        DO      100   I = 1, N
          C1= 0.0
          DO     90   J = 1, N
            C1 = AMAX1(C1,ABS(A(I,J)))
   90     CONTINUE
          C(I) = C1
  100   CONTINUE
        DO      140   I = 1, N
          DO    130   J = 1, N
            A(I,J) = A(I,J)/C(I)
  130     CONTINUE
  140   CONTINUE
C
C Search the pivoting elements
C
      DO      200   J = 1, N-1
        PI1 = 0.0
        DO    150   I = J, N
          PI = ABS(A(INDX(I),J))
          IF (PI.GT.PI1) THEN
            PI1 = PI
            K   = I
          ENDIF
  150   CONTINUE
C
C Eliminate the elements below the diagonal
C
        ITMP    = INDX(J)
        INDX(J) = INDX(K)
        INDX(K) = ITMP
        DO    170   I = J+1, N
          PJ   = A(INDX(I),J)/A(INDX(J),J)
          A(INDX(I),J) = PJ
          DO 160   K = J+1, N
            A(INDX(I),K) = A(INDX(I),K)-PJ*A(INDX(J),K)
  160     CONTINUE
```

```
      170   CONTINUE
      200 CONTINUE
C
C Rescale the coefficients back to the original
C magnitudes
C
        IF(IR) THEN
          DO      240   I = 1, N
            DO    230   J = 1, N
              A(I,J) = C(I)*A(I,J)
      230     CONTINUE
      240   CONTINUE
        ELSE
        ENDIF
C
        RETURN
        END
```

The subroutine above destroys the original matrix. If one wants to save the original matrix, one can just copy it to another matrix before calling the subroutine or introduce another matrix in the subroutine. The determinant of the original matrix can easily be obtained after one has the upper triangular matrix. As we discussed, the procedure used in the partial pivoting Gaussian elimination scheme does not change the value of the determinant except for its sign, which can be fixed with the knowledge of the order of pivoting rows. For an upper triangular matrix, the determinant is given by the product of all its diagonal elements. Therefore, we can obtain the determinant of a matrix as soon as it is transformed into an upper triangular matrix. Here is how one can obtain the determinant of a matrix with the partial pivoting Gaussian elimination scheme.

```
        SUBROUTINE DTRM(A,N,D,INDX)
C
C Subroutine for evaluating the determinant of a matrix
C using the partial pivoting Gaussian elimination
C scheme.
C
        DIMENSION A(N,N),INDX(N)
        LOGICAL IR
C
        IR = .TRUE.
```

```
                CALL ELGS(A,N,IR,INDX)
       C
                D     = 1.0
                DO      100 I = 1, N
                   D = D*A(INDX(I),I)
         100 CONTINUE
       C
                MSGN = 1
                DO      200 I = 1, N
                   DO    150 WHILE (I.NE.INDX(I))
                      MSGN = -MSGN
                      J = INDX(I)
                      INDX(I) = INDX(J)
                      INDX(J) = J
         150     END DO
         200 CONTINUE
       C
                D = MSGN*D
       C
                RETURN
                END
```

We have used the recorded pivoting order in the above subroutine to obtain the correct sign of the determinant.

Similarly, we can solve a linear equation set after having its coefficient matrix transformed into an upper triangular matrix through the Gaussian elimination scheme. The solutions are then obtained with a simple backward substitution scheme,

$$
x_i = \frac{1}{A_{k_i i}^{(n)}} \left(b_{k_i}^{(n)} - \sum_{j=i+1}^{n} A_{k_i j}^{(n)} x_j \right),
\tag{4.34}
$$

for $i = n - 1, n - 2, \ldots, 1$, with

$$
x_n = \frac{b_{k_n}^{(n)}}{A_{k_n n}^{(n)}}.
\tag{4.35}
$$

For example, we have

$$
x_{n-1} = \frac{b_{k_{n-1}}^{(n)} - A_{k_{n-1} n}^{(n)} x_n}{A_{k_{n-1} n-1}^{(n)}}.
\tag{4.36}
$$

We have used k_i as the row index of the pivoting element from the ith column. One can easily show that the above recursion relation does satisfy the linear equation set $\mathbf{A}^{(n)}\mathbf{x} = \mathbf{b}^{(n)}$. The following subroutine is an implementation of the above algorithm.

```
        SUBROUTINE LEGS(A,N,B,X,INDX)
C
C Subroutine for solving the equation A*X = B with the
C partial pivoting Gaussian elimination.
C
        DIMENSION A(N,N),B(N),X(N),INDX(N)
        LOGICAL IR
C
        IR = .FALSE.
        CALL ELGS(A,N,IR,INDX)
C
        DO      100 I = 1, N-1
          DO      90 J = I+1, N
              B(INDX(J)) = B(INDX(J))
     *                     -A(INDX(J),I)*B(INDX(I))
  90      CONTINUE
 100    CONTINUE
C
        X(N) = B(INDX(N))/A(INDX(N),N)
        DO      200 I = N-1, 1, -1
          X(I) = B(INDX(I))
          DO    190 J = I+1, N
              X(I) = X(I)-A(INDX(I),J)*X(J)
 190      CONTINUE
          X(I) =   X(I)/A(INDX(I),I)
 200    CONTINUE
C
        RETURN
        END
```

The inverse of a matrix can be obtained in exactly the same fashion if one realizes that

$$A_{ij}^{-1} = x_i(j), \tag{4.37}$$

for $i = 1, 2, \ldots, n$, where $x_i(j)$ is the solution of the equation $\mathbf{A}\mathbf{x} = \mathbf{b}(j)$ for $j = 1, 2, \ldots, n$. The elements of $\mathbf{b}(j)$ are given by $b_i(j) = \delta_{ij}$, for $i, j = 1, 2, \ldots, n$.

If we expand **b** into a unit matrix, the solution corresponding to each column of the unit matrix forms the corresponding column of the inverse of **A**. With a little modification of the above subroutine for the linear equation set, we obtain the program for matrix inversion. The following subroutine is an implementation of such a scheme.

```
      SUBROUTINE MIGS(A,N,X,INDX)
C
C Subroutine for matrix inversion. A(N,N) is the matrix
C to be inverted, and the inverse is stored in X(N,N)
C in the output.
C
      PARAMETER (NMAX=100)
      DIMENSION A(N,N),X(N,N),INDX(N),B(NMAX,NMAX)
      LOGICAL IR
C
      IF(N.GT.NMAX) STOP 'The matrix dimension is too
C     large.'
C
      DO      20 I = 1, N
        DO    10 J = 1, N
          B(I,J) = 0.0
   10     CONTINUE
   20 CONTINUE
      DO      30 I = 1, N
          B(I,I) = 1.0
   30 CONTINUE
C
      IR = .FALSE.
      CALL ELGS(A,N,IR,INDX)
C
      DO      100 I = 1, N-1
        DO     90 J = I+1, N
          DO 80 K = 1, N
            B(INDX(J),K) = B(INDX(J),K)
     *                          -A(INDX(J),I)*B(INDX(I),K)

   80     CONTINUE
   90     CONTINUE
```

```
      100 CONTINUE
C
          DO      200 I = 1, N
            X(N,I) = B(INDX(N),I)/A(INDX(N),N)
            DO   190 J = N-1, 1, -1
              X(J,I) = B(INDX(J),I)
              DO 180 K = J+1, N
                X(J,I) = X(J,I)-A(INDX(J),K)*X(K,I)
      180       CONTINUE
              X(J,I) =  X(J,I)/A(INDX(J),J)
      190   CONTINUE
      200 CONTINUE
C
          RETURN
          END
```

A scheme related to the Gaussian elimination scheme is called the *LU decomposition scheme*, which decomposes a nonsingular matrix into the product of a lower triangular and an upper triangular matrix. As we discussed, the Gaussian elimination scheme corresponds to a set of $n-1$ matrix operations that can be represented by a matrix multiplication

$$\mathbf{A}^{(n)} = \mathbf{MA}, \tag{4.38}$$

where

$$\mathbf{M} = \mathbf{M}^{(1)}\mathbf{M}^{(2)} \cdots \mathbf{M}^{(n-1)}, \tag{4.39}$$

with each $\mathbf{M}^{(r)}$, for $r = 1, 2, \ldots, n-1$, completing one step of the elimination. It is easy to show that $\mathbf{M}$ is a lower triangular matrix with unit diagonal elements, so Eq. (4.39) is equivalent to

$$\mathbf{A} = \mathbf{LU}, \tag{4.40}$$

with $\mathbf{U}$ an upper triangular matrix and $\mathbf{L}$ a lower triangular matrix. Note that the inverse of a lower triangular matrix is still a lower triangular matrix. The elements in $\mathbf{L}$ and $\mathbf{U}$ can be obtained from comparison of the elements of the product of $\mathbf{L}$ and $\mathbf{U}$ with the elements of $\mathbf{A}$. It is a common choice that the diagonal elements of $\mathbf{U}$ instead of the diagonal elements of $\mathbf{L}$ are set equal to 1. The other elements of $\mathbf{U}$ and $\mathbf{L}$ can easily be obtained from the above equation by comparing corresponding elements on both sides. The result can be cast into two recursion relations,

$$L_{ij} = A_{ij} - \sum_{k=1}^{j-1} L_{ik}U_{kj}, \tag{4.41}$$

$$U_{ij} = \frac{1}{L_{ij}} \left(A_{ij} - \sum_{k=1}^{i-1} L_{ik}U_{kj} \right), \tag{4.42}$$

with $L_{i1} = A_{i1}$ and $U_{1j} = A_{1j}/A_{11}$ as the starting values. Note that **L** is limited to being a lower triangular matrix and **U** is limited to being an upper triangular matrix. In practice, we still need to consider the issue of pivoting, which shows up in the denominator of the expression for U_{ij}. We can manage it in exactly the same way as in the Gaussian elimination scheme. In practice, we can store both **L** and **U** in one matrix, with the lower triangular part for **L** and the rest for **U**. The diagonal elements of **U** are not stored, because they are all equal to 1. The determinant of **A** is given by the product of the diagonal elements of **L** because

$$|\mathbf{A}| = |\mathbf{L}||\mathbf{U}| = \prod_{i=1}^{n} L_{ii}. \tag{4.43}$$

As soon as we obtain **L** and **U** for a given matrix **A**, we can solve the linear equation set $\mathbf{Ax} = \mathbf{b}$ with one set of forward substitutions of the lower triangular matrix and another set of backward substitutions of the upper triangular matrix, because the linear equation set can now be rewritten as

$$\mathbf{Ly} = \mathbf{b}, \tag{4.44}$$

$$\mathbf{Ux} = \mathbf{y}. \tag{4.45}$$

If we compare the elements on both sides of these equations, we have

$$y_i = \frac{1}{L_{ii}} \left(b_i - \sum_{k=1}^{i-1} L_{ik}y_k \right), \tag{4.46}$$

with $y_1 = b_1/L_{11}$ and $i = 2, 3, \ldots, n$. Then we can obtain the solution of the original linear equation set from

$$x_i = y_i - \sum_{k=i+1}^{n} U_{ik}x_k, \tag{4.47}$$

with $x_n = y_n$ and $i = n-1, n-2, \ldots, 1$.

One can also obtain the inverse of a matrix by a method similar to that used in solving the linear equation set by the Gaussian elimination scheme. One can choose a set of vectors with elements $b_i(j) = \delta_{ij}$ with $i, j = 1, 2, \ldots, n$. Here i is used to refer to the element and j is used to refer to a specific vector. The inverse of the matrix $\mathbf{A}$ is then given by $A_{ij}^{-1} = x_i(j)$. Here $x_i(j)$ is the ith element of the solution of $\mathbf{A}\mathbf{x}(j) = \mathbf{b}(j)$. We will demonstrate this scheme in a numerical example in Chapter 6 with a real symmetric tridiagonal matrix. A Fortran subroutine for the LU decomposition can be found in Press et al. (1992).

4.4 Zeros and extremes of a multivariable function

The numerical solution of linear equations discussed in the last section is important in physics and related fields, because many problems can be expressed in terms of linear equations. A few of them will be found in the exercises for this chapter.

However, there is another class of problems that are nonlinear in nature but can be solved iteratively with the linear schemes just developed. Examples include solution of a set of nonlinear multivariable equations and search for the maxima or minima of a multivariable function. In this section, we will show how to extend the matrix methods discussed so far to study nonlinear problems. We will also demonstrate the applications of these numerical schemes with an actual physics problem, the stable geometric configuration of a multicharge system, which is relevant in the study of the geometry of molecules and the formation of solid salts. From the numerical point of view, this is also a problem of searching for the global or local minima on a potential energy surface.

The multivariable Newton method

Nonlinear equations can also be solved with matrix techniques. In Chapter 2, we introduced the Newton method for obtaining the zeros of a single-variable function. Now we would like to extend our study to the solution of a set of multivariable equations. Assume that the equation set is given by

$$\mathbf{f}(\mathbf{x}) = \mathbf{0}, \tag{4.48}$$

with $\mathbf{f} = (f_1, f_2, \ldots, f_n)$ and $\mathbf{x} = (x_1, x_2, \ldots, x_n)$. If the equation has at least one solution $\mathbf{x} = \mathbf{x}_0$, we can perform a Taylor expansion around a neighboring point,

$$\mathbf{f}(\mathbf{x}_0) = \mathbf{f}(\mathbf{x}) + \Delta\mathbf{x} \cdot \nabla\mathbf{f}(\mathbf{x}) + O(\Delta\mathbf{x}^2)$$
$$\simeq \mathbf{f}(\mathbf{x}) + \mathbf{A}(\mathbf{x})\Delta\mathbf{x} \simeq \mathbf{0}, \tag{4.49}$$

with $\Delta\mathbf{x} = \mathbf{x}_0 - \mathbf{x}$ and $\mathbf{A}(\mathbf{x})$ as a partial derivative matrix,

$$A_{ij}(\mathbf{x}) = \frac{\partial f_i(\mathbf{x})}{\partial x_j}. \tag{4.50}$$

The Taylor expansion can be rewritten as

$$\mathbf{x}_0 \simeq \mathbf{x} - \mathbf{A}^{-1}(\mathbf{x})\mathbf{f}(\mathbf{x}), \tag{4.51}$$

which can be used to find the solution of the equation set. Note that the Newton method for single-variable equations discussed in Chapter 2 is a special case with $n = 1$. Here is an outline of the numerical procedure for the multivariable Newton method. We first make an initial guess of the solution $\mathbf{x}$. Then the matrix $\mathbf{A}(\mathbf{x})$ is calculated from the partial derivatives as defined in Eq. (4.50). We can apply any matrix inversion scheme to obtain the inverse of $\mathbf{A}(\mathbf{x})$. The solution is improved with Eq. (4.51). This process can be repeated until the required tolerance is reached.

As with the secant method discussed in Chapter 2, if the expression of the first-order derivative is not easy to obtain, the two-point formula can be used for the partial derivative:

$$A_{ij}(\mathbf{x}) = \frac{f_i(\mathbf{x} + h_j\mathbf{e}_j) - f_i(\mathbf{x})}{h_j}, \tag{4.52}$$

where h_j is the finite interval and $\mathbf{e}_j$ is the unit vector along the direction of x_j. A rule of thumb in practice is to choose

$$h_j \simeq \delta_0 x_j, \tag{4.53}$$

where δ_0 is roughly the square root of the machine tolerance of the specific computer to be used. The machine tolerance is related to the number of bits used in the processor of the computer. For example, the machine tolerance is about $2^{-31} \approx 5 \times 10^{-10}$ for 32-bit computers.

As a matter of fact, either the Newton or the secant method converges locally if the function is smooth enough, and the main computing cost is in the evaluation of the partial derivatives and the inverse of the matrix $\mathbf{A}(\mathbf{x})$ at each iteration.

Extremes of a multivariable function

With the knowledge of the solution of a nonlinear equation set, we can develop numerical schemes to obtain the minima or maxima of a multivariable function. The extremes of a multivariable function $g(\mathbf{x})$ are the solutions of a set of nonlinear equations:

$$\mathbf{f}(\mathbf{x}) = \nabla g(\mathbf{x}) = \mathbf{0}, \tag{4.54}$$

where $\mathbf{f}(\mathbf{x})$ is an n-dimensional vector with each component given by a partial derivative of $g(\mathbf{x})$. In practice, we can use what we have just discussed for the solution of a nonlinear equation set, except that special care is needed to ensure that $g(\mathbf{x})$ will decrease (increase) during the updating steps for a minimum (maximum) of $g(\mathbf{x})$. For example, if we want to obtain a minimum of the function $g(\mathbf{x})$, we can update the position vector $\mathbf{x}$, following Eq. (4.51). To ensure that the updating steps will converge to a minimum of $g(\mathbf{x})$, we can modify the matrix $\mathbf{A}(\mathbf{x})$ from its original form to

$$\mathbf{A}(\mathbf{x}) = \nabla\mathbf{f}(\mathbf{x}) + \mu(\mathbf{x})\mathbf{I}, \tag{4.55}$$

with $\mu(\mathbf{x})$ taken as a positive quantity to ensure that the modified matrix $\mathbf{A}(\mathbf{x})$ is positive definite, that is, $\mathbf{w}^T\mathbf{A}(\mathbf{x})\mathbf{w} \geq 0$ for any $\mathbf{w}$.

A special choice of $\mu(\mathbf{x})$, known as the BFGS (Broyden, 1970; Fletcher, 1970; Goldfarb, 1970; Shanno, 1970) updating scheme, is to have

$$\mathbf{x}_{n+1} = \mathbf{x}_n - \mathbf{A}_n^{-1}\mathbf{f}_n, \tag{4.56}$$

with

$$\mathbf{A}_n = \mathbf{A}_{n-1} + \frac{\mathbf{y}\mathbf{y}^T}{\mathbf{y}^T\mathbf{w}} - \frac{\mathbf{A}_{n-1}\mathbf{w}\mathbf{w}^T\mathbf{A}_{n-1}}{\mathbf{w}^T\mathbf{A}_{n-1}\mathbf{w}}, \tag{4.57}$$

where

$$\mathbf{w} = \mathbf{x}_n - \mathbf{x}_{n-1} \tag{4.58}$$

and

$$\mathbf{y} = \mathbf{f}_n - \mathbf{f}_{n-1}. \tag{4.59}$$

The BFGS scheme ensures that the updating matrix is positive definite, so the search is always moving toward a minimum of the function. This scheme has been very successful in the actual problems and applications. The reason behind its success is still unclear. If one wants to find the maximum of $g(\mathbf{x})$, one can carry out exactly the same steps with $\mathbf{f}(\mathbf{x}) = -\nabla g(\mathbf{x})$. If the gradient is not available analytically, one can use the finite difference instead, which is in the spirit of the secant method. For more details on the optimization of a function, see Dennis and Schnabel (1983). Much work has been done to develop a better numerical method for optimization of a multivariable function in the last few decades. However, the problem of global optimization of a multivariable function with many local minima is still open and may never be solved. For some recent discussions on the problem, see Altschuler et al. (1994; 1995) and Erber and Hockney (1995).

Geometric structures of multicharge clusters

We now turn to a physics problem, the stable geometric structure of a multicharge cluster, which is extremely important in the analysis of small clusters of atoms, ions, and molecules. We will take clusters of Na^+ and Cl^- as illustrative examples.

The geometric structures of n particles with $2 \leq n \leq 6$ are studied. If we take the empirical form of the interaction between two ions in the NaCl crystal,

$$V(r_{ij}) = \eta_{ij}\frac{e^2}{r_{ij}} + \delta_{ij}\alpha e^{-r_{ij}/\rho}, \qquad (4.60)$$

where $\eta_{ij} = -1$ and $\delta_{ij} = 1$ if the particles are of opposite charges; otherwise, $\eta_{ij} = 1$ and $\delta_{ij} = 0$. α and ρ are empirical parameters, which can be determined either from experiments or from first-principles calculations. For solid NaCl (Kittel, 1986, pp. 86–92), $\alpha = 1.09 \times 10^3$ eV and $\rho = 0.321$ Å. We will use these two quantities in what follows.

The function to be optimized is the total interaction potential energy of the system,

$$U(\mathbf{r}_1, \mathbf{r}_2, \ldots, \mathbf{r}_n) = \sum_{i>j}^{n} V(r_{ij}). \qquad (4.61)$$

A local optimal structure of the cluster is obtained when U reaches a local minimum. This minimum can hardly be a global minimum for large clusters, since no relaxation is allowed and a large cluster can have many local minima (Erber and Hockney, 1995). There are $3n$ coordinates as independent variables in $U(\mathbf{r}_1, \mathbf{r}_2, \ldots, \mathbf{r}_n)$. However, since the position of the center of mass and rotation around the center of mass will not change the total potential energy U, we have to remove the center-of-mass motion and the rotational motion around the center of mass during the optimization. We can achieve this by imposing several restrictions on the cluster. First, we fix one of the particles at the origin of the coordinates. This removes three degrees of freedom of the cluster. Second, we restrict another particle on an axis, for example, the x-axis. This removes two more degrees of freedom of the cluster. Finally, we restrict the third particle in a plane, for example, the xy-plane. This removes the last (sixth) degree of freedom fixing the center of mass and making the system rotationless. The potential energy U now only has $3n - 6$ independent variables (coordinates).

We have applied the BFGS scheme to search for the local minima of U for NaCl, Na_2Cl^+, Na_2Cl_2, $Na_3Cl_2{}^+$, and Na_3Cl_3. The stable structures for the first three systems are very simple: a dimer, a straight line, and a square. The stable structures obtained for $Na_3Cl_2{}^+$ and Na_3Cl_3 are shown in Fig. 4.2.

In order to avoid reaching zero separation of any two ions, we have added a term $b(c/r_{ij})^{12}$ in the interaction with $b = 1$ eV and $c = 0.1$ Å, which adds an energy on the order of 10^{-18} eV to the total potential energy. It is worth pointing out that the structure of Na_3Cl_3 is similar to the structure of $(H_2O)_6$, discovered recently by Liu et al. (1996). This might be an indication that water molecules are actually partially charged because of the polarization.

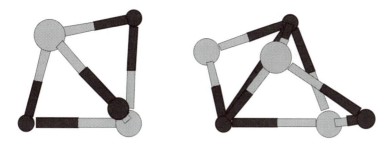

Fig. 4.2. Top view of stable structures of $Na_3Cl_2{}^+$ and Na_3Cl_3.

4.5 Eigenvalue problems

Matrix eigenvalue problems are very important in physics. They are defined by

$$\mathbf{A}\mathbf{x} = \lambda\mathbf{x}, \qquad (4.62)$$

where λ is the eigenvalue corresponding to the eigenvector $\mathbf{x}$ of the matrix $\mathbf{A}$, determined from the secular equation

$$|\mathbf{A} - \lambda\mathbf{I}| = 0, \qquad (4.63)$$

with $\mathbf{I}$ being a unit matrix. For an $n \times n$ matrix, one can have n eigenvalues in general. The eigenvalue problem is quite general in physics, since the matrix in Eq. (4.62) can come from many different problems, for example, the Lagrange equation for the vibrational modes of a large molecule and the Schrödinger equation for $H_3{}^+$ discussed at the beginning of this chapter. In this section, we will discuss the most common and useful methods for obtaining eigenvalues and eigenvectors of the matrix eigenvalue problem defined in Eq. (4.62).

Eigenvalues of a Hermitian matrix

In many problems in physics and related fields, the matrix in question appears to be Hermitian; that is,

$$\mathbf{A}^+ = \mathbf{A}, \qquad (4.64)$$

where $\mathbf{A}^+$ means taking the transpose and complex conjugate of $\mathbf{A}$. The simplicity of the Hermitian eigenvalue problem is due to three important properties associated with a Hermitian matrix:

(1) The eigenvalues of a Hermitian matrix are all real.
(2) The eigenvectors of a Hermitian matrix can be made orthonormal.
(3) A Hermitian matrix can be transformed into a diagonal matrix with the same set of eigenvalues under a similarity transformation of a unitary matrix.

Furthermore, the eigenvalue problem of an $n \times n$ complex Hermitian matrix is equivalent to an eigenvalue problem of a $2n \times 2n$ real symmetric matrix. We can show this easily. A Hermitian matrix $\mathbf{A}$ can be written into the sum of its real and imaginary parts:

$$\mathbf{A} = \mathbf{B} + i\mathbf{C}, \tag{4.65}$$

with $\mathbf{B}$ and $\mathbf{C}$ as the real and imaginary parts of $\mathbf{A}$, respectively. If $\mathbf{A}$ is Hermitian, then $\mathbf{B}$ is a real symmetric matrix and $\mathbf{C}$ is a real skew symmetric matrix; that is,

$$B_{ij} = B_{ji}, \tag{4.66}$$
$$C_{ij} = -C_{ji}. \tag{4.67}$$

If we decompose the eigenvector in a similar fashion, we have $\mathbf{x} = \mathbf{y} + i\mathbf{z}$. The original eigenvalue problem becomes

$$(\mathbf{B} + i\mathbf{C})(\mathbf{y} + i\mathbf{z}) = \lambda(\mathbf{y} + i\mathbf{z}), \tag{4.68}$$

which is equivalent to

$$\begin{pmatrix} \mathbf{B} & -\mathbf{C} \\ \mathbf{C} & \mathbf{B} \end{pmatrix} \begin{pmatrix} \mathbf{y} \\ \mathbf{z} \end{pmatrix} = \lambda \begin{pmatrix} \mathbf{y} \\ \mathbf{z} \end{pmatrix}, \tag{4.69}$$

which is a real symmetric eigenvalue problem with exactly the same set of eigenvalues with a double degeneracy. Therefore we need to solve only the real symmetric eigenvalue problem if the matrix is Hermitian.

The similarity transformation of a real symmetric matrix to a diagonal matrix is orthogonal instead of unitary. This simplifies the procedure considerably. The problem can be simplified further if we separate the procedure into two steps:

(1) Perform a similarity transformation on the real symmetric matrix in Eq. (4.69) with an orthogonal matrix and convert it into a tridiagonal matrix, which has only diagonal elements and the elements next to them nonzero. This is sometimes referred to as a symmetric band matrix with a bandwidth of three.

(2) Solve the eigenvalue problem of the resulting tridiagonal matrix. The similarity transformation preserves the eigenvalues of the original matrix, and the eigenvectors are related by the orthogonal matrix used in the transformation.

The most commonly used method to tridiagonalize a real symmetric matrix is the *Householder method*. Sometimes the method is called the *Givens method*. We give here a brief description of the scheme. The tridiagonalization is achieved with a total of $n - 2$ consecutive transformations, each operating on a row and a column of the matrix. The transformations can be cast into a recursion relation

$$\mathbf{A}^{(k)} = \mathbf{O}_k^T \mathbf{A}^{(k-1)} \mathbf{O}_k, \tag{4.70}$$

for $k = 1, 3, \ldots, n-2$, where $\mathbf{O}_k$ is an orthogonal matrix that works on the row elements with $i = k+2, \ldots, n$ of the kth column and the column elements $j = k+2, \ldots, n$ of the kth row. The recursion begins with $\mathbf{A}^{(0)} = \mathbf{A}$. To be specific, one can write the orthogonal matrix as

$$\mathbf{O}_k = \mathbf{I} - \frac{1}{\eta_k} \mathbf{w}_k \mathbf{w}_k^T, \tag{4.71}$$

where the lth component of the vector $\mathbf{w}_k$ is given by

$$w_k(l) = \begin{cases} 0 & \text{for} \quad l \leq k, \\ A_{kk+1}^{(k-1)} + \alpha_k & \text{for} \quad l = k+1, \\ A_{kl}^{(k-1)} & \text{for} \quad l \geq k+2, \end{cases} \tag{4.72}$$

with

$$\alpha_k = \pm \sqrt{\sum_{l=k+1}^{n} \left[A_{kl}^{(k-1)} \right]^2}, \tag{4.73}$$

and

$$\eta_k = \alpha_k \left[\alpha_k + A_{kk+1}^{(k-1)} \right]. \tag{4.74}$$

Even though the sign of α_k is arbitrary in the above equations, it is always taken to be the same as that of $A_{kk+1}^{(k-1)}$ in practice in order to avoid any possible cancellation, which can make the algorithm ill-conditioned (with a zero denominator in $\mathbf{O}_k$). One can easily show that $\mathbf{O}_k$ as defined above is the desired matrix, which is orthogonal and converts the row elements with $i = k+2, \ldots, n$ for the kth column and the column elements with $j = k+2, \ldots, n$ for the kth row of $\mathbf{A}^{(k-1)}$ to zero just by comparing the elements on both sides of Eq. (4.70). We are not going to provide a subroutine here for the Householder scheme to tridiagonalize a real symmetric matrix, since it is in all standard libraries and reference books, for example, Press et al. (1992).

After one obtains the tridiagonalized matrix, the eigenvalues can be obtained with one of the root search routines discussed in Chapter 2. One can make the observation that the determinant of the tridiagonalized matrix $|\mathbf{A} - \lambda \mathbf{I}| = 0$ is equivalent to a polynomial equation $p_n(\lambda) = 0$. Due to the simplicity of the symmetric tridiagonal matrix, the polynomial $p_n(\lambda)$ can be generated recursively:

$$p_i(\lambda) = (a_i - \lambda) p_{i-1}(\lambda) - b_{i-1}^2 p_{i-2}(\lambda), \tag{4.75}$$

with a_i being the diagonal element A_{ii}, b_i being the off-diagonal element $A_{ii+1} = A_{i+1i}$, and $p_i(\lambda)$ being the polynomial for the submatrix of A_{jk} with $j, k = 1, 2, \ldots, i$. $p_0(\lambda) = 1$ and $p_1(\lambda) = a_1 - \lambda$ can be used to start the recursion. Note that this recursion relation is similar to the one for the orthogonal polynomials discussed in Chapter 2 and can be generated easily with a subroutine.

```
      SUBROUTINE TDPL(A,B,N,X,P)
C
C Subroutine to generate determinant polynomial P_N(X)
C
      DIMENSION P(1:N),A(1:N),B(1:N)
      PO  = 1.0
      IF (N.LT.1) STOP 'The dimension is less than 1.'
      P(1) = A(1)-X
      IF(N.LT.2) RETURN
      P(2) = (A(2)-X)*P(1)-B(1)*B(1)*PO
      IF(N.LT.3) RETURN
      DO    100 I = 2, N-1
         P(I+1) = (A(I+1)-X)*P(I)-B(I)*B(I)*P(I-1)
  100 CONTINUE
      RETURN
      END
```

In principle, one can use any of the root searching routines to find the eigenvalues from the polynomial equation as soon as the polynomial is generated. However, two properties associated with the roots of $p_n(\lambda)$ are useful in developing a fast and accurate subroutine to obtain the eigenvalues of a symmetric tridiagonal matrix.

(1) All the roots of $p_n(\lambda) = 0$ lie in the interval $[-\|\mathbf{A}\|, \|\mathbf{A}\|]$, with $\|\mathbf{A}\|$ being the column modulus of the matrix; that is,

$$\|\mathbf{A}\| = \max \left\{ \sum_{j=1}^{n} |A_{ij}| \right\}, \tag{4.76}$$

for $i = 1, 2, \ldots, n$. One can also use the row modulus of the matrix in the above statement. The row modulus of a matrix is given by

$$\|\tilde{\mathbf{A}}\| = \max \left\{ \sum_{i=1}^{n} |A_{ij}| \right\}, \tag{4.77}$$

for $j = 1, 2, \ldots, n$.

(2) The number of roots for $p_n(\lambda) = 0$ with $\lambda \geq \lambda_0$ is given by the number of agreements of the signs of $p_j(\lambda_0)$ and $p_{j-1}(\lambda_0)$ for $j = 1, 2, \ldots, n$. If any of the polynomials, for example, $p_j(\lambda_0)$, is zero, the sign of the previous polynomial $p_{j-1}(\lambda_0)$ is assigned to that polynomial.

With the help of these properties, one can develop a quite simple but fast algorithm in connection with the bisection method for root searching discussed in Chapter 2 to obtain the eigenvalues of a real symmetric matrix. Let us here outline this algorithm. Consult the relevant sections in Chapter 2 before reading the following discussion.

We first evaluate the column modulus of the matrix with Eq. (4.76). This sets the boundaries for the eigenvalues. Note that each summation in Eq. (4.76) has only three terms, $|A_{ii-1}|$, $|A_{ii}|$, and $|A_{ii+1}|$. We can then start the search for the first eigenvalue in the region of $[-\|\mathbf{A}\|, \|\mathbf{A}\|]$. Since there is a total of n eigenvalues, we need to decide first how many specified eigenvalues are needed. For example, if we are interested only in the ground state – in other words, the lowest eigenvalue – we can design an algorithm to target the lowest eigenvalue only.

A general scheme can be devised and altered, based on the following general procedure. We can bisect the maximum region of $[-\|\mathbf{A}\|, \|\mathbf{A}\|]$ and evaluate the signs of $p_i(0)$ for $i = 1, 2, \ldots, n$. Based on what we have discussed, we will know the number of eigenvalues lying within either $[-\|\mathbf{A}\|, 0]$ or $[0, \|\mathbf{A}\|]$. This includes any possible degeneracy. We can divide the two subintervals into four equal regions and check the signs of the polynomials at each bisected point. This procedure can be continued to narrow down the region where each eigenvalue resides. After l steps of bisection, each eigenvalue sought is narrowed down to a region with a size of $\|\mathbf{A}\|/2^{l-1}$. Note that we can work either on a specific eigenvalue, for example, the one associated with the ground state, or on a group of eigenvalues simultaneously. The error bars are bounded by $\|\mathbf{A}\|/2^{l-1}$ after l steps of bisection. A more realistic estimate of the error bar is obtained from the improvement of a specific eigenvalue at each step of bisection, as in the estimate given in Chapter 2.

Since the Householder scheme for a real symmetric matrix is carried out in two steps, that is, transformation of the original matrix into a tridiagonal matrix followed by solution of the eigenvalue problem of the tridiagonal matrix, one can design different algorithms to achieve each of these two steps separately based on the specific problems studied. Interested readers can find several of them in Wilkinson (1963; 1965). It is worth emphasizing again that a Hermitian matrix eigenvalue problem can be converted into a real symmetric matrix eigenvalue problem with the dimension of the matrix expanded to twice that of the original matrix along each direction. This seems to be a reasonable approach in most cases as long as the size of the matrix is not limited by the available resources.

Eigenvalues of general matrices

Even though most problems in physics are likely to be concerned with Hermitian matrices, there are still problems that require us to deal with general matrices. Here we would like to discuss briefly how to obtain the eigenvalues of a general nondefective matrix. A matrix is nondefective if it can be diagonalized under a matrix transformation and its eigenvectors can form a complete vector space. In this section, we will consider nondefective matrices only, since there is always a well-defined eigenvalue equation $\mathbf{A}\mathbf{x} = \lambda\mathbf{x}$ for a nondefective matrix $\mathbf{A}$ and we hardly encounter defective matrices in physics.

In most cases, we can define a matrix function $f(\mathbf{A})$ with $\mathbf{A}$ playing the same role as x in $f(x)$. For example,

$$f(\mathbf{A}) = e^{-\alpha\mathbf{A}} = \sum_{k=0}^{\infty} \frac{(-\alpha\mathbf{A})^k}{k!} \tag{4.78}$$

is similar to

$$f(x) = e^{-\alpha x} = \sum_{k=0}^{\infty} \frac{(-\alpha x)^k}{k!}, \tag{4.79}$$

except that the power of a matrix is treated as matrix multiplications with $\mathbf{A}^0 = \mathbf{I}$. When the matrix function operates on an eigenvector of the matrix, the matrix can be replaced by the corresponding eigenvalue; that is,

$$f(\mathbf{A})\mathbf{x}_i = f(\lambda_i)\mathbf{x}_i. \tag{4.80}$$

If the function involves the inverse of the matrix and the eigenvalue happens to be zero, one can always add a term $\mu\mathbf{I}$ to the original matrix to remove the singularity. The modified matrix has the eigenvalue $\lambda_i - \mu$ for the corresponding eigenvector $\mathbf{x}_i$. The eigenvalue of the original matrix is recovered with the limit $\mu \to 0$ after the problem is solved. Based on this property of nondefective matrices, we can construct a recursion relation

$$\mathbf{x}^{(k)} = \frac{1}{\sqrt{\mathcal{N}_k}}(\mathbf{A} - \mu\mathbf{I})^{-1}\mathbf{x}^{(k-1)}, \tag{4.81}$$

to extract the eigenvalue that is closest to the parameter μ. Here $k = 1, 2, 3, \ldots$, and $\mathcal{N}_k$ is the normalization constant to ensure that

$$\langle\mathbf{x}^{(k)}\,|\,\mathbf{x}^{(k)}\rangle = (\mathbf{x}^{(k)})^{+}\mathbf{x}^{(k)} = 1. \tag{4.82}$$

To analyze this recursive procedure, let us rewrite the trial state $\mathbf{x}^{(0)}$ as a linear combination of all eigenstates,

$$\mathbf{x}^{(0)} = \sum_{i=1}^{n} a_i^{(0)} \mathbf{x}_i, \tag{4.83}$$

which is always possible, since the eigenstates form a complete vector space. If we substitute the above trial state into the recursion relation, we have

$$\mathbf{x}^{(k)} = \sum_{i=1}^{n} \frac{a_i^{(0)} \mathbf{x}_i}{(\lambda_i - \mu)^k}, \tag{4.84}$$

up to a coefficient. If we normalize the state at each step of iteration, only the state $\mathbf{x}_j$, with the eigenvalue λ_j that is the closest to μ, will survive after a large number of iterations. All other coefficients will vanish as

$$a_i^{(k)} \rightarrow \left(\frac{\lambda_j - \mu}{\lambda_i - \mu} \right)^k \frac{a_i^{(0)}}{a_j^{(0)}} \tag{4.85}$$

for $i \neq j$ after k iterations. However, $a_j^{(k)}$ will grow with k to approach 1. The corresponding eigenvalue is obtained with

$$\lambda_j = \mu + \lim_{k \to \infty} \frac{1}{\sqrt{\mathcal{N}_k}} \frac{x_l^{(k-1)}}{x_l^{(k)}}, \tag{4.86}$$

where l is the index for a specific nonzero component (usually the one with the maximum magnitude) of $\mathbf{x}^{(k)}$ or $\mathbf{x}^{(k-1)}$.

Here $\mathbf{x}^{(0)}$ is a trial state, which should have a nonzero overlap with the eigenvector $\mathbf{x}_j$. One has to be very careful to make sure the overlap is nonzero. Otherwise, one will end up with the eigenvalue that belongs to the eigenvector with a nonzero overlap with $\mathbf{x}^{(0)}$ but still closer to μ than the rest. One way to avoid a zero overlap of $\mathbf{x}^{(0)}$ and $\mathbf{x}_j$ is to generate each component of $\mathbf{x}^{(0)}$ with a random number generator and check each result with at least two different trial states.

The method outlined here is usually called the *inverse iteration method*. It is quite straightforward if one has an efficient algorithm to perform the matrix inversion. One can also rewrite the recursion relation at each step as a linear equation set

$$\sqrt{\mathcal{N}_k}(\mathbf{A} - \mu \mathbf{I})\mathbf{x}^{(k)} = \mathbf{x}^{(k-1)}, \tag{4.87}$$

which can be solved with the Gaussian elimination scheme, for example, from one iteration to another. Note that this iterative scheme also solves the eigenvectors of the matrix at the same time. This is a very useful feature in many applications for which the eigenvectors and eigenvalues are both needed.

Eigenvectors of matrices

We sometimes also need the eigenvectors of an eigenvalue equation. For a nondefective matrix, we can always obtain a complete set of eigenvectors, including the degenerate eigenvalue cases.

First, if the matrix is symmetric, it is much easier to obtain its eigenvectors. In principle, we can transform a real symmetric matrix $\mathbf{A}$ into a tridiagonal matrix $\mathbf{T}$ with a similarity transformation

$$\mathbf{T} = \mathbf{O}^T \mathbf{A} \mathbf{O}, \tag{4.88}$$

with $\mathbf{O}$ being an orthogonal matrix. The eigenvalue equation then becomes

$$\mathbf{A}\mathbf{x} = \lambda \mathbf{x} = \mathbf{O}\mathbf{T}\mathbf{O}^T \mathbf{x}, \tag{4.89}$$

which is equivalent to solving $\mathbf{y}$ from

$$\mathbf{T}\mathbf{y} = \lambda \mathbf{y} \tag{4.90}$$

and then $\mathbf{x}$ from

$$\mathbf{x} = \mathbf{O}\mathbf{y}. \tag{4.91}$$

Thus, we can develop a practical scheme that solves the eigenvectors of $\mathbf{T}$ first and then those of $\mathbf{A}$ by multiplying $\mathbf{y}$ with $\mathbf{O}$, which is a byproduct of any tridiagonalization scheme, such as the Householder scheme.

As we discussed in the preceding section, the recursion relation

$$\sqrt{\mathcal{N}_k}(\mathbf{T} - \mu \mathbf{I})\mathbf{y}^{(k)} = \mathbf{y}^{(k-1)} \tag{4.92}$$

will lead to the eigenvector $\mathbf{y}_j$ with λ_j as the eigenvalue of $\mathbf{T}$ that is closest to μ. Eq. (4.92) can be solved with an elimination scheme. Note that the elimination scheme is now extremely simple, since there are no more than three nonzero elements in each row. In every step of the above recursion, we need to normalize the resulting vector. A numerical example based on the LU decomposition in solving a linear equation set with coefficients in a tridiagonal form is given in detail in Section 6.4.

If the eigenvalue corresponds to more than one eigenvector, we can obtain the other vectors simply by changing the value of μ or the initial guess of $\mathbf{y}^{(0)}$. When all the vectors $\mathbf{x}_k$ corresponding to the same eigenvalue are found, we can transform them into an orthogonal set $\mathbf{z}_k$ with

$$\mathbf{z}_k = \mathbf{x}_k - \sum_{j=1}^{k-1} (\mathbf{z}_j^T \mathbf{x}_k) \mathbf{z}_j, \tag{4.93}$$

which is known as the Gram–Schmidt orthogonalization procedure.

If the tridiagonal matrix is obtained by means of the Householder scheme, we can also obtain the eigenvectors of the original matrix at the same time with

$$\mathbf{x} = \mathbf{O}_1 \mathbf{O}_2 \cdots \mathbf{O}_{(n-2)} \mathbf{y}$$

$$= \left(\mathbf{I} - \frac{1}{\eta_1} \mathbf{w}_1 \mathbf{w}_1^T \right) \cdots \left(\mathbf{I} - \frac{1}{\eta_{n-2}} \mathbf{w}_{n-2} \mathbf{w}_{n-2}^T \right) \mathbf{y}, \tag{4.94}$$

which can be carried out in quite a straightforward fashion if we realize that

$$\left(\mathbf{I} - \frac{1}{\eta_k} \mathbf{w}_k \mathbf{w}_k^T \right) \mathbf{y} = \mathbf{y} - \beta_k \mathbf{w}_k, \tag{4.95}$$

with $\beta_k = \mathbf{w}_k^T \mathbf{y} / \eta_k$. Note that the first k elements in $\mathbf{w}_k$ are all zero.

For a general nondefective matrix, one can also follow a similar scheme to obtain the eigenvectors of the matrix. First one transforms the matrix into an upper Hessenberg matrix. A matrix with all the elements below (above) the first off-diagonal elements equal to zero is called an upper (lower) Hessenberg matrix. This transformation can always be achieved for a nondefective matrix in a similarity transformation with a unitary matrix $\mathbf{U}$ ($\mathbf{U}^{-1} = \mathbf{U}^+$); that is,

$$\mathbf{H} = \mathbf{U}^+ \mathbf{A} \mathbf{U}, \tag{4.96}$$

where $\mathbf{H}$ is an upper Hessenberg matrix and $\mathbf{U}$ is a unitary matrix. There are several methods to reduce a matrix to an upper Hessenberg matrix; they are given in Wilkinson (1965). A typical scheme is similar to the Householder scheme, reducing a symmetric matrix to a triangular matrix so one needs to work on only the lower half of the matrix. We can then solve the eigenvalue problem of the upper Hessenberg matrix first. The eigenvalue problem of an upper (lower) Hessenberg matrix is considerably simpler than that of a general matrix under a typical algorithm, such as the QR algorithm. The so-called QR algorithm decomposes a nonsingular matrix $\mathbf{A}$ into the product of a unitary matrix $\mathbf{Q}$ and an upper triangular matrix $\mathbf{R}$,

$$\mathbf{A} = \mathbf{Q} \mathbf{R}. \tag{4.97}$$

One can construct a series of similarity transformations by alternating the order of $\mathbf{Q}$ and $\mathbf{R}$ to reduce the original matrix to an upper triangular matrix that has the eigenvalues of the original matrix on the diagonal. Assume that $\mathbf{A}^{(0)} = \mathbf{A}$ and

$$\mathbf{A}^{(k)} = \mathbf{Q}_k \mathbf{R}_k; \tag{4.98}$$

then one would have

$$\mathbf{A}^{(k+1)} = \mathbf{R}_k \mathbf{Q}_k = \mathbf{Q}_k^+ \mathbf{A}^{(k)} \mathbf{Q}_k, \tag{4.99}$$

with $k = 0, 1, 2, \ldots$. It is found that this algorithm works best if $\mathbf{A}$ here is a Hessenberg matrix. Taking account of stability and speed considerations, a Householder type of scheme to transform a nondefective matrix into a Hessenberg matrix seems to be an excellent choice (Wilkinson, 1965).

When we have the matrix transformed into the Hessenberg form, we can use the inverse iteration method to obtain the eigenvectors of the Hessenberg matrix $\mathbf{H}$ with

$$\sqrt{\mathcal{N}_k}(\mathbf{H} - \mu\mathbf{I})\mathbf{y}^{(k)} = \mathbf{y}^{(k-1)}, \tag{4.100}$$

and then the eigenvectors of the original matrix $\mathbf{A}$ from

$$\mathbf{x} = \mathbf{U}\mathbf{y}. \tag{4.101}$$

Of course, the normalization of the vector at each iteration is assumed with the normalization constant $\mathcal{N}_k$, with a definition similar to that of $\mathcal{N}_k$ for Eq. (4.82), to ensure the convergence.

4.6 The Faddeev–Leverrier method

A very interesting method developed for matrix inversion and the matrix eigenvalue problem is the *Faddeev–Leverrier method*. The scheme was first discovered by Leverrier in the middle of the last century and later modified by Faddeev (Faddeev and Faddeeva, 1963). Here we give a brief discussion of the method. The characteristic polynomial of the matrix is given by

$$p_n(\lambda) = |\mathbf{A} - \lambda\mathbf{I}| = \sum_{k=0}^{n} a_k \lambda^k, \tag{4.102}$$

where $a_n = \lambda^0 = 1$. Then one can introduce a set of supplementary matrices $\mathbf{S}_k$ with

$$p_n(\lambda)(\lambda\mathbf{I} - \mathbf{A})^{-1} = \sum_{k=0}^{n-1} \lambda^{n-k-1}\mathbf{S}_k. \tag{4.103}$$

If we multiply the above equation by $(\lambda\mathbf{I} - \mathbf{A})$ on both sides, we have

$$\sum_{k=0}^{n} a_k \lambda^k \mathbf{I} = \mathbf{S}_0\lambda^n + \sum_{k=1}^{n-1}(\mathbf{S}_k - \mathbf{A}\mathbf{S}_{k-1})\lambda^{n-k} - \mathbf{A}\mathbf{S}_{n-1}. \tag{4.104}$$

Comparing the coefficients from the same order of λ^l for $l = 0, 1, \ldots, n$ on both sides of the equation, we obtain the recursion for a_{n-k} and $\mathbf{S}_k$,

$$a_{n-k} = -\frac{1}{k}\mathrm{Tr}\,\mathbf{A}\mathbf{S}_{k-1}, \tag{4.105}$$

$$\mathbf{S}_k = \mathbf{A}\mathbf{S}_{k-1} + a_{n-k}\mathbf{I}, \tag{4.106}$$

for $k = 1, 2, \ldots, n$. The recursion is started with $\mathbf{S}_0 = \mathbf{I}$ and ended with a_0. From Eq. (4.103) with $\lambda = 0$, one can easily show that

$$\mathbf{AS}_{n-1} + a_0\mathbf{I} = \mathbf{0}, \tag{4.107}$$

which can be used to obtain the inverse of the matrix $\mathbf{A}$,

$$\mathbf{A}^{-1} = -\frac{1}{a_0}\mathbf{S}_{n-1}. \tag{4.108}$$

Since the recursion relation also generates all the coefficients a_k for $k = 0, 1, \ldots,$ $n - 1$, we can use any root search methods, for example, the bisection method, to obtain all the eigenvalues from the characteristic polynomial $p_n(\lambda) = 0$. This is the same as the situation for the Householder method discussed in the preceding section.

After we have found all the eigenvalues λ_k, we can also obtain the corresponding eigenvectors with the availability of the supplementary matrices $\mathbf{S}_k$. If we define a new matrix

$$\mathbf{X}(\lambda_k) = \sum_{l=1}^{n} \mathbf{S}_{l-1}\lambda_k^{n-l}, \tag{4.109}$$

then the columns of $\mathbf{X}(\lambda_k)$ are the eigenvectors $\mathbf{x}_k$. This can be shown from the limit of the matrix $\mathbf{X}(\lambda)$ with $\lambda = \lambda_k + \delta$. Here $\delta \to 0$ is introduced to make $\mathbf{X}$ a nonsingular matrix. From Eq. (4.103), one can easily show that

$$\mathbf{X}(\lambda) = |\lambda\mathbf{I} - \mathbf{A}|(\lambda\mathbf{I} - \mathbf{A})^{-1}, \tag{4.110}$$

which leads each column of the matrix to the eigenvector $\mathbf{x}_k$ as $\delta \to 0$. The advantage of this scheme over the iterative scheme discussed in Section 4.5 is that one does not need to perform iterations as soon as one has the supplementary matrices $\mathbf{S}_k$. Note that we have not specified anything about the matrix $\mathbf{A}$ except that it is nonsingular. Therefore, the Faddeev–Leverrier method can be used for general matrices as well as symmetric matrices. For more detailed discussions and example programs, see Hostetter, Santina, and D'Carpio-Montalvo (1991, pp. 321–50).

4.7 Electronic structure of atoms

As we have pointed out, there are many applications of matrix operations in physics. We have given a few examples in Section 4.1 and a concrete case study on multicharge clusters in Section 4.4. In this section, we would like to demonstrate how to apply the matrix eigenvalue schemes in the calculation of the electronic structures of many-electron atomic systems within the framework of the Hartree–Fock approximation.

The Schrödinger equation for an N-electron atom is given by

$$\mathcal{H}\Psi(\mathbf{R}) = E\Psi(\mathbf{R}), \tag{4.111}$$

where $\mathbf{R} = (\mathbf{r}_1, \mathbf{r}_2, \ldots, \mathbf{r}_N)$ is a $3N$-dimensional position vector and $\mathcal{H}$ is the Hamiltonian

$$\mathcal{H} = -\frac{\hbar^2}{2m}\sum_{i=1}^{N}\nabla_i^2 - \sum_{i=1}^{N}\frac{Ze^2}{r_i} + \sum_{i>j}\frac{e^2}{|\mathbf{r}_i - \mathbf{r}_j|}. \tag{4.112}$$

Z here is the atomic number and e is the charge of a proton. One cannot obtain an exact solution for the multielectron Schrödinger equation; and approximations have to be made in order to solve the resulting eigenvalue problem.

The so-called Hartree–Fock approximation is to assume that the solution of a system of fermions (electrons in this case) can be written as the determinant of a set of single particle states,

$$\Psi(\mathbf{R}) = \frac{1}{\sqrt{N!}}\begin{vmatrix} \phi_1(\mathbf{r}_1) & \phi_1(\mathbf{r}_2) & \cdots & \phi_1(\mathbf{r}_N) \\ \phi_2(\mathbf{r}_1) & \phi_2(\mathbf{r}_2) & \cdots & \phi_2(\mathbf{r}_N) \\ \vdots & \vdots & \vdots & \vdots \\ \phi_N(\mathbf{r}_1) & \phi_N(\mathbf{r}_2) & \cdots & \phi_N(\mathbf{r}_N) \end{vmatrix}, \tag{4.113}$$

which is commonly known as the Slater determinant. The index used above includes both the spatial and spin indices. In most cases, it is more convenient to write the single particle states as $\phi_{i\sigma}(\mathbf{r})$ with the spin index σ separated from the spatial index i.

In order to have the expectation value of $\mathcal{H}$ optimized (that is, minimized) under the wavefunction $\Psi(\mathbf{R})$, the single particle states have to satisfy

$$\left[-\frac{\hbar^2}{2m}\nabla^2 - \frac{Ze^2}{r} + V_h(\mathbf{r})\right]\phi_{i\sigma}(\mathbf{r})$$

$$-\int V_{x\sigma}(\mathbf{r}', \mathbf{r})\phi_{i\sigma}(\mathbf{r}')\,d\mathbf{r}' = \epsilon_i\phi_{i\sigma}(\mathbf{r}), \tag{4.114}$$

which is known as the Hartree–Fock equation (see Exercise 4.8). Here $V_h(\mathbf{r})$ is the Hartree potential,

$$V_h(\mathbf{r}) = \int \rho(\mathbf{r}')\frac{e^2}{|\mathbf{r} - \mathbf{r}'|}\,d\mathbf{r}', \tag{4.115}$$

with $\rho(\mathbf{r}) = \rho_\uparrow(\mathbf{r}) + \rho_\downarrow(\mathbf{r})$ as the total density of the electrons at $\mathbf{r}$. The exchange interaction $V_{x\sigma}(\mathbf{r}, \mathbf{r}')$ is given by

$$V_{x\sigma}(\mathbf{r}', \mathbf{r}) = \rho_\sigma(\mathbf{r}, \mathbf{r}')\frac{e^2}{|\mathbf{r} - \mathbf{r}'|}, \tag{4.116}$$

where $\rho_\sigma(\mathbf{r}, \mathbf{r}')$ is the density matrix of the electrons, which is given by

$$\rho_\sigma(\mathbf{r}, \mathbf{r}') = \sum_{i=1}^{N_\sigma} \phi_{i\sigma}^+(\mathbf{r})\phi_{i\sigma}(\mathbf{r}'). \tag{4.117}$$

N_σ is the total number of states occupied by the electrons with spin σ. The density of the electrons is the diagonal of the density matrix, that is, $\rho_\sigma(\mathbf{r}) = \rho_\sigma(\mathbf{r}, \mathbf{r})$. ϵ_i above is a parameter introduced during the optimization (minimization) of the expectation value.

The Hartree potential can also be obtained from the solution of the Poisson equation

$$\nabla^2 V_h(\mathbf{r}) = 4\pi e\rho(\mathbf{r}). \tag{4.118}$$

The single-particle wavefunctions in the atomic systems can be assumed to have the form

$$\phi_{i\sigma}(\mathbf{r}) = \frac{1}{r} u_{nl\sigma}(r) Y_{lm}(\theta, \phi), \tag{4.119}$$

where $Y_{lm}(\theta, \phi)$ are the spherical harmonic functions with θ and ϕ as the polar and azimuth angles, respectively. n, l, and m are quantum numbers corresponding to the energy, angular momentum, and z-component of the angular momentum.

We can make the further assumption that the electron density is spherically symmetric and divide the space $[0, r_0]$ into discrete points with an evenly spaced interval h. r_0 is the cutoff in the radial direction, typically a few Bohr radii. If we use a numerical expression for the second-order derivative, for example, the three-point formula, the Hartree–Fock equation for a given l is converted into a matrix equation (see Exercise 4.8)

$$\mathbf{H}\mathbf{u} = \epsilon\mathbf{u}, \tag{4.120}$$

with the diagonal elements of $\mathbf{H}$ given by

$$H_{ii} = -\frac{1}{h^2} + \frac{l(l+1)}{2r_i^2} - \frac{Ze^2}{r_i} + V_h(r_i), \tag{4.121}$$

with the corresponding off-diagonal elements

$$H_{ij} = -\frac{\delta_{ij\pm1}}{2h^2} + hV_{x\sigma}^l(r_i, r_j), \tag{4.122}$$

where $V_{x\sigma}^l(r_i, r_j)$ is the lth component of $V_{x\sigma}(\mathbf{r}_i, \mathbf{r}_j)$ expanded in spherical harmonic functions. We have used $\mathbf{H}$ for the matrix form of $\mathcal{H}$ and $\mathbf{u}$ for the discrete form of the wavefunction $u_{nl\sigma}(r)$. The angular momentum index is suppressed in $\mathbf{H}$ and $\mathbf{u}$ for simplicity of notation. One can easily apply the numerical schemes introduced in the last two sections to solve this matrix eigenvalue problem. The energy levels for different n are obtained for a fixed l. One can, of course, evaluate the density matrix and the Hartree–Fock ground-state energy with the method described here.

4.8 The Lanczos algorithm and the many-body problem

One of the most powerful methods in large-scale matrix computing is the *Lanczos method*, which is an iterative scheme suitable for large, especially sparse (most elements are zero) matrices. The advantage of the Lanczos method is extremely clear when one needs only relatively few eigenvalues and eigenvectors, or when the system is an extremely large and sparse matrix. We will here just sketch a very basic Lanczos algorithm. More elaborate discussions of various Lanczos methods can be found in several specialized books, for example, Wilkinson (1965), Cullum and Willoughby (1985), and Hackbusch (1994).

Assume that the matrix $\mathbf{H}$ is an $n \times n$ real symmetric matrix, and we can tridiagonalize an $m \times m$ subset of $\mathbf{H}$ with

$$\mathbf{O}^T \mathbf{H} \mathbf{O} = \tilde{\mathbf{H}}, \tag{4.123}$$

where $\mathbf{O}$ is an $n \times m$ matrix with its kth column given by

$$\mathbf{v}_k = \frac{\mathbf{u}_k}{\sqrt{\mathcal{N}_k}}, \tag{4.124}$$

for $k = 1, 2, \ldots, m$, where $\mathcal{N}_k = \mathbf{u}_k^T \mathbf{u}_k$ is the normalization constant and the vectors $\mathbf{u}_k$ are generated recursively from an arbitrary vector $\mathbf{u}_1$ as

$$\mathbf{u}_{k+1} = \mathbf{H}\mathbf{v}_k - \alpha_k \mathbf{v}_k - \beta_k \mathbf{v}_{k-1}, \tag{4.125}$$

with $\beta_k = \tilde{H}_{k-1k} = \mathbf{v}_{k-1}^T \mathbf{H} \mathbf{v}_k$ and $\alpha_k = \tilde{H}_{kk} = \mathbf{v}_k^T \mathbf{H} \mathbf{v}_k$. The recursion is started with $\beta_1 = 0$ and $\mathbf{u}_1$ being a selected vector. $\mathbf{u}_1$ can be a normalized vector with each element generated from a uniform random number generator, for example. In principle, the vectors $\mathbf{v}_k$ for $k = 1, 2, \ldots, m$ form an orthonormal set, but in practice one still needs to carry out the Gram–Schmidt orthogonal procedure discussed in Section 4.5 at every step of the recursion to remove the effects due to the rounding errors. One can show that the eigenvalues of the tridiagonal submatrix $\tilde{\mathbf{H}}$ are the approximations of the ones of $\mathbf{H}$ with the largest magnitudes. One can use the standard methods discussed earlier to diagonalize this tridiagonal submatrix $\tilde{\mathbf{H}}$ to obtain its eigenvectors and eigenvalues from $\tilde{\mathbf{H}}\tilde{\mathbf{x}}_k = \lambda_k \tilde{\mathbf{x}}_k$. The eigenvectors $\tilde{\mathbf{x}}_k$ can be used to obtain the approximate eigenvectors of $\mathbf{H}$ with $\mathbf{x}_k \simeq \mathbf{O}\tilde{\mathbf{x}}_k$.

The approximation is improved if one constructs a new initial state $\mathbf{u}_1$ from the eigenvectors $\tilde{\mathbf{x}}_k$ with $k = 1, 2, \ldots, m$, for example, a linear combination,

$$\mathbf{u}_1 = \sum_{k=1}^{m} a_k \tilde{\mathbf{x}}_k, \tag{4.126}$$

and then the recursion is repeated again and again. One can show that this iterative scheme will eventually lead to the m eigenvectors of $\mathbf{H}$ with the largest magnitude eigenvalues. In practice, the selection of the coefficients a_k is rather important in order to have a fast and accurate algorithm. Later in this section we will introduce one of the choices made by Dagotto and Moreo (1985) in the study of the ground state of a quantum many-body system.

One can work out the eigenvalue problem for a specified region of the spectrum of $\mathbf{H}$ with the introduction of the matrix

$$\mathbf{G} = (\mathbf{H} - \epsilon\mathbf{I})^{-1}. \tag{4.127}$$

One can solve $\mathbf{G}$ with the Lanczos algorithm to obtain the eigenvectors with eigenvalues near ϵ. Note that

$$\mathbf{G}\mathbf{x}_k = \frac{1}{\lambda_k - \epsilon}\mathbf{x}_k, \tag{4.128}$$

if $\mathbf{H}\mathbf{x}_k = \lambda_k\mathbf{x}_k$. This is useful if one wants to know about the spectrum at a particular region.

At the beginning of this chapter, we used a many-body Hamiltonian (the Hubbard model) in Eq. (4.13) to describe the electronic behavior of $H_3{}^+$. It is generally believed that the Hubbard model and its variations can describe the majority of highly correlated quantum systems, for example, transition metals, rare earth compounds, conducting polymers, and oxide superconducting materials. One can find good reviews on the Hubbard model in Rasetti (1991). Usually, we want to know the properties of the ground state and the low-lying excited states. For example, if we want to know the ground state and the first excited state of a cluster of N sites with $N_0 < N$ electrons, we can solve the problem with the Lanczos method by taking $m \simeq 10$ and iterating the results a couple of times. Note that the number of many-body states increases very quickly with both N_0 and N. The iteration will converge to the ground state and the first excited state only if the states belong to the ten states with the largest eigenvalue magnitudes. The ground state and first excited state energies will carry the largest magnitude if the chemical potential of the system is set to be zero. We also have to come up with a construction of the next guess of $\mathbf{u}_1$. We can, for example, use

$$\mathbf{u}_1^{(l+1)} = \sum_{k=1}^{5} \mathbf{v}_k^{(l)}. \tag{4.129}$$

A special choice of the iteration scheme for the ground-state properties is given by Dagotto and Moreo (1985). They choose the $(l+1)$th iteration of $\mathbf{v}_1$ as

$$\mathbf{v}_1^{(l+1)} = \frac{1}{\sqrt{1+a^2}} [\mathbf{v}_1^{(l)} + a\mathbf{v}_2^{(l)}], \tag{4.130}$$

where a is determined from the minimization of the expectation value of $\mathbf{H}$ under $\mathbf{v}_1^{(l+1)}$, which gives

$$a = b - \sqrt{1+b^2}, \tag{4.131}$$

where b is expressed in terms of the expectation values of the lth iteration, $\eta_1 = \mathbf{v}_1^T \mathbf{H} \mathbf{v}_1$, $\eta_2 = \mathbf{v}_1^T \mathbf{H}^2 \mathbf{v}_1$, and $\eta_3 = \mathbf{v}_1^T \mathbf{H}^3 \mathbf{v}_1$:

$$b = \frac{\eta_3 - 3\eta_1\eta_2 + 2\eta_1^3}{2(\eta_2 - \eta_1^2)^{3/2}}. \tag{4.132}$$

The second vector

$$\mathbf{v}_2 = \frac{1}{\sqrt{\eta_2 - \eta_1^2}} (\mathbf{H}\mathbf{v}_1 - \eta_1\mathbf{v}_1) \tag{4.133}$$

is also normalized under such a choice of $\mathbf{v}_1$. The advantage of this algorithm is that one needs to store only $\mathbf{v}_1$, $\mathbf{H}\mathbf{v}_1$, and $\mathbf{H}^2\mathbf{v}_1$, three vectors each with n elements, during the iterations. The eigenvalue with the largest magnitude is also given iteratively from

$$\epsilon_1 = \mathbf{v}_1^T \mathbf{H} \mathbf{v}_1 + \frac{a}{\sqrt{\eta_2 - \eta_1^2}}. \tag{4.134}$$

This algorithm is very efficient for calculating the ground-state properties of many-body quantum systems. For more discussions on the method and its recent applications, see Dagotto (1994).

4.9 Random matrices

The distributions of the energy levels of many physical systems have some universal features determined by the fundamental symmetry of the Hamiltonian. This type of feature is usually qualitative. For example, when disorders are introduced into metallic systems, the resistivities will increase and the systems will become insulators if the disorders are strong enough. For each metal, the degree of disorder needed to become an insulator is different. But the general behavior of metallic systems to become insulators under strong disorders is the same. If we want to represent a disordered material with a matrix Hamiltonian, the elements of the matrix have to be randomly selected. The general feature of a physical system is obtained with an ensemble of random matrices satisfying the physical constraints of the system.

Even though the elements of the matrices are random, at least three types of fundamental symmetry can exist in the ensemble of the random matrices for a given physical system. For example, if the system has time reversal symmetry plus rotational invariance for the odd-half-integer spin case, the ensemble is real symmetric, or orthogonal.

If rotational invariance is not present in the odd-half-integer spin case, the ensemble is quaternion real, or symplectic. The general case without time reversal symmetry is described by general Hermitian matrices, or a unitary ensemble.

The structure of the matrix is the other relevant factor that determines the detailed properties of a specific system. Traditionally, the focus of the general properties was on the orthogonal ensemble with real symmetric matrices and a Gaussian distribution for the matrix elements. Efforts made in the last thirty years are described in Mehta (1991). In this section, we will not be able to cover many aspects of random matrices and will only provide a very brief introduction. Interested readers should consult Mehta (1991), Brody et al. (1981), and Bohigas (1991).

One can easily generate a random matrix if the symmetry and the structure of the matrix are specified. The Gaussian orthogonal ensemble of $N \times N$ matrices is specified with a distribution $\mathcal{W}_n(\mathbf{H})$ that is invariant under an orthogonal transformation; that is,

$$\mathcal{W}_N(\mathbf{H}') \, d\mathbf{H}' = \mathcal{W}_N(\mathbf{H}) \, d\mathbf{H}, \tag{4.135}$$

where

$$\mathbf{H}' = \mathbf{O}^T \mathbf{H} \mathbf{O}, \tag{4.136}$$

with $\mathbf{O}$ as an orthogonal matrix. The above condition also implies that $\mathcal{W}_N(\mathbf{H})$ is invariant under the orthogonal transformation, since $d\mathbf{H}$ is invariant. This restricts the distribution to

$$\mathcal{W}_N(\mathbf{H}) = e^{-\mathrm{Tr}\mathbf{H}^2/4\sigma^2}, \tag{4.137}$$

where the trace is given by

$$\mathrm{Tr}\,\mathbf{H}^2 = \sum_{i=1}^{N} H_{ii}^2 + \sum_{i \neq j}^{N} H_{ij}^2. \tag{4.138}$$

The average of the elements $\langle H_{ij} \rangle = 0$ and the variance $\langle H_{ij}^2 \rangle = (1 + \delta_{ij})\sigma^2$. The variance for the off-diagonal elements is only half that of the diagonal elements, because $H_{ij} = H_{ji}$ in symmetric matrices.

One can generate the random matrix ensemble numerically and diagonalize each matrix to obtain the distribution of the eigenvalues and eigenvectors. For example, the Gaussian orthogonal random matrix can be obtained from the following subroutine. The Gaussian random number generator used is the one introduced in Chapter 2.

```
        SUBROUTINE RMSG (N,XS,A)
C
C Subroutine for generating a random matrix in the
C Gaussian orthogonal ensemble with XS as the standard
C deviation of the off-diagonal elements.
C
        DIMENSION A(N,N)
        COMMON /CSEED/ ISEED
C
        DO      100 I = 1, N
          CALL GRNF(G1,G2)
          A(I,I) = SQRT(2.0)*G1*XS
    100 CONTINUE
        DO      300 I = 1, N
          DO    200 J = 1, I-1
            CALL GRNF(G1,G2)
            A(I,J) = G1*XS
            A(J,I) = A(I,J)
    200    CONTINUE
    300 CONTINUE
        RETURN
        END
```

This matrix can then be diagonalized by any method discussed earlier in this chapter. In order to obtain the statistical information, the matrix needs to be generated and diagonalized many times before a smooth and correct distribution can be reached. Based on the distributions of the eigenvalues and the eigenvectors of the random matrix ensemble, we can obtain the correlation functions of the eigenvalues and other important statistical information. Statistical methods and correlation functions are discussed in more detail in Chapter 7 and Chapter 9. The distribution density of the eigenvalues, ρ, in the Gaussian orthogonal ensemble can be obtained analytically, and it is given by a semicircle function

$$\rho(x) = \begin{cases} \sqrt{4 - x^2}/2\pi & \text{for} \quad |x| < 2, \\ 0 & \text{otherwise,} \end{cases} \tag{4.139}$$

which was first derived by Wigner in the 1950s. The variable x represents the eigenvalues measured in $\sigma\sqrt{N}$, and $\rho(x)$ is normalized to 1. The numerical simulations carried out so far seem to suggest that the semicircle distribution is true for any other ensemble as long as the elements satisfy $\langle H_{ij} \rangle = 0$ and $\langle H_{ij}^2 \rangle = \sigma^2$ for $i < j$. We will leave this as an exercise to the readers.

There has been a lot of activity recently in the field of random matrix theory and its applications. Here we just mention a couple of them. Hackenbroich and Weidenmüller (1995) have shown that a general distribution of the form

$$\mathcal{W}_N(\mathbf{H}) = \frac{1}{\mathcal{Z}} e^{-N\mathrm{Tr}\mathbf{V}(\mathbf{H})}, \tag{4.140}$$

with $\mathcal{Z}$ being a normalization constant, would lead to the same correlation functions among the eigenvalues as that of a Gaussian ensemble, in the limit of $N \to \infty$. This is true for any ensemble, orthogonal, symplectic, or unitary. Here $\mathbf{V}(\mathbf{H})$ is a function of $\mathbf{H}$ with the restriction that its eigenvalues are confined within a finite interval with a smooth distribution and that $V(\lambda)$ grows at least linearly as $\lambda \to \infty$. The Gaussian ensemble is a special case with $\mathbf{V}(\mathbf{H}) \propto \mathbf{H}^2$.

Akulin, Bréchignac, and Sarfati (1995) have used the random matrix method in the study of the electronic, structural, and thermal properties of metallic clusters. They have introduced a random interaction V with $\langle V \rangle = 0$ and $\langle V^2 \rangle$ being a function characterized by the electron–electron and electron–ion interactions, electron–phonon interaction, and the shape fluctuation of the cluster. Using this theory, they have predicted deformation transformation in the cluster when the temperature is lowered. This effect is predicted only for open shell clusters and is quite similar to the Jahn–Teller effect, which creates a finite distortion in the lattice in ionic solids due to the electron–phonon interaction.

Exercises

4.1 Find the currents in the unbalanced Wheatstone bridge (Fig. 4.1). Assume that $v_0 = 1.5\,\mathrm{V}$, $r_1 = r_2 = 100\,\Omega$, $r_3 = 150\,\Omega$, $r_x = 120\,\Omega$, $r_a = 1000\,\Omega$, and $r_v = 10\,\Omega$.

4.2 A typical problem in physics is that one usually can calculate physical quantities for a series of finite systems, but ultimately, one would like to know the results for the infinite system. Hulthén (1938) studied the one-dimensional spin-1/2 Heisenberg model

$$\mathcal{H} = \sum_i \mathbf{s}_i \cdot \mathbf{s}_{i+1}$$

and obtained the ground-state energy per site for a series of finite systems, $\epsilon_2 = -2.0000$, $\epsilon_4 = -1.5000$, $\epsilon_6 = -1.4343$, $\epsilon_8 = -1.4128$, and $\epsilon_{10} = -1.4031$. Now assume that the ground-state energy of a finite system is given by

$$\epsilon_n = \epsilon_\infty + \frac{x_1}{n} + \frac{x_2}{n^2} + \cdots + \frac{x_r}{n^r} + \cdots.$$

Truncate the above series at $r = 4$ and find ϵ_∞ by solving the linear equation set numerically.

4.3 Write a subroutine that utilizes the Householder scheme to tridiagonalize a real symmetric matrix.

4.4 Write a subroutine that uses the properties of determinant polynomials of a tridiagonal matrix and the bisection method to solve its first few eigenvalues.

4.5 Find all the 15×15 elements for the Hamiltonian of H_3^+ and use the program developed in Exercise 4.4 to solve it. Compare the numerical results with the analytic results by reducing the matrix into block diagonal form with the largest block being a 2×2 matrix.

4.6 It is of special interest in far infrared spectroscopy to know the vibrational spectrum of a molecule. Study the vibrational modes of Na_2Cl_2 discussed in Section 4.4. Use the empirical interaction potential and associated parameters given in Section 4.4.

4.7 Implement the Faddeev–Leverrier method in a program to obtain the inverse, eigenvalues, and eigenvectors of a general real matrix.

4.8 Derive the Hartree–Fock equation for the atomic systems given in Section 4.7. Show that the matrix form of the Hartree–Fock equation does represent the original equation if the electron density is spherically symmetric, and find the expression for $V_{x\sigma}^l(r_i, r_j)$ in the matrix equation.

4.9 Write a program to generate and diagonalize ensembles of real symmetric matrices with the following distributions:

$$W_N(H_{ij}) = \begin{cases} \frac{1}{2} & \text{for } |H_{ij}| < 1, \\ 0 & \text{otherwise}, \end{cases}$$

$$W_N(H_{ij}) = \frac{1}{2}[\delta(H_{ij} - 1) + \delta(H_{ij} + 1)],$$

$$W_N(H_{ij}) = \frac{1}{\sqrt{2\pi}} e^{-H_{ij}^2/2},$$

for $i \leq j$. Compare the density of eigenvalues with that of the Wigner semicircle.

5

Spectral analysis and Gaussian quadrature

In many cases Nature behaves differently from what our intuition would tell us. It was Fourier who first pointed out that an arbitrary periodic function $f(t)$ with a period T can be decomposed into a summation of simple harmonic terms. Each harmonic term is a periodic function with an angular frequency that is a multiple of the *fundamental angular frequency* $\omega = 2\pi/T$ of the function $f(t)$. The coefficient in each term of the summation is given by an integral of the product of the function and the complex conjugate of that harmonic term.

Assuming that we have a time-dependent function $f(t)$ with a period T, that is, $f(t + T) = f(t)$, the Fourier theorem can be cast into a summation,

$$f(t) = \sum_{n=-\infty}^{\infty} g_n e^{in\omega t}, \tag{5.1}$$

which is commonly known as the Fourier series. Here $\omega = 2\pi/T$ is the fundamental angular frequency and g_n is the Fourier coefficient, which is given by

$$g_n = \frac{1}{T} \int_0^T f(t) e^{-in\omega t} \, dt. \tag{5.2}$$

The Fourier theorem can be derived from the properties of the exponential functions $\phi_n(t) = \exp(in\omega t)/\sqrt{T}$, which form an orthonormal basis set in the region of one period of $f(t)$; that is,

$$\int_{t_0}^{t_0+T} \phi_m^+(t)\phi_n(t) \, dt = \langle m \mid n \rangle = \delta_{mn}, \tag{5.3}$$

where t_0 is an arbitrary starting point, $\phi_m^+(t)$ is the complex conjugate of $\phi_m(t)$, and δ_{nm} is the Kronecker δ-function,

$$\delta_{mn} = \begin{cases} 1 & \text{for} \quad m = n, \\ 0 & \text{for} \quad m \neq n. \end{cases} \tag{5.4}$$

The Fourier coefficient g_m is then obtained with Eq. (5.1) multiplied by $\exp(-im\omega t)$ on both sides of the equation and integrated over one period of $f(t)$. We can always take $t_0 = 0$ for convenience, since it is arbitrary.

5.1 The Fourier transform and orthogonal functions

One can generalize the Fourier theorem to a nonperiodic function defined in a region of $x \in [a, b]$ if one has a complete basis set of orthonormal functions $\phi_n(x)$ with

$$\int_a^b \phi_n^+(x)\phi_m(x)\,dx = \langle n \mid m \rangle = \delta_{nm}. \tag{5.5}$$

For any arbitrary function $f(x)$ defined in the region of $x \in [a, b]$, one can always write

$$f(x) = \sum_n g_n \phi_n(x) \tag{5.6}$$

if the function is square integrable, that is,

$$\int_a^b |f(x)|^2 dx < \infty, \tag{5.7}$$

and has consistent boundary values with $\phi_n(x)$; that is, $f(a)$ and $f(b)$ can be produced by the series in Eq. (5.6). The summation in the series is over all possible states, and the coefficient g_n is given by

$$g_n = \int_a^b f(x)\phi_n^+(x)\,dx = \langle n \mid f \rangle. \tag{5.8}$$

One can obtain the continuous Fourier transform by restricting the Fourier series to the region of $t \in [-T/2, T/2]$ and then extending the period T to infinity. We need to redefine $n\omega \to \omega$ and $\sum_n \to d\omega/\sqrt{2\pi}$. Then the summation becomes an integral,

$$f(t) = \frac{1}{\sqrt{2\pi}} \int_{-\infty}^{\infty} g(\omega)e^{i\omega t}\,d\omega, \tag{5.9}$$

which is commonly known as the Fourier integral. The Fourier coefficient is then given by

$$g(\omega) = \frac{1}{\sqrt{2\pi}} \int_{-\infty}^{\infty} f(t)e^{-i\omega t}\,dt. \tag{5.10}$$

Equations (5.9) and (5.10) define an integral transform and its inverse, which are commonly known as the Fourier transform and the inverse Fourier transform. One can show that the two equations above are consistent, that is, that one can obtain the second equation by multiplying the first equation by $\exp(-i\omega' t)$ and then integrating it. One will need to use the Dirac δ-function during this process. The Dirac δ-function is defined by

$$\delta(\omega - \omega') = \frac{1}{2\pi} \int_{-\infty}^{\infty} e^{i(\omega - \omega')t} \, dt, \qquad (5.11)$$

with

$$\int_a^b f(\omega)\delta(\omega - \omega') \, d\omega = \begin{cases} f(\omega') & \text{for } \omega' \in [a, b], \\ 0 & \text{otherwise.} \end{cases} \qquad (5.12)$$

The Fourier transform can also be applied to other types of variables or to higher dimensions. For example, the Fourier transform for a function $f(\mathbf{r})$ in three dimensions is given by

$$f(\mathbf{r}) = \frac{1}{(2\pi)^{3/2}} \int g(\mathbf{k}) e^{i\mathbf{k}\cdot\mathbf{r}} \, d\mathbf{k}, \qquad (5.13)$$

with the Fourier coefficient

$$g(\mathbf{k}) = \frac{1}{(2\pi)^{3/2}} \int f(\mathbf{r}) e^{-i\mathbf{k}\cdot\mathbf{r}} \, d\mathbf{r}. \qquad (5.14)$$

Note that both of the above integrals are three dimensional. The space defined by $\mathbf{k}$ is usually called the *momentum space*.

5.2 The discrete Fourier transform

The Fourier transform of a specific function is usually necessary in the analysis of experimental data, since it is often difficult to establish a clear physical picture just from the raw data taken from an experiment. Numerical procedures for the Fourier transform are inevitable in physics and other scientific fields. Usually we have to deal with a large number of data points, and the speed of the algorithm for the Fourier transform becomes a very important issue. We will first discuss a straightforward scheme for the one-dimensional case in this section to illustrate some basic aspects of the discrete Fourier transform (DFT). In the next section, we will outline a fast algorithm for the discrete Fourier transform, commonly known as the fast Fourier transform (FFT).

As we pointed out earlier, the one-dimensional Fourier transform is defined by Eq. (5.9) and Eq. (5.10); as always, we need to convert the continuous variables into discrete ones before we develop a numerical procedure.

Let us consider $f(t)$ as a time-dependent physical quantity obtained from experimental measurements. If the measurements are conducted between $t = 0$ and $t = T$, $f(t)$ is nonzero only for $t \in [0, T]$. To simplify our problem, we can assume that the data are taken at evenly spaced points with each interval $\tau = T/(N - 1)$.

The initial time is set at $t = 0$. The corresponding frequency ω is then discrete too, with an interval $\nu = 2\pi/T$. Here N is the total number of data points.

The discrete Fourier transform with such a data set can then be expressed in terms of a summation,

$$f_n = \frac{1}{\sqrt{N}} \sum_{k=0}^{N-1} g_k e^{i2\pi nk/N}, \tag{5.15}$$

with the Fourier coefficients given by

$$g_k = \frac{1}{\sqrt{N}} \sum_{n=0}^{N-1} f_n e^{-i2\pi nk/N}, \tag{5.16}$$

where we have used our convention that $f_n = f(t = n\tau)$ and $g_k = g(\omega = k\nu)$. One can show that the above two summations are consistent, meaning that the Fourier transform of the Fourier coefficients will give the exact values of $f(t)$ at $t = 0$, $\tau, \ldots, (N-1)\tau$. However, this inverse Fourier transform does not ensure smoothness in the recovered data function.

Note that the exponential functions in the series form a discrete basis set of orthogonal functions, that is,

$$\frac{1}{N} \sum_{n=0}^{N-1} e^{i2\pi n(k-l)} = \delta_{kl}. \tag{5.17}$$

Before we introduce the fast Fourier transform, let us examine how one can implement the discrete Fourier transform of Eq. (5.16) in a straightforward manner. We can separate the real (Re) and imaginary (Im) parts of the coefficients,

$$\mathrm{Re}(g_k) = \frac{1}{\sqrt{N}} \sum_{n=0}^{N-1} \left[\cos \frac{2\pi nk}{N} \mathrm{Re}(f_n) + \sin \frac{2\pi nk}{N} \mathrm{Im}(f_n) \right], \tag{5.18}$$

$$\mathrm{Im}(g_k) = \frac{1}{\sqrt{N}} \sum_{n=0}^{N-1} \left[\cos \frac{2\pi nk}{N} \mathrm{Im}(f_n) - \sin \frac{2\pi nk}{N} \mathrm{Re}(f_n) \right], \tag{5.19}$$

for convenience. Note that it is not necessary to separate them in practice even though most people would do so. The convenience is that one needs to deal only with real numbers after the real and imaginary parts of the coefficients are separated.

The following program is an implementation of Eqs. (5.18) and (5.19) with the real and imaginary parts of the coefficients separated.

```
          PROGRAM DFT_EXAMPLE
     C
     C Example of the discrete Fourier transform with
     C function f(x) = x(1-x) in [0,1]. The inverse
```

```
C transform is also performed for comparison.
C
      PARAMETER (N=128,M=8)
      DIMENSION FR(N),FI(N),GR(N),GI(N)
C
      F0 = 1.0/SQRT(FLOAT(N))
      H  = 1.0/(N-1)
C
      DO     150  I = 1, N
        X    = H*(I-1)
        FR(I) = X*(1.0-X)
        FI(I) = 0.0
  150 CONTINUE
C
      CALL DFT (FR,FI,GR,GI,N)
      DO     200  I = 1, N
        GR(I) = F0*GR(I)
        GI(I) = F0*GI(I)
  200 CONTINUE
C
C     Perform inverse Fourier transform
C
      DO     300  I = 1, N
        GI(I) = -GI(I)
  300 CONTINUE
      CALL DFT (GR,GI,FR,FI,N)
      DO     400  I = 1, N
        FR(I) =  F0*FR(I)
        FI(I) = -F0*FI(I)
  400 CONTINUE
      WRITE (6,999) (H*(I-1),FR(I),I=1,N,M)
      WRITE (6,999)  H*(N-1),FR(N)
      STOP
  999 FORMAT (2F16.8)
      END
C
      SUBROUTINE DFT (FR,FI,GR,GI,N)
C
C Subroutine to perform the discrete Fourier transform
C with FR and FI as the real and imaginary parts of the
C input and GR and GI as the corresponding parts of the
```

```
C output.
C
      DIMENSION FR(N),FI(N),GR(N),GI(N)
      PI = 4.0*ATAN(1.0)
      X  = 2*PI/N
C
      DO      150  I = 1, N
        GR(I) = 0.0
        GI(I) = 0.0
        DO    100  J = 1, N
          Q      = X*(J-1)*(I-1)
          GR(I) = GR(I)+FR(J)*COS(Q)+FI(J)*SIN(Q)
          GI(I) = GI(I)+FI(J)*COS(Q)-FR(J)*SIN(Q)
100     CONTINUE
150 CONTINUE
      RETURN
      END
```

The above program will take an amount of computing time that is proportional to N^2. We have used $f(x) = x(1-x)$ as the data function to test the subroutine in the region of $x \in [0, 1]$. This program demonstrates how one can perform the discrete Fourier transform and its inverse, that is, to obtain g_k from f_n and f_n from g_k in exactly the same manner. Note that we do not have the factor $1/\sqrt{N}$ in the subroutine, which is a common practice in designing a Fourier transform subroutine.

As we have pointed out, the inverse Fourier transform provides exactly the same values as the original data at the lattice points. The result of the inverse of the discrete Fourier transform from the above program and the original data are shown in Fig. 5.1. Inaccuracy can still appear at these lattice points, because rounding errors occur during computing. A more important issue here is the accuracy of the recovered data function at places other than the lattice points. When the Fourier transform or its inverse is performed, we can have only a discrete set of data points. So an interpolation is inevitable if we want to know any value that is not at the lattice points. We can increase the number of points to reduce the errors in the interpolation of the data, but we also have to watch the growth of the rounding errors with the number of points used.

5.3 The fast Fourier transform

As one can see clearly, the straightforward discrete Fourier transform algorithm used in the last section is very inefficient, since the computing time needed is

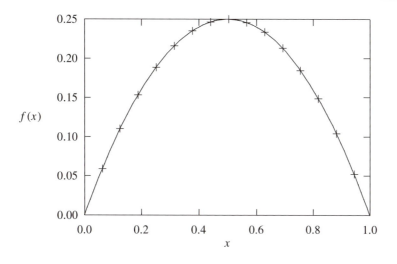

Fig. 5.1. This figure shows the function $f(x) = x(1 - x)$ (line) and the corresponding values from the inverse transform of its Fourier coefficients ($+$) calculated with the program DFT_EXAMPLE.

proportional to N^2. In order to solve this problem, many people have come up with the idea that is now known as the fast Fourier transform algorithm. The key element of the fast Fourier transform is to rearrange the terms in the series and have the summation performed in a hierarchical manner. For example, one can perform a series of pairwise additions to accomplish the summation if the number of data points is of the power of 2; that is, $N = 2^M$, where M is an integer.

The idea of the fast Fourier transform was conceived a long time ago, even before any computer was built. As mentioned in Chapter 1, Gauss considered the idea of the fast Fourier transform algorithm and published his work in neoclassical Latin (Gauss, 1886; Goldstine, 1977, pp. 249–53). But no one really noticed Gauss's idea or made any connection with modern computation. The fast Fourier transform algorithm was formally discovered and put into practice by Cooley and Tukey (1965). Here we will give a brief outline of their idea.

The simplest fast Fourier transform algorithm is accomplished with the observation that one can separate the odd and even terms in the Fourier transform:

$$g_k = \sum_{n=0}^{N/2-1} f_{2n} e^{-i2\pi(2n)k/N} + \sum_{n=0}^{N/2-1} f_{2n+1} e^{-i2\pi(2n+1)k/N},$$

$$= x_k + y_k e^{-i2\pi k/N}, \tag{5.20}$$

where

$$x_k = \sum_{n=0}^{N/2-1} f_{2n} e^{-i2\pi nk/(N/2)} \tag{5.21}$$

and

$$y_k = \sum_{n=0}^{N/2-1} f_{2n+1} e^{-i2\pi nk/(N/2)}. \tag{5.22}$$

Here we have ignored the factor $1/\sqrt{N}$, which can always be added in the program that calls the fast Fourier transform subroutine. What we have done is to rewrite the Fourier transform with a summation of N terms into two summations, each of $N/2$ terms. This process can be carried on further and further until eventually we have only two terms in each summation if $N = 2^M$, with M being an integer. There is one more symmetry between g_k for $k < N/2$ and g_k for $k \geq N/2$, that is,

$$g_k = x_k + w^k y_k, \tag{5.23}$$

$$g_{k+N/2} = x_k - w^k y_k, \tag{5.24}$$

where $w = \exp(-2\pi i/N)$ and $k = 0, 1, \ldots, N/2 - 1$. This is quite important in practice, since now we need to perform the transform only up to $k = N/2 - 1$, and the Fourier coefficients with higher k are obtained with the above equation at the same time.

There are two important ingredients in the fast Fourier transform algorithm. After the summation is decomposed M times, we need to add individual data points in pairs. However, due to the sorting of the odd and even terms in every level of decomposition, the points in each pair at the first level of additions can be very far apart in the original data string. However, Cooley and Tukey (1965) find that if one records the data string index with a binary number, a *bit-reversed order* will put each pair of data points next to each other for the summations at the first level. Let us take a set of 16 data points $f_0, f_1, \ldots, f_{15}$ as an example. If we record them with a binary index, we have $0000, 0001, 0010, 0011, \ldots, 1111$, for all the data points. Bit-reversed order is achieved if we reverse the order of each binary string. For example, the bit-reversed order of 1000 is 0001. So the order of the data after bit reversal is $f_0, f_8, f_4, f_{12}, \ldots, f_3, f_{11}, f_7, f_{15}$. Then the first level of additions is performed between f_0 and f_8, f_4 and $f_{12}, \ldots, f_3$ and f_{11}, and f_7 and f_{15}. After bit reversal, Eqs. (5.23) and (5.24) can be applied repeatedly with additions of pairs that are 2^{l-1} spaces apart. Here l indicates the level of additions; for example, the first set of additions corresponds to $l = 1$. At each level of additions, two data points are created from each pair. Note that w^k is associated with the second term in the equation.

With the fast Fourier transform algorithm, the computing time needed for a large set of data points is tremendously reduced. One can see this by examining the calculation steps needed in the transform. Assume that $N = 2^M$, so that after bit reversal, one needs to perform M levels of additions and $N/2$ additions at each level. A careful analysis will show that the total computing time in the fast Fourier transform algorithm is proportional to $N \log_2 N$ instead of to N^2, as is the case for the straightforward discrete Fourier transform. Similar algorithms can be devised for N of the power of 4, 8, and so forth.

Many versions and variations of the fast Fourier transform programs are available in Fortran (Burrus and Parks, 1985) and in other programming languages. One of the earliest Fortran programs of the fast Fourier transform was written by Cooley, Lewis, and Welch (1969). Now many computers come with a fast Fourier transform library, which is usually written in machine language and tailored specifically for the architecture of the system.

Most subroutines for the fast Fourier transform are written with complex variables. However, it is easier just to deal with real variables. One can always separate the real and imaginary parts in Eqs. (5.20) through (5.24), as discussed earlier. The following subroutine is such an example. However, it is more concise for programming to have both input and result in a complex form. If the data are real, one can simply remove the lines related to the imaginary part of the input data.

```
      SUBROUTINE FFT (AR,AI,N,M)
C
C The fast Fourier transform subroutine with N = 2**M.
C AR and AI are the real and imaginary parts of the data
C in the input and of the corresponding Fourier
C coefficients in the output.
C
      DIMENSION AR(N),AI(N)
C
      PI = 4.0*ATAN(1.0)
      N1 = 2**M
      IF(N1.NE.N) STOP 'Indices do not match.'
C
C Rearrange the data to the bit-reversed order
C
      L = 1
      DO      150  K = 1, N-1
        IF(K.LT.L) THEN
          A1    = AR(L)
          A2    = AI(L)
          AR(L) = AR(K)
          AI(L) = AI(K)
```

```
          AR(K) = A1
          AI(K) = A2
        ENDIF
        J   = N/2
        DO      100  WHILE (J.LT.L)
          L = L-J
          J = J/2
100     END DO
        L = L+J
150 CONTINUE
C
C Perform additions at all the levels with bit-reversed
C data
C
      L2 = 1
      DO      200  L = 1, M
        Q   =  0.0
        L1  =  L2
        L2  =  2*L1
        DO    190  K = 1, L1
          U   =  COS(Q)
          V   = -SIN(Q)
          Q   =  Q+PI/L1
          DO  180  J = K, N, L2
            I     =  J+L1
            A1    =  AR(I)*U-AI(I)*V
            A2    =  AR(I)*V+AI(I)*U
            AR(I) =  AR(J)-A1
            AR(J) =  AR(J)+A1
            AI(I) =  AI(J)-A2
            AI(J) =  AI(J)+A2
180       CONTINUE
190     CONTINUE
200 CONTINUE
      RETURN
      END
```

As one can see from the above subroutine, rearranging the data points into bit-reversed order is nontrivial. However, it can still be understood if one examines the part of the program iteration by iteration with a small N. To convince oneself that the above subroutine is a correct implementation of the $N = 2^M$ case, the reader can take $N = 16$ as an example and then work out the terms by hand.

A very important issue here is how much time the fast Fourier transform can save. On a scalar machine, it is quite significant. However, on a vector machine, the saving is somewhat restricted. A vector processor can have the inner loop in the discrete Fourier transform performed in parallel within each clock cycle, and this makes the computing time for the straightforward discrete Fourier transform proportional to N^α, with α somewhere between 1 and 2. In real problems, α may be a little bit higher than 2 for a scalar machine. However, the advantage of vectorization in the fast Fourier transform is not as significant as in the straightforward discrete Fourier transform. So in general, one may need to examine the problem under study, and the fast Fourier transform is certainly an important tool, since the available computing resources are always limited.

5.4 The power spectrum of a driven pendulum

We discussed in Chapter 3 the fact that a damped pendulum under a driving force can exhibit either periodic or chaotic behavior. One way to analyze the dynamics of a nonlinear system is to study its power spectrum.

The power spectrum of a dynamical variable is defined as the square of the modulus of the Fourier coefficient of the time dependence of the variable,

$$S(\omega) = |g(\omega)|^2, \tag{5.25}$$

where $g(\omega)$ is the Fourier transform of a dynamic variable $x(t)$,

$$g(\omega) = \frac{1}{\sqrt{2\pi}} \int_{-\infty}^{\infty} x(t) e^{-i\omega t} \, dt. \tag{5.26}$$

We discussed in the preceding section how to achieve a fast Fourier transform numerically. The evaluation of the power spectrum of a time-dependent variable then becomes straightforward.

The damped pendulum under a driving force is described by

$$\frac{dy_1}{dt} = y_2, \tag{5.27}$$

$$\frac{dy_2}{dt} = -qy_2 - \sin y_1 + b \cos \omega_0 t, \tag{5.28}$$

where $y_1(t) = \theta(t)$ is the angle between the pendulum and the vertical, $y_2(t) = d\theta(t)/dt$ is its corresponding angular velocity, q is the measure of the damping, and b and ω_0 are the amplitude and angular frequency of the external driving force. We already showed in Chapter 3 how to solve this problem numerically with the fourth-order Runge–Kutta algorithm. The power spectrum of the time-dependent angle and the time-dependent angular velocity can be obtained easily by performing a discrete Fourier transform with the fast Fourier transform algorithm. We show the numerical results in Figs. 5.2(a) and (b), corresponding to the power spectra of the time-dependent angle $\theta(t)$ in periodic and chaotic regions, respectively.

We have taken 65,536 data points with 192 points in one period of the driving force. The initial condition, $y_1(0) = 0$ and $y_2(0) = 2$, is used in generating the data. In Fig. 5.2, we have shown the first 2,048 points of the power spectra. As one can see from the figures, the power spectrum for periodic motion has separated sharp peaks with δ-function-like distributions. The broadening of the major peak at the angular frequency of the driving force, ω_0, is due to rounding errors. One can reduce the broadening by increasing the number of points used. The power

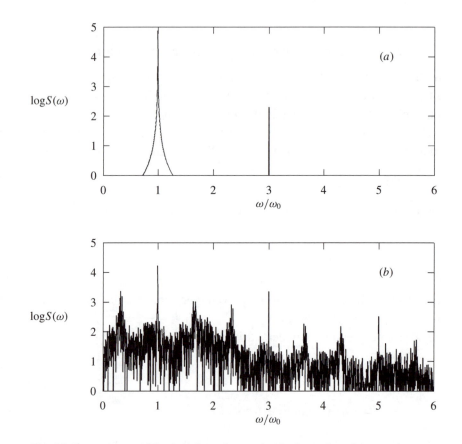

Fig. 5.2. Power spectra of the time-dependent angle of a damped pendulum under a driving force, (*a*) for periodic behavior with $\omega_0 = 2/3$, $q = 0.5$, and $b = 0.9$, and (*b*) for chaotic behavior with $\omega_0 = 2/3$, $q = 0.5$, and $b = 1.15$.

spectrum for the chaotic case is quite irregular, but still has noticeable peaks at the positions of the periodic case. This is because the contribution at the angular frequency of the driving force is still high, even though the system is chaotic. The peaks at multiples of ω_0 are due to relatively large contributions at the higher frequencies with $\omega = n\omega_0$ in the Fourier coefficient.

5.5 The Fourier transform in higher dimensions

The Fourier transform can be performed in a very straightforward manner in higher dimensions if one realizes that one can transform each coordinate as if it were a one-dimensional problem, with all other coordinate indices held constant.

Let us take the two-dimensional case as an example. Assume that the data are from a rectangular domain with N_1 mesh points in one direction and N_2 in the other, so the total number of points is $N = N_1 N_2$. The discrete Fourier transform is then given by

$$
g_{kl} = \frac{1}{\sqrt{N}} \sum_{n=0}^{N_1-1} \sum_{m=0}^{N_2-1} f_{nm} e^{-2\pi i(kn/N_1 + lm/N_2)}
$$

$$
= \frac{1}{\sqrt{N_1}} \sum_{n=0}^{N_1-1} e^{-2\pi kni/N_1} \frac{1}{\sqrt{N_2}} \sum_{m=0}^{N_2-1} f_{nm} e^{-2\pi lmi/N_2}. \tag{5.29}
$$

Thus, we can carry out the transform first for all the terms under index m with a fixed n and then carry out the transform for all the terms under index n with a fixed l. The procedure can be seen easily from the subroutine given below.

```
      SUBROUTINE FFT2D (FR,FI,N1,N2,M1,M2)
C
C Subroutine for the two-dimensional fast Fourier
C transform with N = N1*N2 and N1 = 2**M1 and
C N2 = 2**M2.
C
      DIMENSION FR(N1,N2), FI(N1,N2)
C
C Transformation under the second index
C
      DO        100  I = 1, N1
        CALL FFT (FR(I,1),FI(I,1),N2,M2)
  100 CONTINUE
```

```
C
C Transformation under the first index
C
      DO          250  J = 1, N2
          CALL FFT (FR(1,J),FI(1,J),N1,M1)
  200 CONTINUE
      RETURN
      END
```

We have used the one-dimensional fast Fourier transform twice, once for each of the two indices. This procedure works well if the boundary of the data is rectangular.

5.6 Wavelet analysis

Wavelet analysis was first introduced by Haar (1910) but not recognized as a powerful mathematical tool until about a decade ago. It was Morlet et al. (1982a; 1982b) who first used wavelet analysis in seismic data analysis. The wavelet transform contains spectral information at a different scale and a different location of the data stream, for example, if we want to know the intensity for a signal around a specific frequency and time. This is in contrast to the Fourier analysis, in which a specific transform coefficient contains the information of a specific scale or frequency from the entire data space without referring to the difference from one location to another in the data stream. The wavelet method is extremely powerful in the analysis of short time signals, transient data, or complex patterns. The development and applications of wavelet analysis in the last decade have shown that many more applications will emerge in the near future. Here we will give just a brief introduction to the subject and point out its potential applications in computational physics. More details on the method and some initial applications can be found in several monographs (Chui, 1992; Daubechies, 1992; Meyer, 1993; Young, 1993; Holschneider, 1996) and collections (Combes, Grossmann, and Tchanmitchian, 1990; Chui, Montefusco, and Puccio, 1994; Foufoula–Georgiou and Kumar, 1994).

It is desirable in many applications that the local structure in a set of data points can be amplified and analyzed. This is because we may not be able to obtain the data through the whole space or on a specific scale of particular interest. Sometimes we also want to filter out the noise around the boundaries of the data. A windowed Fourier transform can be formulated to select the information from the data at a specific location. One can define

$$g(\omega, \tau) = \frac{1}{\sqrt{2\pi}} \int_{-\infty}^{\infty} f(t)w(t - \tau)e^{-i\omega t} dt \tag{5.30}$$

as the windowed Fourier transform of the function $f(t)$ under the window function $w(t - \tau)$. The window function $w(t - \tau)$ is used here to extract information about $f(t)$ in the neighborhood of $t = \tau$. $w(t)$ is commonly chosen to be a real, even function with

$$\int_{-\infty}^{\infty} w^2(t)\, dt = 1. \tag{5.31}$$

Typical window functions such as the triangular window function

$$w(t) = \begin{cases} \mathcal{N}(1 - |t|/\sigma) & \text{for} \quad |t| < \sigma, \\ 0 & \text{otherwise,} \end{cases} \tag{5.32}$$

and the Gaussian window function

$$w(t) = \mathcal{N}e^{-(t/\sigma)^2}, \tag{5.33}$$

are often used in data analysis. Here $\mathcal{N}$ is the normalization constant and σ is the measure of the window width. As soon as σ is selected, $\mathcal{N}$ can be evaluated with Eq. (5.31). From the definition, one can also recover the data points from their Fourier coefficients through the inverse transform with the same window function:

$$f(t) = \frac{1}{\sqrt{2\pi}} \int_{-\infty}^{\infty} g(\omega, \tau) w(\tau - t) e^{i\omega t}\, d\omega\, d\tau. \tag{5.34}$$

As in the conventional Fourier transform, the inner product, that is, the integral over the square of the data, is equal to the integral over the square of its transform coefficient,

$$\int_{-\infty}^{\infty} |f(t)|^2 dt = \int_{-\infty}^{\infty} |g(\omega, \tau)|^2\, d\omega\, d\tau. \tag{5.35}$$

The advantage of the windowed Fourier transform is that it tunes the data with the window function so that we can obtain the local structure of the data or suppress unwanted effects in the data string. However, the transform treats the whole data space uniformly and would not be able to distinguish detailed structures of the data at different scales.

Continuous wavelet transform

One can obtain information about a set of data locally and also at different scales through wavelet analysis. The continuous wavelet transform of a function $f(t)$ is defined through the integral

$$g(\lambda, \tau) = \int_{-\infty}^{\infty} f(t) u_{\lambda\tau}^{+}(t)\, dt, \tag{5.36}$$

where $u_{\lambda\tau}^{+}(t)$ is the complex conjugate of the *wavelet*

$$u_{\lambda\tau}(t) = \frac{1}{\sqrt{|\lambda|}} u\left(\frac{t-\tau}{\lambda}\right), \tag{5.37}$$

with $\lambda \neq 0$ as the *dilate* and τ as the *translate* of the wavelet transform. λ and τ are usually chosen to be real. There are some constraints on the selection of a meaningful wavelet $u(t)$. For example, in order to have the inverse transform defined, we must have

$$\mathcal{Z} = \int_{-\infty}^{\infty} \frac{1}{|\omega|} |\tilde{u}(\omega)|^2 d\omega < \infty, \tag{5.38}$$

where $\tilde{u}(\omega)$ is the Fourier transform of $u(t)$. The above condition is equivalent to

$$\tilde{u}(0) = \int_{-\infty}^{\infty} u(t)\, dt = 0, \tag{5.39}$$

if $u(t)$ is square integrable and decays as $t \to \pm\infty$. The wavelet $u(t)$ is usually normalized with

$$\langle u \mid u \rangle = \int_{-\infty}^{\infty} |u(t)|^2 dt = 1 \tag{5.40}$$

for convenience. Typical examples of continuous wavelets are the modified Gaussian wavelet

$$u(t) = \frac{2}{\sqrt{3}}(1 - \pi t^2)e^{-\pi t^2/2}, \tag{5.41}$$

whose Fourier transform is given by $\tilde{u}(\omega) = 2\omega^2 \exp(-\omega^2/2\pi)/\sqrt{2\pi^3}$, and the Morlet wavelet

$$u(t) = e^{ivt - \pi t^2/2}, \tag{5.42}$$

whose Fourier transform is $\tilde{u}(\omega) = \exp[-(\omega - v)^2/2\pi]/\sqrt{\pi}$. The Morlet wavelet satisfies Eq. (5.39) approximately when v is large. From the definition of the continuous wavelet transform and the constraints on the selection of the wavelets $u(t)$, we can show that the data function can be recovered from the inverse transform

$$f(t) = \frac{1}{\mathcal{Z}} \int_{-\infty}^{\infty} \frac{1}{\lambda^2} g(\lambda, \tau) u_{\lambda\tau}(t)\, d\lambda\, d\tau. \tag{5.43}$$

One can also show that the wavelet transform satisfies an identity similar to that in the Fourier transform; that is,

$$\int_{-\infty}^{\infty} |f(t)|^2 \, dt = \frac{1}{\mathcal{Z}} \int_{-\infty}^{\infty} \frac{1}{\lambda^2} |g(\lambda\tau)|^2 \, d\lambda\, d\tau. \tag{5.44}$$

The physical meaning of the variables λ and τ can be interpreted geometrically with a two-dimensional graph. Grossmann, Kronland–Martinet, and Morlet (1989) have discussed this graphic interpretation of the wavelet transform. Mathematically, the continuous wavelet transform is easier to deal with than the discrete wavelet transform. However, most data obtained are discrete in nature. More important, it is much easier to implement the data analysis numerically if the transform is defined with discrete variables.

Discrete and orthonormal wavelet transform

A discrete version of the wavelet transform can be introduced similarly to the discrete Fourier transform by taking $\lambda = 1/a^j$ and $\tau = k\lambda$, with j being an integer. The most popular discrete wavelet transform is to take $a = 2$ and k to be an integer; the wavelet is then given by

$$u_{jk}(t) = 2^{j/2} u(2^j t - k). \tag{5.45}$$

A proper choice of $u_{jk}(t)$ can ensure that $u_{ij}(t)$ forms a complete orthonormal basis set (Daubechies, 1992). Then we can express a function $f(t)$ in terms of the wavelet in a series,

$$f(t) = \sum_{j=-\infty}^{\infty} \sum_{k=-\infty}^{\infty} c_{jk} u_{jk}^+(t), \tag{5.46}$$

with c_{jk} derived from the inner product of $f(t)$ and $u_{jk}(t)$. In practice, u_{ij} is chosen from a set of dual orthogonal functions; that is,

$$\langle u_{jk} \mid u_{lm} \rangle = \int_{-\infty}^{\infty} u_{jk}^+(t) u_{lm}(t)\, dt = \delta_{jl}\delta_{km}, \tag{5.47}$$

which leads to the coefficients in the expansion of $f(t)$ as

$$c_{jk} = \int_{-\infty}^{\infty} f(t) u_{jk}(t)\, dt = \langle f \mid u_{jk} \rangle. \tag{5.48}$$

This is quite similar to the Fourier series for a periodic function, except that we are dealing with two indices here. Note that the factor $2^{j/2}$ is introduced to ensure that $\langle u_{jk} \mid u_{jk} \rangle = 1$. In the case of a discrete data set, the inner product above should be interpreted as a summation with

$$c_{jk} = \sum_i f(t_i) u_{jk}(t_i) = \langle f \mid u_{jk} \rangle, \tag{5.49}$$

which is similar to the discrete Fourier transform.

As discovered by Mallat (1989), one does not need to directly generate u_{jk}, and the related integrals or summations for c_{jk} can be obtained through a series of matrix multiplications. A numerical algorithm known as the multiresolution decomposition scheme has been devised to combine the wavelet generation and transform. This so-called *tree algorithm* relates the data $f(t)$ and its wavelet transform coefficients by a set of consecutive matrix multiplications. In order to establish this algorithm, we need to know how to generate a wavelet numerically. Here we will discuss only the iterative generation of the wavelet; for more discussions on other methods, see Strang (1989), which also forms a very concise introduction to the subject. The dual orthogonal requirement of Eq. (5.47) drastically restricts the choices of suitable functions as appropriate wavelets. Daubechies (1988) devised a dyadic recursive scheme to generate the wavelets through a *scaling function* $v(t)$. The scaling function satisfies the *dilation equation*

$$v(t) = \sum_{k=0}^{N-1} d_k v(2t - k),$$
(5.50)

where d_k is a function of k, which is obtained from the constraints placed on $v(t)$. The corresponding wavelet is generated with the same set of coefficients,

$$u(t) = \sum_{k=0}^{N-1} (-1)^k d_k v(2t + k - N + 1).$$
(5.51)

Here N has to be an even integer and a different value of N gives a different wavelet basis set. The Daubechies wavelets are localized in the data space and semilocalized in the transformed space.

The scaling function has to satisfy a set of constraints so that the associated wavelets form an orthonormal and complete basis set. The dilation equation and $u(t)$ defined in Eq. (5.51) lead to the scaling and translating relations of the wavelets defined in Eq. (5.45) automatically. The scaling function itself is normalized,

$$\int_{-\infty}^{\infty} v(t) \, dt = 1.$$
(5.52)

Note that the same integral for the wavelet $u(t)$ vanishes. This normalization condition sets up the first equation for the coefficients d_k in the dilation equation,

$$\sum_{k=0}^{N-1} d_k = 2.$$
(5.53)

In order to have the corresponding wavelets form a dual orthogonal basis set, the scaling function is orthogonal to its translate,

$$\langle v_{jk} \mid v_{jl} \rangle = \delta_{kl},$$
(5.54)

where

$$v_{jk}(t) = 2^{j/2} v(2^j t - k).$$
(5.55)

One can show that the above requirement for the scaling function ensures all the conditions needed for the corresponding wavelets to form an orthogonal and complete basis set.

Before we discuss how to achieve the wavelet transform, we would like to mention how to obtain the coefficients d_k and then perform the summations of Eqs. (5.50) and (5.51). We have already mentioned that the normalization of the scaling function requires that the summation of all the coefficients be equal to 2. From the orthonormal conditions of $v_{ij}(x)$ and the dual orthogonal conditions of $u_{ij}(x)$, we have

$$\sum_{k=0}^{N-1} d_k^+ d_{k+2l} = 2\delta_{l0}, \qquad (5.56)$$

for $l = 0, 1, \ldots, N/2 - 1$, where d_k^+ is the complex conjugate of d_k. For the case of real coefficients, $d_k^+ = d_k$. The above $N/2$ equations cannot uniquely determine all N coefficients d_k. However, if we want to reproduce polynomials up to the $(N/2 - 1)$th order, we need all the moments under the wavelet to be zero (Strang, 1989),

$$\sum_{k=0}^{N-1} (-1)^k k^l d_k = 0, \qquad (5.57)$$

for $l = 0, 1, \ldots, N/2 - 1$, which supplies another $N/2$ equations. Note that the case of $l = 0$ in either Eq. (5.56) or Eq. (5.57) can be derived from Eq. (5.53) and other given relations. So Eqs. (5.53), (5.56), and (5.57) all together provide N independent equations for N coefficients d_k, and therefore they can be uniquely determined for a given N. For example, if we choose $N = 4$, we have

$$
\begin{aligned}
d_0 &= (1 + \sqrt{3})/4; \quad & d_1 &= (3 + \sqrt{3})/4; \\
d_2 &= (3 - \sqrt{3})/4; \quad & d_3 &= (1 - \sqrt{3})/4,
\end{aligned}
\qquad (5.58)
$$

which is commonly known as the D4 wavelet, discovered by Daubechies (1988). After one has all the coefficients d_k, one can easily generate the scaling function and wavelet iteratively. For example, if one starts from a box function $v^{(-1)}(t) = 1$ for $0 < t \leq 1$, and otherwise $v^{(-1)}(t) = 0$, the scaling function $v(t)$ can be obtained through the recursion relation (Strang, 1989)

$$v^{(n)}(t) = \sum_{k=0}^{N-1} d_k v^{(n-1)}(2t - k), \qquad (5.59)$$

for $n = 0, 1, 2, \ldots$, and we can generate $u(t)$ similarly through Eq. (5.51). The recursion needs to be performed many times until the difference reaches the finest scale in the data. Each recursion can be performed as a matrix multiplication.

However, one does not need to generate the scaling function or the wavelet in order to accomplish the discrete wavelet transform. With the scaling function and wavelet defined with Eq. (5.59), the transform of Eq. (5.46) can be rewritten into two parts,

$$f(t) = \sum_{k=-\infty}^{\infty} b_k v(t-k) + \sum_{j=0}^{\infty} \sum_{k=-\infty}^{\infty} c_{jk} u_{jk}(t), \tag{5.60}$$

where b_k is a set of new coefficients that comes from the summation of the terms from $j = -\infty$ to $j = -1$ under the definition of the scaling function and the wavelet. Now we can use the properties of the dilation equation to accomplish the multiresolution decomposition.

As we mentioned in Section 3.6, any one-dimensional finite region $[a, b]$ can be transformed into the region $[0, 1]$. So we can always assume that a finite set of data lies in $[0, 1]$. For the convenience of the transform, we can divide the region $[0, 1]$ into $N = 2^M$ evenly spaced points, with M an integer. Then let us assume that we have the data function $f(t_i)$, with $i = 1, 2, \ldots, 2^M$.

The discrete wavelet expansion can be written in a quite compact form if we cut the functions $v(t-k)$ and $u(2^j t - k)$ into segments in the region $[0, 1]$, $[1, 2]$, $\ldots$, and so forth. We can translate all the segments into the region $[0, 1]$ and then add them all up. For example, if we perform such an operation on a function $g(t)$, we have

$$g(t) = \sum_{l=-\infty}^{\infty} g(t-l) \tag{5.61}$$

for $0 < t \leq 1$, which is similar to the dilation equation without rescaling.

If this operation is applied to both $v(t-k)$ and $u(2^j t - k)$, we have

$$f(t) = b_0 v(t) + \sum_{j=0}^{\infty} \sum_{k=0}^{2^j-1} 2^{j/2} b_{2^j+k} u(2^j t - k), \tag{5.62}$$

which forms the basis of the tree algorithm of Mallat (1989) for the multiresolution decomposition scheme. The coefficients are now formally given by

$$b_0 = \int_0^1 f(t) v(t)\, dt \tag{5.63}$$

and

$$b_{2^j+k} = \int_0^1 f(t) u(2^j t - k)\, dt, \tag{5.64}$$

with $v(t)$ and $u(2^j t - k)$ as the summation of all the segments of the scaling function and the wavelet in the region $[0, 1]$ and all their translates. Note that the orthonormal conditions for $v_{jk}(t)$ and $u_{jk}(t)$ are still the same, except now the region of integration is $[0, 1]$. One can also show that if the scaling function is generated from a square box with $t \in [0, 1]$, the segment-added function $v(t) \equiv 1$ for $t \in [0, 1]$.

In the fast Fourier transform, when we have $N = 2^M$ data points in $f(t)$, we have then M levels of pairwise additions in the transform. In the discrete wavelet transform, this aspect is similar and we have M levels of resolutions. In other words, we have $N = 2^M$ coefficients b_k in the transform. As one can see from the dilation equation, the scaling function at the half-integer $t = k/2^j$ can be generated from the function at the integer $t = j$, and this iteration can be repeated to obtain all the points $t_k = k/2^M$, with $k = 1, 2, \ldots, 2^M$. Note that the $t = 0$ point is equivalent to the $t = 1$ point after the functions are wrapped up with the segment additions. So the scaling function will have its value at all the 2^M points generated after M iterations provided that $v(t)$ is a box function at the beginning of the iteration. This generation process can be cast into M matrix multiplications. Here we outline the procedure of the tree algorithm of Mallat (1989). Interested readers should consult the original work of Mallat and the discussion by Strang (1989).

First one has to decide what type of wavelet is to be used, that is, to determine the coefficients d_k. One can, for example, use the D4 wavelet discussed earlier in this section. Symbolically, we can write the data as

$$f(t) = b_0 v(t) + \sum_{j=0}^{M-1} f^{(j)}(t), \tag{5.65}$$

with

$$f^{(j)}(t) = \sum_{k=0}^{2^j-1} b_{2^j+k} 2^{j/2} u(2^j t - k). \tag{5.66}$$

It turns out that the above transform can be cast into a series of matrix multiplications and all the matrices involved can be generated recursively with permutation operations (Mallat, 1989). The inverse transform can also be cast into a series of matrix multiplications, with each matrix as the inverse of the corresponding matrix in the original transform. We will not go into further detail here but direct interested readers to Mallat (1989) and Strang (1989). Readers are encouraged to explore this area further if they are interested in wavelet analysis. Good discussions on the numerical aspects, including some computer programs associated with MATLAB or *Mathematica*, can be found in Newland (1993) and Rowe and Abbott (1995).

Like the discrete Fourier transform, the discrete wavelet transform can also be used in higher-dimensional spaces. Interested readers can find discussions of the two-dimensional wavelet transform in Newland (1993).

5.7 Special functions

The solutions of a special set of differential equations can be expressed in terms of polynomials. In some cases, each polynomial has an infinite number of terms. For example, the Schrödinger equation for a particle in a central potential $V(r)$ is

$$-\frac{\hbar^2}{2m}\nabla^2\Psi(\mathbf{r}) + V(r)\Psi(\mathbf{r}) = \epsilon\Psi(\mathbf{r}), \tag{5.67}$$

with

$$\nabla^2 = \frac{1}{r^2}\frac{\partial}{\partial r}r^2\frac{\partial}{\partial r} + \frac{1}{r^2\sin\theta}\frac{\partial}{\partial\theta}\sin\theta\frac{\partial}{\partial\theta} + \frac{1}{r^2\sin^2\theta}\frac{\partial^2}{\partial\phi^2} \tag{5.68}$$

in spherical coordinates. One can assume that the solution $\Psi(\mathbf{r})$ is of the form

$$\Psi(r,\theta,\phi) = R(r)Y(\theta,\phi), \tag{5.69}$$

and then the equation becomes

$$\frac{1}{R}\frac{d}{dr}r^2\frac{dR}{dr} + \frac{2mr^2}{\hbar^2}[\epsilon - V(r)] = -\frac{1}{Y}\left(\frac{1}{\sin\theta}\frac{\partial}{\partial\theta}\sin\theta\frac{\partial Y}{\partial\theta} + \frac{1}{\sin^2\theta}\frac{\partial^2 Y}{\partial\phi^2}\right)$$
$$= \lambda, \tag{5.70}$$

where λ is an introduced parameter to be determined by solving the eigenvalue problem. One can further assume that $Y(\theta,\phi) = \Theta(\theta)\Phi(\phi)$ and that the equation for $\Theta(\theta)$ is

$$\frac{1}{\sin\theta}\frac{d}{d\theta}\sin\theta\frac{d\Theta}{d\theta} + \left(\lambda - \frac{m^2}{\sin^2\theta}\right)\Theta = 0, \tag{5.71}$$

with the corresponding equation

$$\Phi''(\phi) = -m^2\Phi(\phi) \tag{5.72}$$

for $\Phi(\phi)$. Here m is another introduced parameter and has to be an integer, since $\Phi(\phi)$ has to be a periodic function of ϕ, that is, $\Phi(\phi+2\pi) = \Phi(\phi)$. Note that $\Phi(\phi) = A\exp(im\phi) + B\exp(-im\phi)$ is the solution of the differential equation for $\Phi(\phi)$.

In order to have the solution of $\Theta(\theta)$ to be finite everywhere with $\theta \in [0,\pi]$, we have to have $\lambda = l(l+1)$, with l being a positive integer. Such solutions $\Theta(\theta) = P_l^m(\cos\theta)$ are called the associated Legendre polynomials,

$$P_l^m(x) = (1-x^2)^{m/2}\frac{d^m}{dx^m}P_l(x), \tag{5.73}$$

with $P_l(x)$ being the Legendre polynomials that satisfy the recursion relation

$$(l+1)P_{l+1}(x) = (2l+1)x P_l(x) - l P_{l-1}(x), \tag{5.74}$$

starting with $P_0(x) = 1$ and $P_1(x) = x$. We can then obtain all $P_l(x)$ from Eq. (5.74) with $l = 2, 3, 4, \ldots$. One can easily show that $P_l(x)$ is the solution of the equation

$$\frac{d}{dx}(1-x^2)\frac{d}{dx}P_l(x) + l(l+1)P_l(x) = 0 \tag{5.75}$$

and is a polynomial of order l in the region $[-1, 1]$, as we discussed in Chapter 4 in the calculation of the electronic structure of atoms.

Legendre polynomials are useful in almost every subfield of physics and engineering where partial differential equations involving spherical coordinates need to be solved. For example, the electrostatic potential of a charge distribution $\rho(\mathbf{r})$ can be written as

$$\Phi(\mathbf{r}) = \int \frac{\rho(\mathbf{r}') d\mathbf{r}'}{|\mathbf{r} - \mathbf{r}'|}, \tag{5.76}$$

where $1/|\mathbf{r} - \mathbf{r}'|$ can be expanded with the application of Legendre polynomials:

$$\frac{1}{|\mathbf{r} - \mathbf{r}'|} = \sum_{l=0}^{\infty} \frac{r_<^l}{r_>^{l+1}} P_l(\cos\theta), \tag{5.77}$$

where $r_<$ $(r_>)$ is the smaller (greater) value of r and r' and θ is the angle between $\mathbf{r}$ and $\mathbf{r}'$. So if we can generate $P_l(\cos\theta)$, we can evaluate the electrostatic potential with any given charge distribution $\rho(\mathbf{r})$ term by term.

We can generate the numerical values of the Legendre polynomials from the recursion relation given in Eq. (5.74). Here is a simple subroutine that generates the Legendre polynomials up to a maximum l for any given x.

```
        SUBROUTINE LGND(LMAX,X,P)
C
C Subroutine to generate the Legendre polynomials P_l(X)
C for l = 0,1,...,LMAX with a given x.
C
        DIMENSION P(0:LMAX)
        P(0) = 1.0
        P(1) = X
        DO 100 L = 1, LMAX-1
          P(L+1) = ((2.0*L+1)*X*P(L)-L*P(L-1))/(L+1)
    100 CONTINUE
        RETURN
        END
```

Legendre polynomials form a set of orthogonal basis functions in the region $[-1, 1]$, which can be used to achieve the least-squares approximations, as discussed in Chapter 2. If we choose

$$U_l(x) = \sqrt{\frac{2l+1}{2}} P_l(x),$$ (5.78)

one can easily show that $U_l(x)$ satisfy

$$\int_{-1}^{1} U_l(x) U_{l'}(x)\, dx = \delta_{ll'}$$ (5.79)

and can be used to express any function $f(x)$ in the region $[-1, 1]$ as

$$f(x) = \sum_{l=0}^{\infty} a_l U_l(x)$$ (5.80)

with

$$a_l = \int_{-1}^{1} f(x) U_l(x)\, dx.$$ (5.81)

This is very similar to the Fourier transform. In fact, almost every aspect associated with the Fourier transform can be generalized to other orthogonal functions. There is a whole class of orthogonal polynomials that are similar to the Legendre polynomials and can be generated in exactly the same fashion we have just described. Detailed discussions of these orthogonal polynomials can be found in Hochstrasser (1965).

In the case of cylindrical coordinates, the equation governing the Laplace operator in the radial direction is the so-called Bessel equation

$$\frac{d^2 J(x)}{dx^2} + \frac{1}{x}\frac{d J(x)}{dx} + \left(1 - \frac{v^2}{x^2}\right) J(x) = 0,$$ (5.82)

where v is a parameter and the solution of the equation is called the Bessel function of order v. Here v can be an integer or a fraction. The recursion relation for the Bessel functions is

$$J_{v\pm1}(x) = \frac{2v}{x} J_v(x) - J_{v\mp1}(x),$$ (5.83)

which can be used in a similar fashion as for the Legendre polynomials to generate $J_{v\pm1}(x)$ from $J_v(x)$ and $J_{v\mp1}(x)$.

Bessel functions can be further divided into two types, depending on their asymptotic behavior. We call the ones with finite values as $x \to 0$ the first kind and denote them by $J_v(x)$; the ones that diverge as $x \to 0$ are called the second kind and denoted by $Y_v(x)$. We will consider only the case where v is an integer.

In practice, there are two more problems in generating Bessel functions numerically. Bessel functions have an infinite number of terms in series representation, so
it is difficult to initiate the recursion relation numerically, and the Bessel function
of the second kind, $Y_\nu(x)$, increases exponentially with ν when $\nu > x$.

These problems can be resolved if we carry out the recursion relation forward
for $Y_\nu(x)$ and backward for $J_\nu(x)$. We can use several properties to initiate the
recursion. For $J_\nu(x)$, we can set the first two points as $J_N(x) = 0$ and $J_{N-1}(x) = 1$
in the backward recursion. The functions generated can be rescaled afterward
with

$$R_N = J_0(x) + 2J_2(x) + \cdots + 2J_N(x), \tag{5.84}$$

since $R_\infty = 1$ for the actual Bessel functions and

$$\lim_{\nu \to \infty} J_\nu(x) = 0. \tag{5.85}$$

For $Y_\nu(x)$, we can use the values obtained for $J_0(x), J_1(x), \ldots, J_N(x)$ to initiate
the first two points,

$$Y_0(x) = \frac{2}{\pi}[\ln(x/2) + \gamma]J_0(x) - \frac{4}{\pi}\sum_{k=1}^{\infty}(-1)^k\frac{J_{2k}(x)}{k}, \tag{5.86}$$

and

$$Y_1(x) = \frac{1}{J_0(x)}\left[J_1(x)Y_0(x) - \frac{2}{\pi x}\right], \tag{5.87}$$

in the forward recursion. Here γ is the Euler constant,

$$\gamma = \lim_{N \to \infty}\left[\sum_{k=1}^{N}\frac{1}{k} - \ln N\right] = 0.5772156649\ldots. \tag{5.88}$$

The summation in Eq. (5.86) is truncated at $2k = N$, which should not introduce
too much error into the functions, since $J_\nu(x)$ exponentially decreases with ν for
$\nu > x$. The following subroutine is an implementation of the above schemes for
generating the first and second kind of Bessel functions.

```
      SUBROUTINE BSSL (BJ,BY,N,X)
C
C Subroutine to generate J_n(x) and Y_n(x) with a
C given x up to n = NMAX - NTEL.
C
      PARAMETER(NMAX=30,NTEL=5)
      DIMENSION BJ(0:N),BY(0:N),B1(0:NMAX)
C
      IF(N.GT.(NMAX-NTEL)) STOP 'N is too large.'
C
      PI    = 4.0*ATAN(1.0)
      GAMMA = 0.5772156649
C
      B1(NMAX)   = 0.0
      B1(NMAX-1) = 1.0
C
C Generating J_n(x)
C
      SUM = 0.0
      DO   100  I  = NMAX-1, 1, -1
        B1(I-1) = 2*I*B1(I)/X-B1(I+1)
        IF(MOD(I,2).EQ.0) SUM = SUM+2.0*B1(I)
  100 CONTINUE
      SUM = SUM + B1(0)
C
      DO   200  I  = 0, N
        BJ(I) = B1(I)/SUM
  200 CONTINUE
C
C Generating Y_n(x) starts here
C
      SUM1 = 0.0
      DO   300  K  = 1, NMAX/2
         SUM1 =SUM1+(-1)**K*B1(2*K)/K
  300 CONTINUE
C
      SUM1  = -4.0*SUM1/(PI*SUM)
      BY(0) = 2.0*(ALOG(X/2.0)+GAMMA)*BJ(0)/PI+SUM1
      BY(1) = (BJ(1)*BY(0)-2.0/(PI*X))/BJ(0)
C
```

```
      DO    400   I  = 1, N-1
         BY(I+1) = 2*I*BY(I)/X-BY(I-1)
  400 CONTINUE
C
      RETURN
      END
```

If one needs to generate only $J_\nu(x)$, the lines associated with $Y_\nu(x)$ can be deleted from the above subroutine. Note that the variable x needs to be smaller than the maximum ν in order to have the value of $J_\nu(x)$ be accurate.

5.8 Gaussian quadrature

We can use special functions to construct numerical integration quadratures that automatically minimize the possible errors due to the deviation of approximate functions from the data, the same concept we used in Chapter 2 to obtain the approximation of a function by orthogonal polynomials. In fact, all special functions form vector spaces.

An integral defined in the region $[a, b]$ can be written as

$$I = \int_a^b w(x)f(x)\,dx, \tag{5.89}$$

where $w(x)$ is the weight of the integral, which has exactly the same meaning as the weight of the orthogonal functions defined in Chapter 2. Examples of $w(x)$ will be given in this section.

Now we divide the region $[a, b]$ into N points ($N - 1$ slices) and approximate the integral as

$$I \simeq \sum_{n=1}^{N} w_n f(x_n), \tag{5.90}$$

with x_n and w_n determined according to two criteria, the simplicity of the expression and the accuracy of the approximation.

The expression in Eq. (5.90) is quite general and includes the simplest quadratures introduced in Chapter 2 with $w(x) = 1$. For example, if we take

$$x_n = a + \frac{n-1}{N-1}(b-a) \tag{5.91}$$

for $n = 1, 2, \ldots, N$ and

$$w_n = \frac{1}{N(1 + \delta_{n1} + \delta_{nN})}, \tag{5.92}$$

we recover the trapezoid rule. The Simpson rule is recovered if we take a more complicated w_n with

$$
w_n = \begin{cases} 1/3N & \text{for} \quad n = 1 \text{ or } N, \\ 1/N & \text{for} \quad n = \text{even number}, \\ 2/3N & \text{for} \quad n = \text{odd number}, \end{cases} \tag{5.93}
$$

but still the same set of x_n with $w(x) = 1$.

The so-called Gaussian quadrature is constructed from a set of orthogonal polynomials $\phi_l(x)$ with

$$
\int_a^b \phi_l(x) w(x) \phi_{l'}(x) \, dx = \langle \phi_l \mid \phi_{l'} \rangle = \mathcal{N}_l \delta_{ll'}, \tag{5.94}
$$

where the definition of each quantity is exactly the same as in Section 2.1. One can show that to choose x_n to be the roots of $\phi_N(x) = 0$ and to choose

$$
w_n = \frac{-a_N \mathcal{N}_N}{\phi_N'(x_n)\phi_{N+1}(x_n)}, \tag{5.95}
$$

with $n = 1, 2, \ldots, N$, ensure that the error in the quadrature is given by

$$
\Delta I = \frac{\mathcal{N}_N}{A_N^2 (2N)!} f^{(2N)}(x_0), \tag{5.96}
$$

where x_0 is a value of $x \in [a, b]$, A_N is the coefficient of the x^N term in $\phi_N(x)$, and $a_N = A_{N+1}/A_N$. We can use any kind of orthogonal polynomials for this purpose. For example, with the Legendre polynomials, we have $a = -1$, $b = 1$, $w(x) = 1$, $a_N = (2N + 1)/(N + 1)$, and $\mathcal{N}_N = 2/(2N + 1)$.

Another set of very useful orthogonal polynomials is the Chebyshev polynomials, defined in the region $[-1, 1]$. For example, the recursion relation for the Chebyshev polynomials of the first kind is

$$
T_{n+1}(x) = 2x T_n(x) - T_{n-1}(x), \tag{5.97}
$$

starting with $T_0(x) = 1$ and $T_1(x) = x$. One can easily show that

$$
\int_{-1}^1 T_n(x) w(x) T_m \, dx = \delta_{mn}, \tag{5.98}
$$

with $w(x) = 1/\sqrt{1 - x^2}$. If we write an integral with the same weight into a series,

$$
\int_{-1}^1 \frac{f(x) \, dx}{\sqrt{1 - x^2}} \simeq \sum_{k=1}^N w_k f(x_k), \tag{5.99}
$$

we have

$$
x_k = \cos \frac{(2k - 1)\pi}{2N} \tag{5.100}
$$

and $w_k = \pi/N$, which are extremely easy to use. Note that we can translate a given integration region $[a, b]$ to another, for example, $[0, 1]$ or $[-1, 1]$, through a linear coordinate transformation, as discussed in Section 3.6.

Exercises

5.1 Run the discrete Fourier transform program and the fast Fourier transform program on a computer and establish the time dependence on the number of points for both of them. Will it be different if the processor is a vector processor, that is, if a segment of the inner loop is performed in one clock cycle?

5.2 Analyze the power spectrum of the Duffing model from Exercise 3.4. What is the most significant difference between the Duffing model and the driven pendulum with damping?

5.3 Write a subroutine that generates the Chebyshev polynomials of the first kind discussed in Section 5.8.

5.4 The Laguerre polynomials form an orthogonal basis set in the region $[0, \infty]$ and satisfy the following recursion relation

$$(n + 1)L_{n+1}(x) = (2n + 1 - x)L_n(x) - nL_{n-1}(x),$$

starting with $L_0(x) = 1$ and $L_1(x) = 1 - x$. (a) Write a subroutine that generates the Laguerre polynomials for a given x and maximum n. (b) Show that the weight of the polynomials $w(x) = \exp(-x)$, and that the normalization factor $\mathcal{N}_n = \int_0^\infty w(x)L_n^2(x)dx = 1$. (c) If we want to construct the Gaussian quadrature for the integral

$$I = \int_0^\infty e^{-x} f(x)\, dx = \sum_{n=1}^{N} w_n f(x_n) + \Delta I,$$

with x_n as the nth root of $L_N(x) = 0$, show that w_n is given by

$$w_n = \frac{(N!)^2 x_n}{(N + 1)^2 L_{N+1}^2(x_n)}.$$

This quadrature is extremely useful in statistical physics, where the integrals are typically weighted with an exponential function.

5.5 The Hermite polynomials are another set of important orthogonal polynomials in the region $[-\infty, \infty]$ and satisfy the following recursion relation

$$H_{n+1}(x) = 2x H_n(x) - 2n H_{n-1}(x),$$

starting with $H_0(x) = 1$ and $H_1(x) = 2x$. (a) Write a subroutine that generates the Hermite polynomials for a given x and maximum n. (b) Show that the weight of the polynomials $w(x) = \exp(-x^2)$, and that the normalization factor $\mathcal{N}_n = \int_{-\infty}^{\infty} w(x)H_n^2(x)dx = \sqrt{\pi}2^n n!$. (c) Now if we want to construct the Gaussian quadrature for the integral

$$I = \int_{\infty}^{\infty} e^{-x^2} f(x)\,dx = \sum_{n=1}^{N} w_n f(x_n) + \Delta I,$$

with x_n as the nth root of $H_N(x) = 0$, show that w_n is given by

$$w_n = \frac{2^{N-1}N!\sqrt{\pi}}{N^2 H_{N-1}^2(x_n)}.$$

Many integrals involving the Gaussian distribution can take advantage of this quadrature.

5.6 The integral expressions for the Bessel functions of the first and second kinds of order zero are given by

$$J_0(x) = \frac{2}{\pi} \int_0^{\infty} \sin(x \cosh t)\,dt,$$

$$Y_0(x) = -\frac{2}{\pi} \int_0^{\infty} \cos(x \cosh t)\,dt,$$

with $x > 0$. Calculate these two expressions numerically and compare them with the values generated with the subroutine in Section 5.7.

5.7 Demonstrate numerically that for a small angle θ but large l,

$$P_l(\cos \theta) \simeq J_0(l\theta).$$

5.8 In quantum scattering, when the incident particle has very high kinetic energy, the cross section

$$\sigma = 2\pi \int_0^{\pi} \sin \theta |f(\theta)|^2\,d\theta$$

has a very simple form with

$$f(\theta) \simeq -\frac{2m}{q\hbar^2} \int_0^{\infty} \sin(qr)V(x)r\,dr,$$

where $q = |\mathbf{k}_f - \mathbf{k}_i| = 2k \sin(\theta/2)$ and $\mathbf{k}_i$ and $\mathbf{k}_f$ are initial and final momenta of the particle with the same magnitude k. (a) Show that the above $f(\theta)$ is the dominant term as $k \to \infty$. (b) If the particle is an electron and the scattering potential is an ionic potential

$$V(r) = \frac{Ze^2}{r}e^{-r/r_0},$$

write a program to evaluate the cross section with the Gaussian quadrature from the Laguerre polynomials for the integrals. Here Z, e, and r_0 are the number of protons, the proton charge, and the screening length. Use $Z = 2$ and $r_0 = a_0/4$ with a_0 being the Bohr radius as a testing example.

5.9 Another approximation in quantum scattering, called the Eikonal approximation, is valid for high-energy and small-angle scattering. The scattering cross section can be written as

$$\sigma = 8\pi \int_0^\infty b \sin^2[\alpha(b)]\, db,$$

with

$$\alpha(b) = -\frac{m}{2\hbar^2 k} \int_{-\infty}^\infty V(b, x)\, dx,$$

where b is the impact parameter and x is the coordinate of the particle along the impact direction with the scattering center at $x = 0$. (a) Show that the Eikonal approximation is valid for high-energy and small-angle scattering. (b) If the scattering potential is

$$V(r) = -V_0 e^{-(r/r_0)^2},$$

calculate the cross section with the Gaussian quadratures of Hermite and Laguerre polynomials for the integrals. Here V_0 and r_0 are given parameters.

6

Partial differential equations

In Chapter 3, we discussed numerical methods for solving initial-value and boundary-value problems given in the form of differential equations with one independent variable, that is, ordinary differential equations. Many of those methods can be generalized to study differential equations involving more than one independent variable, that is, partial differential equations. Many physics problems are given in the form of a second-order partial differential equation, either elliptic, parabolic, or hyperbolic.

6.1 Partial differential equations in physics

In this chapter, we discuss numerical schemes for solving partial differential equations. There are several types of partial differential equations in physics. For example, the Poisson equation

$$\nabla^2 \phi(\mathbf{r}) = -4\pi\rho(\mathbf{r}) \tag{6.1}$$

for the electrostatic potential $\phi(\mathbf{r})$, at the position $\mathbf{r}$, with a charge distribution $\rho(\mathbf{r})$, is a typical elliptic equation. The diffusion equation

$$\frac{\partial n(\mathbf{r}, t)}{\partial t} - \nabla \cdot D(\mathbf{r})\nabla n(\mathbf{r}, t) = S(\mathbf{r}, t) \tag{6.2}$$

for the concentration of $n(\mathbf{r}, t)$, at the position $\mathbf{r}$ and time t, with a source $S(\mathbf{r}, t)$, is a typical parabolic equation. Here $D(\mathbf{r})$ is the diffusion coefficient at the position $\mathbf{r}$. The wave equation

$$\frac{1}{c^2}\frac{\partial^2 u(\mathbf{r}, t)}{\partial t^2} - \nabla^2 u(\mathbf{r}, t) = R(\mathbf{r}, t) \tag{6.3}$$

for the displacement $u(\mathbf{r}, t)$, at the position $\mathbf{r}$ and time t, with a source $R(\mathbf{r}, t)$, is a typical hyperbolic equation. The time-dependent Schrödinger equation

$$-\frac{\hbar}{i}\frac{\partial \Psi(\mathbf{r}, t)}{\partial t} = \mathcal{H}\Psi(\mathbf{r}, t) \tag{6.4}$$

is for a quantum system described by the Hamiltonian $\mathcal{H}$. The Schrödinger equation looks like a diffusion equation with an imaginary time.

All the above equations are linear if the sources and other quantities in the equations are not related to the solutions. There are also nonlinear equations in physics, which rely heavily on numerical solutions. For example, the equations for fluid dynamics,

$$\frac{\partial \mathbf{v}}{\partial t} = -(\mathbf{v} \cdot \nabla)\mathbf{v} - \frac{1}{\rho}\nabla P + \eta \nabla^2 \mathbf{v}, \tag{6.5}$$

$$\frac{\partial \rho}{\partial t} + \nabla \cdot \rho \mathbf{v} = 0, \tag{6.6}$$

$$f(P, \rho) = 0, \tag{6.7}$$

require numerical solutions under most conditions. Here the first equation is the Navier–Stokes equation, with $\mathbf{v}$ being the velocity, ρ being the density, P being the pressure, and η being the kinematic viscosity of the fluid. The Navier–Stokes equation can be derived from the Newton equation for a small element in the fluid. The second equation is the continuity equation, which is the result of mass conservation. The third equation is the equation of state, which can also involve temperature as another variable in many cases.

In this chapter, we will first outline an analytic scheme that can simplify numerical tasks drastically in many cases and is commonly known as the separation of variables. The basic idea of the separation of variables is to reduce a partial differential equation to several ordinary differential equations. The separation of variables is the standard method for analytic solutions of most partial differential equations. Even for the numerical study of partial differential equations, due to the limitation of the computing resources, the separation of variables can serve the purpose of simplifying problems to the point where they can be solved with the available computing resources.

6.2 Separation of variables

Before we introduce numerical schemes for partial differential equations, we would like to briefly discuss an analytic method, that is, the *separation of variables*, in solving partial differential equations. In many cases, a combination of the separation of variables and a numerical scheme will increase the speed of computing and the accuracy of the solution. The combination of analytic and numerical methods becomes critical in cases where the memory and speed of the available computing resources are limited. This section is far from being a complete discussion; interested readers should consult a standard textbook, such as Courant and Hilbert (1989).

Here we will just illustrate how the method works. Each variable (time or one of the spatial coordinates) is isolated from the rest, and the ordinary equation for each variable is solved before they are combined into the general solution. Boundary and initial conditions are then used to determine the general parameters in the solution, which appear as coefficients or eigenvalues.

We will take the diffusion equation in an isotropic, three-dimensional, and infinite space, with a time-dependent source and a zero initial value, as an example. The diffusion equation with a time-dependent source and a constant diffusion coefficient is given by

$$\frac{\partial n(\mathbf{r}, t)}{\partial t} - D\nabla^2 n(\mathbf{r}, t) = S(\mathbf{r}, t). \tag{6.8}$$

The initial value $n(\mathbf{r}, 0) = 0$ will be considered first. One can view the source as an infinite number of impulsive sources, each within a time interval $d\tau$, given by $S(\mathbf{r}, \tau)\delta(t - \tau)d\tau$. Then the solution $n(\mathbf{r}, t)$ is the superposition of the solution in each time interval $d\tau$, with

$$n(\mathbf{r}, t) = \int_0^t \eta(\mathbf{r}, t; \tau) \, d\tau, \tag{6.9}$$

where $\eta(\mathbf{r}, t; \tau)$ satisfies

$$\frac{\partial \eta(\mathbf{r}, t; \tau)}{\partial t} - D\nabla^2 \eta(\mathbf{r}, t; \tau) = S(\mathbf{r}, \tau)\delta(t - \tau), \tag{6.10}$$

with $\eta(\mathbf{r}, 0; \tau) = 0$. The impulse equation above is equivalent to a homogeneous equation

$$\frac{\partial \eta(\mathbf{r}, t; \tau)}{\partial t} - D\nabla^2 \eta(\mathbf{r}, t; \tau) = 0, \tag{6.11}$$

with the initial condition $\eta(\mathbf{r}, \tau; \tau) = S(\mathbf{r}, \tau)$. Thus we have transformed the original inhomogeneous equation with a zero initial value into an integral of a function that is a solution of a homogeneous equation with a nonzero initial value. We can generally write the solution of the inhomogeneous equation as the sum of a solution of the corresponding homogeneous equation and a particular solution of the inhomogeneous equation with the given initial condition satisfied. So if we find a way to solve the corresponding homogeneous equation with a nonzero initial value, we find the solution of the general problem.

Now let us turn to the separation of variables for a homogeneous diffusion equation with a nonzero initial value. We will assume that the space is an isotropic, three-dimensional, and infinite space, in order to simplify our discussion. Other finite boundary situations can be solved in a similar manner. We will also suppress the parameter τ to simplify our notation. The solution of $\eta(\mathbf{r}, t) = \eta(\mathbf{r}, t; \tau)$ can be assumed to be

$$\eta(\mathbf{r}, t) = X(\mathbf{r})T(t), \tag{6.12}$$

where $X(\mathbf{r})$ is a function of $\mathbf{r}$ only and $T(t)$ is a function of t only. If we substitute this assumed solution into the equation, we have

$$XT' - DT\nabla^2 X = 0, \tag{6.13}$$

or

$$\frac{T'}{DT} = \frac{\nabla^2 X}{X} = -k^2, \tag{6.14}$$

where $k = |\mathbf{k}|$ is an introduced parameter that will be determined later. Now we have two separated equations,

$$\nabla^2 X(\mathbf{r}) + k^2 X(\mathbf{r}) = 0, \tag{6.15}$$

and

$$T'(t) + \omega T(t) = 0, \tag{6.16}$$

which are related by $\omega = Dk^2$. The spatial equation is now a standard eigenvalue problem with k^2 as the eigenvalues to be determined by the specific boundary condition. Since we have assumed that the space is an isotropic, three-dimensional, and infinite space, the solution is given by plane waves,

$$X(\mathbf{r}) = Ce^{i\mathbf{k}\cdot\mathbf{r}}, \tag{6.17}$$

where C is a general constant. Similarly, the time-dependent equation for $T(t)$ is a standard initial-value problem with the solution

$$T(t) = A(\mathbf{k})e^{-\omega(t-\tau)}, \tag{6.18}$$

where $\omega = Dk^2$ and $A(\mathbf{k})$ is a coefficient to be determined by the initial condition of the problem. If we combine the spatial solution and the time solution, we have

$$\eta(\mathbf{r}, t) = \int A(\mathbf{k})e^{i\mathbf{k}\cdot\mathbf{r}-\omega(t-\tau)} \, d\mathbf{k}, \tag{6.19}$$

and we can obtain $A(\mathbf{k})$ from the initial value of $\eta(\mathbf{r}, \tau) = S(\mathbf{r}, \tau)$ with the Fourier transform of the above equation:

$$A(\mathbf{k}) = \frac{1}{(2\pi)^{3/2}} \int S(\mathbf{r}, \tau)e^{-i\mathbf{k}\cdot\mathbf{r}} \, d\mathbf{r}. \tag{6.20}$$

If we substitute the above result for $A(\mathbf{k})$ into the integral for $\eta(\mathbf{r}, t)$ and complete the integration over $\mathbf{k}$, we obtain

$$\eta(\mathbf{r}, t) = \frac{1}{\sqrt{4\pi Dt}} \int S(\mathbf{r}', \tau) \exp\left[-\frac{(\mathbf{r} - \mathbf{r}')^2}{4Dt}\right] d\mathbf{r}' \tag{6.21}$$

which can be substituted into the integral expression of $n(\mathbf{r}, t)$. We then reach the final solution of the original problem,

$$n(\mathbf{r}, t) = \int_0^t d\tau \int S(\mathbf{r}', \tau) \frac{1}{\sqrt{4\pi D(t - \tau)}} \exp\left[-\frac{(\mathbf{r} - \mathbf{r}')^2}{4D(t - \tau)}\right] d\mathbf{r}'. \tag{6.22}$$

This integral can be evaluated numerically if the form of $S(\mathbf{r}, t)$ is given. One should realize that if the space is not infinite but confined by some boundary, the above closed form of the solution will be in a different but similar form, since the idea of the Fourier transform is quite general in the sense of the spatial eigenstates. In most cases, the spatial eigenstates form an orthogonal basis set which can be used for a general Fourier transform, as discussed in Chapter 5. It is worth pointing out that the above procedure is also similar to the Green's function method, that is, transforming the problem into a problem of Green's function and expressing the solution in a convolution integral of Green's function and the source. For more discussions on the Green's function method, see Courant and Hilbert (1989).

For the wave equation with a spatial and time-dependent source, we can follow more or less the same steps to complete the separation of variables. The general solution for a nonzero initial displacement and/or velocity is a linear combination of the solution of the homogeneous equation $u_0(\mathbf{r}, t)$ and a particular solution of the inhomogeneous equation, $u_p(\mathbf{r}, t)$. For the homogeneous equation

$$\nabla^2 u_0(\mathbf{r}, t) = \frac{1}{c^2} \frac{\partial^2 u_0(\mathbf{r}, t)}{\partial t^2}, \tag{6.23}$$

we can assume that $u_0(\mathbf{r}, t) = X(\mathbf{r})T(t)$, and then we have $T''(t) = -\omega_k^2 T(t)$ and $\nabla^2 X(\mathbf{r}) = -k^2 X(\mathbf{r})$, with $\omega_k^2 = c^2 k^2$. If we solve the eigenvalue problem for the spatial part, we have the general solution as

$$u_0(\mathbf{r}, t) = \sum_{\mathbf{k}} (A_{\mathbf{k}} e^{-i\omega_k t} + B_{\mathbf{k}} e^{i\omega_k t}) u_{\mathbf{k}}(\mathbf{r}), \tag{6.24}$$

where $u_{\mathbf{k}}(\mathbf{r})$ is the eigenstate of the equation for $X(\mathbf{r})$ with the eigenvalue $k^2 = |\mathbf{k}|^2$. The particular solution $u_p(\mathbf{r}, t)$ can be obtained by performing a Fourier transform of time on both sides of Eq. (6.3). Then we have

$$\nabla^2 \tilde{u}_p(\mathbf{r}, \omega) + \frac{\omega^2}{c^2} \tilde{u}_p(\mathbf{r}, \omega) = -\tilde{R}(\mathbf{r}, \omega), \tag{6.25}$$

where $\tilde{u}_p(\mathbf{r}, \omega)$ and $\tilde{R}(\mathbf{r}, \omega)$ are the Fourier transforms of $u_p(\mathbf{r}, t)$ and $R(\mathbf{r}, t)$, respectively. We can, in principle, expand $\tilde{u}_p(\mathbf{r}, \omega)$ and $\tilde{R}(\mathbf{r}, \omega)$ in terms of the eigenstates $u_\mathbf{k}(\mathbf{r})$ as

$$\tilde{u}_p(\mathbf{r}, \omega) = \sum_\mathbf{k} c_\mathbf{k} u_\mathbf{k}(\mathbf{r}) \tag{6.26}$$

and

$$\tilde{R}(\mathbf{r}, \omega) = \sum_\mathbf{k} d_\mathbf{k} u_\mathbf{k}(\mathbf{r}), \tag{6.27}$$

which give

$$c_\mathbf{k} = \frac{c^2}{\omega_k^2 - \omega^2} d_\mathbf{k} \tag{6.28}$$

from Eq. (6.25). If we put all these together, we have the general solution for the inhomogeneous equation,

$$u(\mathbf{r}, t) = u_0(\mathbf{r}, t) + \frac{1}{\sqrt{2\pi}} \int_{-\infty}^{\infty} d\omega \, e^{-i\omega t} \sum_\mathbf{k} \frac{d_\mathbf{k}}{k^2 - \omega^2/c^2}, \tag{6.29}$$

with $d_\mathbf{k}$ given by

$$d_\mathbf{k} = \int \tilde{R}(\mathbf{r}, \omega) u_\mathbf{k}^+(\mathbf{r}) \, d\mathbf{r}, \tag{6.30}$$

which is a function of ω. We have assumed that $u_\mathbf{k}(\mathbf{r})$ are normalized and form an orthogonal basis set. Here $u_\mathbf{k}^+(\mathbf{r})$ is the complex conjugate of $u_\mathbf{k}(\mathbf{r})$. The initial condition can now be used to determine $A_\mathbf{k}$ and $B_\mathbf{k}$.

The useful result of the separation of variables is that we may then need to deal with an equation or equations involving only a single variable, so that all the methods we discussed in Chapter 3 become applicable.

6.3 Discretization of the equation

The essence of all numerical schemes in solving differential equations is in the discretization of the continuous variables, that is, the spatial coordinates and time. If we use the rectangular coordinate system, we have to discretize $\partial A(\mathbf{r}, t)/\partial r_i$, $\partial^2 A(\mathbf{r}, t)/\partial r_i \partial r_j$, $\partial A(\mathbf{r}, t)/\partial t$, and $\partial^2 A(\mathbf{r}, t)/\partial^2 t$, where r_i is either x, y, or z. Sometimes we also need to deal with a situation similar to having a spatially dependent diffusion constant, for example, to discretize $\nabla \cdot D(\mathbf{r})\nabla A(\mathbf{r}, t)$. Here $A(\mathbf{r}, t)$ is a generic function from one of the equations discussed at the beginning of this chapter. We can apply the numerical schemes developed in Chapter 2 for first-order and/or second-order derivatives for all the partial derivatives. Typically, one uses the two-point or three-point formula for the first-order partial derivatives

$$\frac{\partial A(\mathbf{r}, t)}{\partial t} = \frac{A(\mathbf{r}, t_{n+1}) - A(\mathbf{r}, t_n)}{\tau} \tag{6.31}$$

or

$$\frac{\partial A(\mathbf{r}, t)}{\partial t} = \frac{A(\mathbf{r}, t_{n+1}) - A(\mathbf{r}, t_{n-1})}{2\tau}. \tag{6.32}$$

One can also use the three-point formula for the second-order derivatives

$$\frac{\partial^2 A(\mathbf{r}, t)}{\partial^2 t} = \frac{A(\mathbf{r}, t_{n+1}) - 2A(\mathbf{r}, t_n) + A(\mathbf{r}, t_{n-1})}{\tau^2}. \tag{6.33}$$

Here $t_n = t$ and $\tau = t_{n+1} - t_n = t_n - t_{n-1}$. The same formulas can be applied to the spatial variables as well, for example,

$$\frac{\partial A(\mathbf{r}, t)}{\partial x} = \frac{A(x_{n+1}, y, z, t) - A(x_{n-1}, y, z, t)}{2h_x}, \tag{6.34}$$

with $h_x = x_{n+1} - x_n$, and

$$\frac{\partial^2 A(\mathbf{r}, t)}{\partial^2 x} = \frac{A(x_{n+1}, y, z, t) - 2A(x_n, y, z, t) + A(x_{n-1}, y, z, t)}{h_x^2}. \tag{6.35}$$

The partial differential equations can then be solved at discrete points.

However, when we need to deal with a discrete mesh that is not uniform or with an inhomogeneity such as $\nabla \cdot D(\mathbf{r})\nabla A(\mathbf{r}, t)$, we may need to introduce some other discretization scheme. Typically, one can construct a functional from the field of the solution in an integral form. The differential equation is recovered from functional variation. This is in the same spirit as deriving the Lagrange equation from the optimization of the action integral. The advantage here is that one can discretize the integral first and then carry out the functional variation at the lattice points. The corresponding difference equation results. Let us take the one-dimensional Poisson equation, with a spatially dependent dielectric constant,

$$\frac{d}{dx}\epsilon(x)\frac{d\phi(x)}{dx} = -4\pi\rho(x) \tag{6.36}$$

for $x \in [0, 1]$, as an example. For simplicity, let us take the homogeneous Dirichlet boundary condition $\phi(0) = \phi(1) = 0$ here. We can construct a functional

$$U = \int_0^1 \left\{ \frac{1}{2}\epsilon(x)\left[\frac{d\phi(x)}{dx}\right]^2 - 4\pi\rho(x)\phi(x) \right\} dx, \tag{6.37}$$

which leads to the Poisson equation if we take

$$\frac{\delta U}{\delta \phi} = 0. \tag{6.38}$$

If we use the three-point formula for the first-order derivative in the integral and the trapezoid rule to convert the integral into a summation, we have

$$U \simeq \frac{1}{2h} \sum_k \epsilon_{k-1/2}(\phi_k - \phi_{k-1})^2 - 4\pi h \sum_k \rho_k \phi_k \qquad (6.39)$$

as the discrete representation of the functional. We have used $\phi_k = \phi(x_k)$ for notation convenience and $\epsilon_{k-1/2} = \epsilon(x_{k-1/2})$ as an approximation of $\epsilon(x)$ in the interval $[x_{k-1}, x_k]$. Now if we treat each discrete variable ϕ_k as an independent variable, the functional variation in Eq. (6.38) becomes a partial derivative

$$\frac{\partial U}{\partial \phi_k} = 0. \qquad (6.40)$$

We then obtain the desired difference equation

$$(\epsilon_{k+1/2} + \epsilon_{k-1/2})\phi_k - \epsilon_{k+1/2}\phi_{k+1} - \epsilon_{k-1/2}\phi_{k-1} = 4\pi h^2 \rho_k, \qquad (6.41)$$

which is a discrete representation of the original Poisson equation. This scheme is extremely useful when the geometry of the system is not rectangular or when the parameters in the equation are not constants. For example, we may want to study the Poisson equation in a cylindrically or spherically symmetric case, or diffusion of some contaminated materials underground where we have a spatially dependent diffusion coefficient. The stationary one-dimensional diffusion equation is equivalent to the one-dimensional Poisson equation if we replace $\epsilon(x)$ with $D(x)$ and $4\pi\rho(x)$ with $S(x)$ in Eqs. (6.36) through (6.41). The scheme can also be generalized for the time-dependent diffusion equation by replacing the source term $h^2 S_k$ with $h^2\{S_k(t_n) - [n_k(t_{n+1}) - n_k(t_n)]/\tau\}$. For higher spatial dimensions, the integral is a multidimensional integral. The aspects of time dependence and higher dimensions will be discussed in more detail later in this chapter.

Sometimes it is more convenient to deal with the physical quantities only at the lattice points. Equation (6.41) can be modified into such a form, with the quantity midway between two neighboring points replaced by the average of two lattice points. For example, the difference equation in Eq. (6.41) can be modified into

$$4\epsilon_k \phi_k - (\epsilon_{k+1} + \epsilon_k)\phi_{k+1} - (\epsilon_{k-1} + \epsilon_k)\phi_{k-1} = 8\pi h^2 \rho_k, \qquad (6.42)$$

which does not sacrifice too much accuracy if $\epsilon(x)$ is a slow-varying function of x.

6.4 The matrix method for difference equations

When a partial differential equation is discretized and given in a difference equation form, we can generally solve it with the matrix methods, which we introduced in Chapter 4. Recall that we outlined how to solve the Hartree–Fock equation for atomic systems with the matrix method in Chapter 4 as a special example. In general, when we have a differential equation with the form

$$\mathcal{L}\psi(\mathbf{r}, t) = f(\mathbf{r}, t), \tag{6.43}$$

where $\mathcal{L}$ is a linear operator of the spatial and time variables, $\psi(\mathbf{r}, t)$ is the physical quantity to be solved, and $f(\mathbf{r}, t)$ is the source, we can always discretize the equation and put it in the form of a matrix equation

$$\mathbf{A}\mathbf{x} = \mathbf{b}, \tag{6.44}$$

where $\mathbf{A}$ is from the discretization of the operator $\mathcal{L}$, $\mathbf{x}$ is the discrete values of ψ excluding the boundary points, and $\mathbf{b}$ is the known vector constructed from the discrete values of $f(\mathbf{r}, t)$ and the boundary points of ψ. The time variable is usually separated with the Fourier transform first unless it has to be dealt with at different spatial points. Situations like this will be discussed in Section 6.7.

For example, the difference equations for the one-dimensional Poisson equation obtained in the preceding section can be cast into such a form. For the difference equation given in Eq. (6.42), we have

$$A_{ij} = \begin{cases} 4\epsilon_i & \text{for} \quad i = j, \\ -(\epsilon_i + \epsilon_{i+1}) & \text{for} \quad i = j - 1, \\ -(\epsilon_i + \epsilon_{i-1}) & \text{for} \quad i = j + 1, \\ 0 & \text{otherwise}, \end{cases} \tag{6.45}$$

and

$$b_i = 8\pi h^2 \rho_i, \tag{6.46}$$

with $x_i = \phi_i$ of Eq. (6.42). This matrix representation of the difference equation resulting from the discretization of a differential equation is very general.

Let us illustrate this method further by studying an interesting example. Consider a situation in which a person is sitting on a long bench supported at both ends. Assume that the length of the bench is L. If we want to know the displacement of the bench at every point, we can establish the equation for the curvature at different locations from Newton's equation for each tiny slice of the bench. Then we obtain

$$YI \frac{d^2 u(x)}{dx^2} = f(x), \tag{6.47}$$

where $u(x)$ is the curvature (that is, the inverse of the effective radius of the curve at x); Y is Young's modulus of the bench; $I = t^3 w/3$ is a geometric constant, with t being the thickness and w being the width of the bench; and $f(x)$ is the force density on the bench. Since both ends are fixed, the curvatures there are zero; that is, $u(0) = u(L) = 0$. If we discretize the equation with evenly spaced intervals, that is, $x_0 = 0, x_1 = h, \ldots, x_{n+1} = L$, we have

$$-2u_i + u_{i+1} + u_{i-1} = \frac{h^2 f_i}{YI} \tag{6.48}$$

for $i = 0, 1, \ldots, n + 1$, which is equivalent to

$$\begin{pmatrix} 2 & -1 & \cdots & \cdots & 0 \\ -1 & 2 & -1 & \cdots & 0 \\ \vdots & \vdots & \vdots & \vdots & \vdots \\ 0 & \cdots & 0 & -1 & 2 \end{pmatrix} \begin{pmatrix} u_1 \\ u_2 \\ \vdots \\ u_n \end{pmatrix} = \begin{pmatrix} b_1 \\ b_2 \\ \vdots \\ b_n \end{pmatrix}, \tag{6.49}$$

with $u_0 = u_{n+1} = 0$ as required by the boundary condition and $b_i = -h^2 f_i / YI$. One can easily solve the problem with the Gaussian elimination scheme or the LU decomposition scheme. The LU decomposition scheme in this case is extremely simple, since the matrix is a real, symmetric, and tridiagonal matrix with all the diagonal elements the same and all the nonzero off-diagonal elements the same as well.

Let us develop the LU decomposition for a general tridiagonal matrix **A**. We have

$$\mathbf{A} = \mathbf{LU}, \tag{6.50}$$

with

$$A_{ij} = \begin{cases} a_i & \text{for} \quad i = j, \\ e_i & \text{for} \quad i = j - 1, \\ c_i & \text{for} \quad i = j + 1, \\ 0 & \text{otherwise.} \end{cases} \tag{6.51}$$

L here is a lower triangular matrix with

$$L_{ij} = \begin{cases} w_i & \text{for} \quad i = j, \\ v_i & \text{for} \quad i = j + 1, \\ 0 & \text{otherwise,} \end{cases} \tag{6.52}$$

and **U** is an upper triangular matrix with

$$U_{ij} = \begin{cases} 1 & \text{for} \quad i = j, \\ u_i & \text{for} \quad i = j - 1, \\ 0 & \text{otherwise.} \end{cases} \tag{6.53}$$

The elements in **L** and **U** can be related easily to a_i, e_i, and c_i if we multiply **L** and **U** and compare element by element. We have

$$v_i = c_i, \tag{6.54}$$

$$u_i = e_i / w_i, \tag{6.55}$$

$$w_i = a_i - v_i u_{i-1}, \tag{6.56}$$

with $w_1 = a_1$. The solution of the linear equation $\mathbf{Ax} = \mathbf{b}$ can be obtained by one forward substitution and one backward substitution, since

$$\mathbf{Ax} = \mathbf{LUx} = \mathbf{Ly} = \mathbf{b}, \tag{6.57}$$

with $\mathbf{Ux} = \mathbf{y}$. We can first solve $\mathbf{Ly} = \mathbf{b}$ and then $\mathbf{Ux} = \mathbf{y}$ with

$$y_i = (b_i - v_i y_{i-1})/w_i, \tag{6.58}$$

$$x_i = y_i - u_i x_{i+1}, \tag{6.59}$$

where $y_1 = b_1/w_1$ and $x_n = y_n$.

For the specific problem discussed here, we have $e_i = c_i = 1$ and $a_i = -2$. Assume that the force distribution on the bench is given by

$$f(x) = \begin{cases} f_0[e^{-(x-L/2)^2/x_0^2} - e^{-1}] + \rho g & \text{for} \quad |x - L/2| \le x_0, \\ \rho g & \text{otherwise,} \end{cases} \tag{6.60}$$

with $f_0 = 200$ N/m and $x_0 = 0.25$ m. We can assume the bench has a length of 3.0 m, a width of 0.20 m, a thickness of 0.030 m, a linear density of $\rho = 3.0$ kg/m, and a Young's modulus of 1.0×10^9 N/m^2. $g = 9.80$ m/s^2 is the gravitational acceleration due to the weight of the bench.

The following program is a realization of the algorithm described above with the given specific force distribution and parameters for the situation of a person sitting on a bench. We have used SI units in the program. Since all the nonzero off-diagonal elements are the same, so are the diagonal elements, and we do not need to store them as arrays. Instead, they are stored as simple constants.

```
        PROGRAM BENCH
C
C This program solves the problem of a person sitting on
C a bench as described in this section.
C
        PARAMETER (N=99)
        DIMENSION B(N),X(N),Y(N),W(N),U(N)
C
        XL  =  3.0
        H   =  XL/(N+1)
        H2  =  H*H
        Y0  =  1.0E09*0.03**3*0.20/3.0
        X0  =  0.25
        RHO =  3.0
        G   =  9.8
```

```
      D   =   2.0
      E   = -1.0
      E0  =   1.0/EXP(1.0)
C
C Find elements in L and U
C
      W(1) =  D
      U(1) =  E/D
      DO    100 I = 2, N
        W(I) = D-E*U(I-1)
        U(I) = E/W(I)
  100 CONTINUE
C
C Assign the array B
C
      DO    200 I = 1, N
        XD   =  H*I
        B(I) = -H2*RHO*G
        IF (ABS(XD-XL/2.0).LT.X0) THEN
          B(I) = B(I)-H2*F0*(EXP(-((XD-XL/2)/X0)**2)-E0)
        ELSE
        ENDIF
        B(I) = B(I)/Y0
  200 CONTINUE
C
C Find the solution of the curvature
C
      Y(1) = B(1)/W(1)
      DO   300 I = 2, N
        Y(I) = (B(I)-Y(I-1))/W(I)
  300 CONTINUE
C
      X(N) = Y(N)
      DO   400 I = N-1,1,-1
        X(I) = Y(I)-U(I)*X(I+1)
  400 CONTINUE
      WRITE(6,9999) (X(I),I=1,N)
      STOP
 9999 FORMAT(F20.10)
      END
```

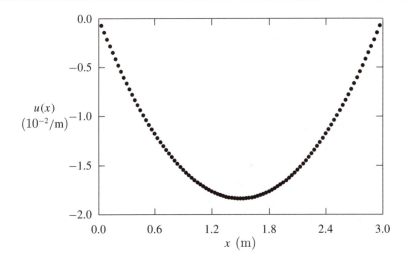

Fig. 6.1. The numerical result of the curvature of the bench, evaluated with the program BENCH.

The numerical result for the curvature of the bench calculated with the above program is illustrated in Fig. 6.1. Under these realistic parameters, the bench does not deviate from the original position very much. Even at the middle of the bench, the effective radius for the curve is still about 50 m.

An interesting aspect of this problem is that when the equation is discretized, the difference equation is extremely simple, a real, symmetric, and tridiagonal matrix. As one would expect, all one-dimensional problems with the same mathematical structure can be solved in the same fashion – for example, the one-dimensional Poisson equation or the stationary one-dimensional diffusion equation. We will show later that equations with higher spatial dimensions and with time dependence can be solved in a similar manner. One needs to note that the boundary condition as well as the coefficient matrix can become more complicated. However, if we split the coefficient matrix among the coordinates correctly, we can still preserve its tridiagonal nature, at least, in each step along each coordinate direction. As long as the coefficient matrix is tridiagonal, we can use the simple LU decomposition outlined above to obtain the solution of the linear equation set. Sometimes one may also need to solve the linear problem with the full matrix, which is no longer tridiagonal. Then the Gaussian elimination scheme introduced in Chapter 4 or the general LU decomposition outlined there can be used.

6.5 The relaxation method

As we have discussed, a functional can be constructed for the purpose of discretizing a differential equation. The procedure for reaching the differential equation is the optimization of the functional. In most cases, the optimization is equivalent to the minimization of the functional. A numerical scheme can therefore be devised to find the numerical solution iteratively, as long as the functional associated with the equation decreases with each iteration. The true solution is obtained when the functional reaches its minimum. This numerical scheme, formally known as the *relaxation method*, is extremely powerful in solving elliptic equations, including the Poisson equation, the stationary diffusion equation, and the spatial part of the wave equation after separation of variables.

Let us first examine the stationary one-dimensional diffusion equation

$$-\frac{d}{dx}\left[D(x)\frac{dn(x)}{dx}\right] = S(x), \tag{6.61}$$

which can be written into a discrete form

$$n_i = \frac{1}{D_{i+1/2} + D_{i-1/2}}(D_{i+1/2}n_{i+1} + D_{i-1/2}n_{i-1} + h^2 S_i), \tag{6.62}$$

as discussed in Section 6.3. It is the above equation that gives us the basic idea of the relaxation method: A guessed solution that satisfies the required boundary condition is gradually modified to satisfy the difference equation within the given tolerance. So the key is to establish an updating scheme that can modify the guessed solution gradually toward the correct direction, namely, the direction with the functional minimized. In practice, one uses the following updating scheme:

$$n_i^{(l+1)} = (1 - p)n_i^{(l)} + pn_i, \tag{6.63}$$

where $n_i^{(l)}$ is the solution of the *l*th iteration at the lattice point i and n_i is given from Eq. (6.62) with the terms on the right-hand side calculated with $n_i^{(l)}$. Here p is an adjustable parameter restricted in the region $p \in [0, 2]$. The reason for such a restriction on p is that the procedure has to ensure the optimization of the functional during the iterations, and this is equivalent to having the solution of Eq. (6.63) approach the true solution of the diffusion equation defined in Eq. (6.61). One can find more discussion on the quantitative aspect of p in Young and Gregory (1988, pp. 1026–39). The points at the boundaries can be updated under the constraints of the boundary condition. Later we will discuss how to achieve this numerically. The following subroutine performs such an iteration once and can be called by other parts of a complete program to solve the equation with the relaxation method.

```
        SUBROUTINE RLXN (FN,DN,S,N,P,H)
    C
    C Subroutine performing one iteration of relaxation for
```

```
C the stationary one-dimensional diffusion equation. DN
C is the diffusion coefficient shifted one half step
C toward x = 0.
C
      DIMENSION FN(N),DN(N),S(N)
      H2 = H*H
      Q  = 1.0-P
      DO 100 I = 2, N-1
        FN(I) = Q*FN(I)+P*(D(I+1)*FN(I+1)
     *            +D(I)*FN(I-1)+H2*S(I))/(D(I+1)+D(I))
  100 CONTINUE
      RETURN
      END
```

Note that we have used the updated n_{i-1} on the right-hand side of the relaxation scheme, which is usually more efficient but less stable. Note that we can also use $D_{i-1/2} \simeq (D_i + D_{i-1})/2$, $D_{i+1/2} \simeq (D_i + D_{i+1})/2$, and $D_{i+1/2} + D_{i-1/2} \simeq 2D_i$ in Eq. (6.62) if the diffusion coefficient is given only at the lattice points.

The points at the boundaries are not updated in the subroutine. For the Dirichlet boundary condition, one can just keep them constant. For the Neumann boundary condition, one can use either a four-point or a six-point formula for the first-order derivative to update the solution at the boundary. For example, if we have

$$\frac{dn(x,t)}{dx}\bigg|_{x=0} = 0, \tag{6.64}$$

we can update the first point n_0 by setting

$$n_0 = \frac{1}{3}(4n_1 - n_2), \tag{6.65}$$

which is the result of the four-point formula of the first-order derivative at $x = 0$,

$$\frac{dn(x)}{dx}\bigg|_{x=0} \simeq \frac{1}{6}(-2n_{-1} - 3n_0 + 6n_1 - n_2) = 0, \tag{6.66}$$

with $n_{-1} = n_1$ as the result of the zero derivative. A partial derivative is dealt with in the same manner. Higher accuracy can be achieved if we use a formula with more points. We have to use the even-numbered point formulas, since we want to have n_0 in the expression.

Now let us turn to the two-dimensional case, and we will use the Poisson equation

$$\nabla^2\phi(\mathbf{r}) = -4\pi\rho(\mathbf{r}) = -s(\mathbf{r}) \tag{6.67}$$

as an example. Here $s(\mathbf{r})$ is introduced for convenience. Now if we consider the case with a rectangular boundary, we have

$$\frac{\phi_{i+1j} + \phi_{i-1j} - 2\phi_{ij}}{h_x^2} + \frac{\phi_{ij+1} + \phi_{ij-1} - 2\phi_{ij}}{h_y^2} = -s_{ij}, \tag{6.68}$$

where i and j are used for the x and y coordinates and h_x and h_y are the intervals along the x and y directions, respectively. We can rearrange the above equation so that the value of the solution at a specific point is given by the corresponding values at the neighboring points,

$$\phi_{ij} = \frac{1}{2(1+\alpha)} [\phi_{i+1j} + \phi_{i-1j} + \alpha(\phi_{ij+1} + \phi_{ij-1}) + h_x^2 s_{ij}], \tag{6.69}$$

with $\alpha = (h_x/h_y)^2$. So the solution is obtained if the values of ϕ_{ij} at all the lattice points satisfy the above equation and the boundary condition. The relaxation scheme is based on the fact that the solution of the equation is approached iteratively. One first guesses a solution that satisfies the boundary condition. Then one can update or improve the guessed solution with

$$\phi_{ij}^{(l+1)} = (1-p)\phi_{ij}^{(l)} + p\phi_{ij}, \tag{6.70}$$

where p is an adjustable parameter on the order of 1. $\phi_{ij}^{(l)}$ is the result of the lth iteration, and ϕ_{ij} is solved from Eq. (6.69) with $\phi_{ij}^{(l)}$ used on the right-hand side.

The second part of the above equation can be viewed as the correction to the solution, since the second part is obtained through the differential equation. The choice of p will determine the speed of convergence. If p is selected outside the allowed range by the specific geometry and discretization, the algorithm will be unstable. Usually, the optimized p is between 0 and 2. In practice, one can find the optimized p easily by just running the program for a few iterations, say, ten, with different values of p. The convergence can be analyzed easily with the result from iterations of each choice of p. Mathematically, one can place an upper limit on p but in practice, this is not really necessary, since it is much easier to test the choice of p numerically.

6.6 Groundwater dynamics

Groundwater dynamics is very rich because of the complexity of the underground structures. For example, a large piece of rock may modify the speed and direction of flow drastically. The dynamics of groundwater is of importance in the construction of any underground structure.

In this section, since we just want to illustrate the power of the method, we will confine ourselves to a relatively simple case: a two-dimensional aquifer with a rectangular geometry of dimensions $L_x \times L_y$. The steady groundwater flow is described by the so-called Darcy's law

$$\mathbf{q} = -\kappa \nabla \phi, \tag{6.71}$$

where $\mathbf{q}$ is the specific discharge vector (the flux density), which is a measure of the volume of the fluid passing through a unit section perpendicular to the velocity in a unit of time. The average velocity of the flow at a specific point, $\mathbf{v}$, can be related to the specific discharge by $\mathbf{v} = \mathbf{q}/\beta$, with β being the porosity, which is the percentage of the empty space (voids) on a cross section perpendicular to the flow. κ is the hydraulic conductivity, which in general is a 3×3 matrix depending on the specific porous medium and carries a unit of velocity. ϕ is referred to as the head, which is a measure of the height of the water in a standpipe from the datum (zero point). The head at elevation z can be related to the hydraulic pressure P at the reference elevation ($z = 0$) as

$$\phi = \frac{P}{\rho g} + z, \tag{6.72}$$

where ρ is the density of the fluid and g is the gravitational acceleration. If we combine the above equation with the continuity equation for a steady flow

$$\nabla \cdot \mathbf{q}(\mathbf{r}) = -\beta \frac{\partial \rho}{\partial t} = -N(\mathbf{r}), \tag{6.73}$$

with $N(\mathbf{r})$ being the given rate of infiltration, a function of spatial variables only, we obtain a generalized Poisson equation

$$\nabla \cdot \kappa \nabla \phi(\mathbf{r}) = -N(\mathbf{r}), \tag{6.74}$$

which describes the groundwater dynamics in most cases. When there is no infiltration and the system is isotropic and homogeneous, the above equation reduces to its simplest form

$$\nabla^2 \phi(\mathbf{r}) = 0, \tag{6.75}$$

which is a Laplace equation for the head.

Now let us demonstrate how one can solve the groundwater dynamics problem by assuming that we are dealing with a rectangular geometry with nonuniform conductivity and nonzero infiltration. Boundary conditions play a significant role in determining the behavior of groundwater dynamics. We discretize the equation with the scheme discussed in Section 6.3, and the equation becomes

$$\phi_{ij} = \frac{1}{4(1+\alpha)\kappa_{ij}}\{(\kappa_{i+1j} + \kappa_{ij})\phi_{i+1j} + (\kappa_{ij} + \kappa_{i-1j})\phi_{i-1j}$$

$$+ \alpha[(\kappa_{ij+1} + \kappa_{ij})\phi_{ij+1} + (\kappa_{ij} + \kappa_{ij-1})\phi_{ij-1}] + 2h_x^2 N_{ij}\}, \qquad (6.76)$$

where $\alpha = (h_x/h_y)^2$. The above difference equation can form the basis for an iterative approach with

$$\phi_{ij}^{(l+1)} = (1-p)\phi_{ij}^{(l)} + p\phi_{ij}, \qquad (6.77)$$

where $\phi_{ij}^{(l)}$ is the value of the lth iteration and ϕ_{ij} is the value evaluated from Eq. (6.76), with the right-hand side evaluated with $\phi_{ij}^{(l)}$. We would like to illustrate this procedure by studying an actual example, with $L_x = 1,000$ m, $L_y = 500$ m, $\kappa(x, y) = \kappa_0 + ay$ with $\kappa_0 = 1.0$ m/s and $a = -0.04$ s^{-1}, and $N(x, y) = 0$. The boundary condition is $\partial\phi/\partial x = 0$ at $x = 0$ and $x = 1,000$ m, $\phi = \phi_0$ at $y = 0$, and $\phi = \phi_0 + b\cos(x/L_x)$ at $y = 500$ m, with $\phi_0 = 200$ m and $b = -20$ m. The four-point formula for the first-order derivative is used to ensure zero partial derivatives for two of the boundaries. The following program is an implementation of the relaxation method applied to this groundwater dynamics problem.

```
      PROGRAM G_WATER
C
C This program solves the groundwater dynamics problem
C in rectangular geometry through the relaxation method.
C
      PARAMETER (NX=1001,NY=501,ITMX=5)
      DIMENSION PHI(NX,NY),CK(NX,NY),SN(NX,NY)
C
      PI = 4.0*ATAN(1.0)
      A0 = 1.0
      B0 = -0.04
      H0 = 200.0
      CH = -20.0
      SX = 1000.0
      SY = 500.0
      HX = SX/(NX-1)
      HY = SY/(NY-1)
      P  = 0.5
```

```
C
C Set up the boundary condition and initial guess
C
      DO    150  I = 1, NX
        X = (I-1)*HX
        DO  100   J = 1, NY
          Y = (J-1)*HY
          CK(I,J)  = A0+B0*Y
          PHI(I,J) = H0+CH*COS(PI*X/SX)*Y/SY
          SN(I,J)  = 0.0
  100   CONTINUE
  150 CONTINUE
C
      DO 200 ISTP = 1, ITMX
C
C Ensure the boundary conditions with the four-point
C formula
C
      DO  180 J = 1, NY
        PHI(1,J)  =(4.0*PHI(2,J)-PHI(3,J))/3.0
        PHI(NX,J) =(4.0*PHI(NX-1,J)-PHI(NX-2,J))/3.0
  180   CONTINUE
C
      CALL RX2D (PHI,CK,SN,NX,NY,P,HX,HY)
  200 CONTINUE
C
      DO   400  I = 1, NX, 100
        X = (I-1)*HX
        DO 390   J = 1, NY, 50
          Y = (J-1)*HY
          WRITE (6,9999) X,Y,PHI(I,J)
  390   CONTINUE
  400 CONTINUE
 9999 FORMAT (3F14.6)
      STOP
      END
C
      SUBROUTINE RX2D (F,D,S,NX,NY,P,HX,HY)
C
C Subroutine performing one iteration of the relaxation
```

```
C for the two-dimensional Poisson equation.
C
      DIMENSION F(NX,NY),D(NX,NY),S(NX,NY)
      H2  = HX*HX
      A   = H2/(HY*HY)
      Q   = 1.0-P
      DO   200 I = 2, NX-1
        DO 100 J = 2, NY-1
          E       = 4.*(1.0+A)*D(I,J)
          C1      = D(I+1,J)+D(I,J)
          C2      = D(I-1,J)+D(I,J)
          C3      = A*(D(I,J+1)+D(I,J))
          C4      = A*(D(I,J-1)+D(I,J))
          F(I,J)  = Q*F(I,J)+P*(C1*F(I+1,J)+C2*F(I-1,J)
     *              +C3*F(I,J+1)+C4*F(I,J-1)+2.0
     *              *H2*S(I,J))/E
100     CONTINUE
200   CONTINUE
      RETURN
      END
```

Note that we have used the four-point formula to update the boundary points to ensure zero partial derivatives there. One can use the six-point formula to improve the accuracy if needed. The output of the above program is shown in Fig. 6.2.

The transient state of groundwater dynamics involves the time variable; as we will discuss in the next two sections. Dynamics involving time can be solved

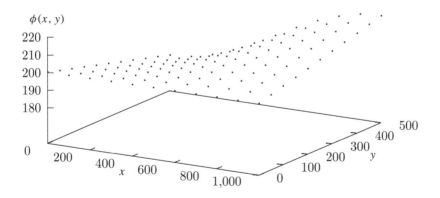

Fig. 6.2. The head in the groundwater dynamics problem obtained from the program G_WATER. Here x, y, and $\phi(x, y)$ are all plotted in meters.

with the combination of the discrete scheme discussed in the last two sections for the spatial variables and the scheme discussed in Chapter 3 for the initial-value problems. It is interesting that time evolution is almost identical to the iterations of the relaxation scheme. A more stable scheme requires the use of the tridiagonal matrix scheme discussed in Section 6.4.

6.7 Initial-value problems

A typical initial-value problem can be either the time-dependent diffusion equation or the time-dependent wave equation. Some initial-value problems are nonlinear equations, such as the equation for a stretched elastic string or the Navier–Stokes equation in fluid dynamics. One can, in most cases, apply the Fourier transform for the time variable of the equation to reduce it to a stationary equation, which can be solved by the relaxation method discussed earlier in this chapter. Then the time dependence can be obtained with an inverse Fourier transform after the solution of the corresponding stationary case is obtained. For equations with higher-order time derivatives, we can also redefine the derivatives as new variables in order to convert the equations to ones with only first-order time derivatives, as we did in Chapter 3. For example, we can redefine the first-order time derivative in the wave equation, that is, the velocity

$$v(\mathbf{r}, t) = \frac{\partial u(\mathbf{r}, t)}{\partial t}, \tag{6.78}$$

into a new variable. Then we have two coupled first-order equations,

$$\frac{\partial u(\mathbf{r}, t)}{\partial t} = v(\mathbf{r}, t), \tag{6.79}$$

$$\frac{1}{c^2} \frac{\partial v(\mathbf{r}, t)}{\partial t} = \nabla^2 u(\mathbf{r}, t) + R(\mathbf{r}, t), \tag{6.80}$$

which describe the same physics as the original second-order wave equation. Note that the above equation set now has a mathematical structure similar to that of a first-order equation such as the diffusion equation

$$\frac{\partial n(\mathbf{r}, t)}{\partial t} = \nabla \cdot D(\mathbf{r}) \nabla n(\mathbf{r}, t) + S(\mathbf{r}, t). \tag{6.81}$$

This means we can develop numerical schemes for equations with first-order time derivatives only. In the case of higher-order time derivatives, we will always introduce new variables. In the next chapter, however, we will introduce some numerical algorithms designed to solve second-order differential equations directly. As one can see, after discretization of the spatial variables, we have practically the same initial-value problem as that discussed in Chapter 3. However, there is one more complication. The specific scheme used to discretize the spatial variables as well as the time variable will certainly affect the stability and accuracy of the solution. Even though it is not the goal here to analyze all aspects of various algorithms, we will still make a comparison among the most popular algorithms with some actual examples in physics and discuss specifically the relevant aspect of the instability with the spatial and time intervals adopted.

In order to analyze the stability of the problem, let us first take the one-dimensional diffusion equation

$$\frac{\partial n(x, t)}{\partial t} = D \frac{\partial^2 n(x, t)}{\partial^2 x} + S(x, t), \tag{6.82}$$

as an example. If we discretize the first-order time derivative by means of the two-point formula with an interval τ and the second-order spatial derivative by means of the three-point formula with an interval h, we obtain the Euler method for the equation

$$n_i(t + \tau) = n_i(t) + \gamma[n_{i+1}(t) + n_{i-1}(t) - 2n_i(t)] + \tau S_i(t), \tag{6.83}$$

with $\gamma = D\tau/h^2$. So the problem is solved if we know the initial value $n(x, 0)$ and the source $S(x, t)$. However, this algorithm is unstable if γ is significantly larger than $1/2$. One can show this by examining the case with $x \in [0, 1]$ and $n(0, t) = n(L, t) = 0$. For detailed discussions, see Young and Gregory (1988, pp. 1078–84).

A better scheme is the Crank–Nicolson method, which modifies the Euler method by using the average of the second-order spatial derivative and the source at t and $t + \tau$ on the right side of the equation; that is,

$$n_i(t + \tau) = n_i(t) + \frac{1}{2}\{[H_i n_i(t) + \tau S_i(t)]$$

$$+ [H_i n_i(t + \tau) + \tau S_i(t + \tau)]\}, \tag{6.84}$$

where we have used

$$H_i n_i(t) = \gamma[n_{i+1}(t) + n_{i-1}(t) - 2n_i(t)] \tag{6.85}$$

to simplify the notation. The implicit iterative scheme in Eq. (6.84) can be rewritten into the form

$$(2 - H_i)n_i(t + \tau) = (2 + H_i)n_i(t) + \tau[S_i(t) + S_i(t + \tau)], \tag{6.86}$$

which has all the unknown terms at $t + \tau$ on the left. More importantly, Eq. (6.86) is a linear equation set with a symmetric tridiagonal coefficient matrix, which can easily be solved as discussed in Section 6.4 for the problem of a person sitting on a bench. One can also show that the algorithm is stable for any γ and converges as

$h \to 0$, and that the error in the solution is on the order of h^2 (Young and Gregory, 1988).

However, the above tridiagonal matrix will not hold if the system is in a higher-dimensional space. There are two ways to deal with this problem in practice. One can use the same recursive scheme to solve the problem with some other method to solve the linear system, such as the Gaussian elimination scheme or a general LU decomposition scheme discussed in Chapter 4. A more practical approach is to deal with each spatial coordinate separately. For example, if we are dealing with the two-dimensional diffusion equation, we have

$$H_{ij}n_{ij}(t) = (H_i + H_j)n_{ij}(t), \tag{6.87}$$

with

$$H_i n_{ij}(t) = \gamma_x [n_{i+1j}(t) + n_{i-1j}(t) - 2n_{ij}(t)], \tag{6.88}$$

$$H_j n_{ij}(t) = \gamma_y [n_{ij+1}(t) + n_{ij-1}(t) - 2n_{ij}(t)]. \tag{6.89}$$

Here $\gamma_x = D\tau/h_x^2$ and $\gamma_y = D\tau/h_y^2$. The decomposition of H_{ij} into H_i and H_j can be used to take one half of each time step along the x direction and the other half along the y direction; that is,

$$(2 - H_j)n_{ij}\left(t + \frac{\tau}{2}\right) = (2 + H_j)n_{ij}(t) + \frac{\tau}{2}\left[S_{ij}(t) + S_{ij}\left(t + \frac{\tau}{2}\right)\right], \quad (6.90)$$

and

$$(2 - H_i)n_{ij}(t + \tau) = (2 + H_i)n_{ij}\left(t + \frac{\tau}{2}\right)$$
$$+ \frac{\tau}{2}\left[S_{ij}\left(t + \frac{\tau}{2}\right) + S_{ij}(t + \tau)\right], \tag{6.91}$$

which is a combined implicit scheme developed by Peaceman and Rachford. As one can see, in each of the above two steps, one has a tridiagonal coefficient matrix. One still has to be careful in using the Peaceman–Rachford algorithm. Even though it has the same accuracy as the Crank–Nicolson algorithm, the convergence in the Peaceman–Rachford algorithm can sometimes be very slow in practice. One should compare the Peaceman–Rachford algorithm with the Crank–Nicolson method in the test runs before deciding which one to be used in the productive runs of a specific problem. One can easily monitor the convergence of the algorithm by changing h_x and h_y. For more discussions and comparisons on various algorithms, see Young and Gregory (1988).

6.8 Temperature field of nuclear waste storage facilities

The temperature increase around nuclear waste rods is not only an interesting physics problem but also an important safety question that has to be addressed before building any nuclear waste storage. The typical plan for nuclear waste storage is an underground facility with the waste rods arranged in an array. In this section, we will study a very simple case, the temperature around a single rod.

The problem can be described by the thermal diffusion equation in cylindrical coordinates with a source,

$$\frac{1}{\kappa}\frac{\partial T(r,t)}{\partial t} - \frac{1}{r}\frac{\partial}{\partial r}r\frac{\partial T(r,t)}{\partial r} = S(r,t), \tag{6.92}$$

where κ is a system-dependent parameter and the temperature far from the source is equal to the environment temperature. We have made several assumptions. The rods are very long, so the temperature field is a two-dimensional field. The heat discharge is assumed to be a radial function only. For the purpose of the numerical evaluation, we assume that

$$S(r,t) = \begin{cases} \frac{T_0}{a^2}e^{-t/\tau_0} & \text{for } r \leq a, \\ 0 & \text{otherwise,} \end{cases} \tag{6.93}$$

where T_0, a, and τ_0 are system-dependent parameters. If we take advantage of the cylindrical symmetry of the problem, the discretized equation along the radial direction is equivalent to a one-dimensional diffusion equation with a spatially dependent diffusion coefficient $D \propto r$; that is,

$$(2 - H_i)T(t + \tau) = (2 + H_i)T(t) + \tau\kappa[S_i(t) + S_i(t + \tau)], \tag{6.94}$$

with

$$H_i T(t) = \frac{\tau\kappa}{r_i h^2}[r_{i+1/2}T_{i+1}(t) + r_{i-1/2}T_{i-1}(t) - 2r_i T_i(t)]. \tag{6.95}$$

The numerical problem is now more complicated than the constant diffusion coefficient case, but the matrix involved is still tridiagonal. One can solve the problem iteratively with the simple method described in Section 6.4 by performing an LU decomposition first and then one forward substitution followed by a backward substitution.

Now let us see an actual numerical example. Assume that the radius of the rod a is 25 cm, parameter κ is 1.0×10^6 cm^2/s, $T_0 = 10$ K, and $\tau_0 = 100$ years. The initial condition is given by $T(r,0) = 300$ K. The following program puts everything together for the temperature field around a nuclear waste rod.

```
      PROGRAM N_T_FIELD
C
C This program calculates the time-dependent temperature
C field around a nuclear waste rod in a two-dimensional
C model.
C
      PARAMETER (M1=1001,M=2001,N=19)
      DIMENSION A(N),B(N),C(N),S(M,N),X(M,N+1),Y(N),
     *          G(N),W(N),V(N),U(N)
C
      TL  = 1.0
      HT  = 1.0/(M1-1)
      TC  = 1.0
      XK  = 3153.6
      RA  = 25.0
      RB  = 100.0
      H   = RB/(N+1)
      H2  = H*H
      T0  = 10.0
      S0  = HT*XK*T0/RA**2
      CS  = HT*XK/H2
C
      DO   90  I = 1, M
        DO  80  J = 1, N+1
          X(I,J) = 0.0
   80   CONTINUE
   90 CONTINUE
C
      DO   100  I = 1, N
        R = H*I
        A(I) =   2.0*(1.0+CS)*R
        B(I) = -(1.0+0.5/I)*CS*R
        C(I) = -(1.0-0.5/I)*CS*R
        IF(R.LT.RA) THEN
          S(1,I) = S0*R
        ENDIF
  100 CONTINUE
C
C Assign the source of the radiation heat
C
```

```
        DO    200   I = 2, M
         T    = HT*(I-1)
          DO    150   J = 1, N
            R = H*J
             IF(R.LT.RA) THEN
               S(I,J) = R*S0/EXP(T/TC)
             ENDIF
   150    CONTINUE
   200 CONTINUE
C
        DO    1000  I = 2, M
C
C Find the values from the last time step
C
          A0 = 2.0*(1.0-CS)*H
          B0 = (1.0+0.5)*CS*H
          C0 = (1.0-0.5)*CS*H
          G(1) = A0*X(I-1,1)
       *         +B0*X(I-1,2)+S(I-1,1)+S(I,1)
          DO    300   J = 2, N
            R  = J*H
            A0 = 2.0*(1.0-CS)*R
            B0 = (1.0+0.5/J)*CS*R
            C0 = (1.0-0.5/J)*CS*R
            G(J) = A0*X(I-1,J)+B0*X(I-1,J+1)
       *           +C0*X(I-1,J-1)+(S(I-1,J)+S(I,J))
   300    CONTINUE
C
C Find the elements in L and U
C
          W(1) = A(1)
          V(1) = C(1)
          U(1) = B(1)/W(1)
          DO    400 J = 2, N
            V(J) = C(J)
            W(J) = A(J)-V(J)*U(J-1)
            U(J) = B(J)/W(J)
   400    CONTINUE
C
C Find the solution of the temperature
```

```
C
          Y(1) = G(1)/W(1)
          DO    500 J = 2, N
            Y(J) = (G(J)-V(J)*Y(J-1))/W(J)
  500     CONTINUE
C
          X(I,N) = Y(N)
          DO    600 J = N-1,1,-1
            X(I,J) = Y(J)-U(J)*X(I,J+1)
  600     CONTINUE
 1000 CONTINUE
          I    = M
          DO    800 J  =  1, N+1
           R = J*H
           WRITE(6,9999) R,X(I,J)
  800     CONTINUE
  900 CONTINUE
      STOP
 9999 FORMAT(2F16.8)
      END
```

The parameters used in the above programs are not from actual storage units. However, if one wants to study the real situation, one needs only to modify the

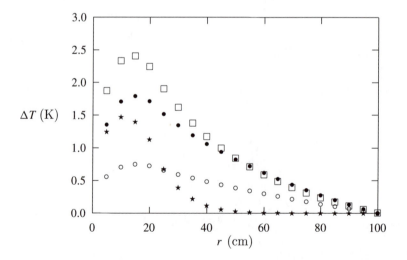

Fig. 6.3. Temperature change around a nuclear waste rod calculated with the program N_T_FIELD. Stars are for $t = 5$ years, squares are for $t = 50$ years, solid dots are for $t = 100$ years, and circles are for $t = 200$ years.

parameters for the actual system and environment. The numerical scheme and program are quite applicable to realistic situations. The output of the program N_T_FIELD is shown in Fig. 6.3. Note that the maximum change of the temperature (the peak) increases with time initially and then decreases afterward. The largest change of the temperature over time and the rate of the temperature change are critical in designing a safe and practical nuclear waste storage.

Exercises

6.1 Develop a numerical scheme that solves the Poisson equation

$$\nabla^2 \phi(r, \theta) = -4\pi \rho(r, \theta)$$

in polar coordinates. Assume that the geometry of the boundary is a circular ring with the potential at the inner radius, $\phi(a, \theta)$, and outer radius, $\phi(b, \theta)$, given. Test the scheme with some special choice of the boundary values.

6.2 If the charge distribution in the Poisson equation is spherically symmetric, derive the difference equation for the potential along the radius. Test the algorithm with $\rho(r) = \rho_0 \exp(-r/r_0)$.

6.3 Derive the relaxation scheme for a three-dimensional system with rectangular boundaries. Analyze the choice of p in a program for the Poisson equation with constant potentials at the boundaries.

6.4 Modify the program for the groundwater dynamics problem given in Section 6.5 to study the general case of the transient state, that is, the case where the time derivative of the head is no longer zero. Apply the program to study the stationary case as given in Section 6.5 and analyze the evolution of the solution with time.

6.5 Write a program that solves the wave equation of a finite string with both ends fixed. Assume that the initial displacement and velocity are given. Test the program with some specific choice of the initial condition.

6.6 Develop a numerical scheme for the time-dependent Schrödinger equation with the Crank–Nicolson method. Consider only the one-dimensional case and test it by applying it to a square well with an initial state being a Gaussian wave packet coming from the left.

6.7 Obtain the algorithm for solving the three-dimensional wave equation with 1/3 of the time step applied to each coordinate direction. Test the algorithm with the equation under the homogeneous Dirichlet boundary condition. Take the initial condition as $u(\mathbf{r}; 0) = 0$ and $v(\mathbf{r}; 0) = \sin(\pi x/L_x) \sin(\pi y/L_y) \sin(\pi z/L_z)$, where L_x, L_y, and L_z are the lengths of the box along the three directions.

7

Molecular dynamics simulations

Most physical systems are collections of interacting objects. For example, a drop of water is a collection of more than 10^{21} water molecules, and a galaxy can be a collection of millions of stars. In general, there is no analytical solution for an interacting system with more than two objects. One can solve the problem of a two-body system, such as the Earth–Sun system, analytically, but not a three-body system, such as the Moon–Earth–Sun system. The situation is similar in quantum mechanics, in that one can obtain the energy levels of the hydrogen atom (one electron and one proton) analytically, but not the helium atom (two electrons and a nucleus). Numerical techniques beyond what we have discussed so far are needed to study systems with more than two interacting objects, or so-called many-body systems. Of course, there is a distinction between three-body systems such as the Moon–Earth–Sun system and a more complicated system, such as a drop of water. Statistical mechanics has to be applied to the latter.

7.1 General behavior of a classical system

In this chapter, we would like to introduce a class of simulation techniques called *molecular dynamics*, which solves the dynamics of a classical many-body system described by a many-body Hamiltonian $\mathcal{H}$. From Hamilton's principle, the generalized coordinates $\mathbf{q}_i$ and generalized momenta $\mathbf{p}_i$ satisfy Hamilton's equations,

$$\dot{\mathbf{q}}_i = \frac{\partial \mathcal{H}}{\partial \mathbf{p}_i}, \tag{7.1}$$

$$\dot{\mathbf{p}}_i = -\frac{\partial \mathcal{H}}{\partial \mathbf{q}_i}, \tag{7.2}$$

which are valid for any given Hamiltonian, including cases in which the system can exchange energy or heat with the environment. For an N-body system with a pairwise interaction potential $V(\mathbf{r}_{ij})$ and an external potential $U_{\text{ext}}(\mathbf{r}_i)$, the Hamiltonian is

$$\mathcal{H} = \sum_{i=1}^{N} \frac{\mathbf{p}_i^2}{2m_i} + \sum_{i>j}^{N} V(\mathbf{r}_{ij}) + \sum_{i=1}^{N} U_{\text{ext}}(\mathbf{r}_i), \qquad (7.3)$$

where m_i and $\mathbf{p}_i$ are the mass and momentum of the ith particle. Hamilton's equations for each coordinate and its corresponding momentum are then given by

$$\dot{\mathbf{r}}_i = \frac{\mathbf{p}_i}{m_i}, \qquad (7.4)$$

$$\dot{\mathbf{p}}_i = \mathbf{f}_i, \qquad (7.5)$$

with

$$\mathbf{f}_i = -\nabla_i U_{\text{ext}}(\mathbf{r}_i) - \sum_{j \neq i}^{N} \nabla_i V(\mathbf{r}_{ij}) \qquad (7.6)$$

as the force exerted on the ith particle. The methods for solving Newton's equation discussed in Chapter 1 and Chapter 3 can be used to solve the above equation set. However, those methods are not as practical as the ones about to be discussed, in terms of the speed and accuracy of the computation and given the statistical nature of large systems. In this chapter, we would like to discuss several commonly used molecular dynamics simulation schemes and offer a few examples.

Before we go into the numerical schemes and actual physical systems, we need to discuss several issues. There are several ways to simulate a many-body system. Most simulations are done either through a stochastic process, such as the Monte Carlo simulation, which will be discussed in Chapter 9, or through a deterministic process, such as a molecular dynamics simulation. Some numerical simulations are performed in a hybridized form of both, for example, Langevin dynamics, which is similar to molecular dynamics except for the presence of a random dissipative force, and Brownian dynamics, which is performed under the condition that the acceleration is balanced out by the drifting and random dissipative forces. We will not discuss the Langevin or Brownian dynamics in this book; but interested readers can find some detailed discussions in Heermann (1986).

Another issue is the distribution function of the system. In statistical mechanics, each special environment is dealt with by way of a special ensemble. For example, for an isolated system we use the microcanonical ensemble, which assumes a constant total energy, number of particles, and volume. A system in good contact with a thermal bath is dealt with using the canonical ensemble, which assumes a constant temperature, number of particles, and volume (or pressure). For any ensemble, the system is described by a probability function $\mathcal{W}(\mathbf{R})$. The statistical average of a physical quantity A is then given by

$$\langle A \rangle = \frac{1}{\mathcal{Z}} \int A(\mathbf{R}) \mathcal{W}(\mathbf{R}) \, d\mathbf{R}, \tag{7.7}$$

where $\mathcal{Z}$ is the partition function of the system,

$$\mathcal{Z} = \int \mathcal{W}(\mathbf{R}) \, d\mathbf{R}. \tag{7.8}$$

Note that $\mathbf{R}$ here is an arbitrary point in the phase space, which includes all the coordinates and momenta of the particles in the system in most cases. The ensemble average given above is equivalent to the time average

$$\langle A \rangle = \lim_{\tau \to \infty} \frac{1}{\tau} \int_0^\tau A(t) \, dt, \tag{7.9}$$

if the system is *ergodic*; that is, every possible state is accessed with an equal probability. Since molecular dynamics simulations are deterministic in nature, almost all physical quantities are obtained through a time average. Sometimes the average over all the particles is also needed. For example, the average kinetic energy of the system can be obtained from any ensemble average, and the result is given by the partition theorem

$$\langle E_k \rangle = \left\langle \sum_{i=1}^N \frac{\mathbf{p}_i^2}{2m_i} \right\rangle = \frac{G}{2} k_B T, \tag{7.10}$$

where G is the total number of degrees of freedom and T is the temperature in the system and k_B is the Boltzmann constant. For a very large system, $G \simeq 3N$, since each particle has three degrees of freedom. In molecular dynamics simulations, the average kinetic energy of the system is obtained through

$$\langle E_k \rangle = \frac{1}{M} \sum_{j=1}^M E_k(t_j), \tag{7.11}$$

where M is the total number of data points taken at different time steps and $E_k(t_j)$ is the kinetic energy of the system at time t_j. If the system is ergodic, the time average is equivalent to the ensemble average. The temperature T of the simulated system is then given by the average kinetic energy with the application of the partition theorem, $T = 2\langle E_k \rangle / G k_B$.

7.2 Basic methods for many-body systems

In general, we can define an n-body density function

$$\rho_n(\mathbf{r}_1, \mathbf{r}_2, \ldots, \mathbf{r}_n) = \frac{1}{\mathcal{Z}} \frac{N!}{(N-n)!} \int \mathcal{W}(\mathbf{R}) \, d\mathbf{r}_{n+1} \, d\mathbf{r}_{n+2} \cdots d\mathbf{r}_N, \tag{7.12}$$

with the particle density $\rho(\mathbf{r}) = \rho_1(\mathbf{r})$ as the special case of $n = 1$. The two-body density function is related to the *pair-distribution function* $g(\mathbf{r}, \mathbf{r}')$ through

$$\rho_2(\mathbf{r}, \mathbf{r}') = \rho(\mathbf{r})g(\mathbf{r}, \mathbf{r}')\rho(\mathbf{r}'), \tag{7.13}$$

and one can easily show that

$$\rho_2(\mathbf{r}, \mathbf{r}') = \langle \hat{\rho}(\mathbf{r})\hat{\rho}(\mathbf{r}')\rangle - \delta(\mathbf{r} - \mathbf{r}')\rho(\mathbf{r}), \tag{7.14}$$

where the first term is the so-called *density–density correlation function*. $\hat{\rho}(\mathbf{r})$ is the density operator, defined as

$$\hat{\rho}(\mathbf{r}) = \sum_{i=1}^{N} \delta(\mathbf{r} - \mathbf{r}_i). \tag{7.15}$$

The density of the system is given by the average of the density operator,

$$\rho(\mathbf{r}) = \langle \hat{\rho}(\mathbf{r})\rangle. \tag{7.16}$$

If the density of the system is nearly a constant, the expression for $g(\mathbf{r}, \mathbf{r}')$ can be reduced to a much simpler form with

$$g(\mathbf{r}) = \frac{1}{\rho N}\left\langle \sum_{i\neq j}^{N} \delta(\mathbf{r} - \mathbf{r}_{ij})\right\rangle, \tag{7.17}$$

where ρ is the average density from the points $\mathbf{r}' = 0$ and $\mathbf{r}$. If the angular distribution is not the information needed, we can take the angular average to obtain the *radial distribution function*

$$g(r) = \frac{1}{4\pi}\int g(\mathbf{r})\sin\theta\, d\theta\, d\phi, \tag{7.18}$$

where θ and ϕ are angles in the spherical coordinate system. The pair-distribution or radial distribution function is related to the *static structure factor* through the Fourier transform,

$$g(\mathbf{r}) - 1 = \frac{1}{(2\pi)^3\rho}\int [S(\mathbf{k}) - 1]e^{i\mathbf{k}\cdot\mathbf{r}}\, d\mathbf{k}, \tag{7.19}$$

$$S(\mathbf{k}) - 1 = \rho\int [g(\mathbf{r}) - 1]e^{-i\mathbf{k}\cdot\mathbf{r}}\, d\mathbf{r}. \tag{7.20}$$

The angular average of $S(\mathbf{k})$ is given by

$$S(k) = 4\pi\rho\int_0^{\infty} \frac{\sin kr}{kr}[g(r) - 1]r^2\, dr. \tag{7.21}$$

The structure factor of a system can be measured with the light- or neutron-scattering experiment.

The behavior of the pair-distribution function can provide a lot of information regarding the translational nature of the particles in the system. For example, a solid structure would have a pair-distribution function with sharp peaks at the distances of nearest neighbors, next nearest neighbors, and so forth. If the system is a liquid, the pair-distribution will still have some broad peaks at the average distances of nearest neighbors, next nearest neighbors, and so forth, but the feature fades away after several peaks.

If the bond orientational order is important, one can also define an orientational correlation function

$$g_n(\mathbf{r}, \mathbf{r}') = \langle q_n(\mathbf{r}) q_n(\mathbf{r}') \rangle, \tag{7.22}$$

where $q_n(\mathbf{r})$ is a quantity associated with the orientation of a specific bond. Detailed discussions on orientational order can be found in Strandburg (1992).

Here we would like to discuss how one can calculate $\rho(\mathbf{r})$ and $g(r)$ in a numerical simulation. The density at a specific point is given by

$$\rho(\mathbf{r}) \simeq \frac{\langle N(\mathbf{r}, \Delta r) \rangle}{\Omega(\mathbf{r}, \Delta r)}, \tag{7.23}$$

where $\Omega(\mathbf{r}, \Delta r)$ is the volume of a sphere centered at $\mathbf{r}$ with a radius Δr and $N(\mathbf{r}, \Delta r)$ is the number of particles in the volume. Note that we may need to adjust the radius Δr to have a smooth and realistic density distribution $\rho(\mathbf{r})$ for a specific system. The average is taken over the time steps.

Similarly, we can obtain the radial distribution function numerically. We need to measure the radius r from the position of a specific particle $\mathbf{r}_i$, and then the radial distribution function $g(r)$ is the probability of another particle's showing up at a distance r. Numerically, we have

$$g(r) \simeq \frac{\langle \Delta N(r, \Delta r) \rangle}{\rho \Delta \Omega(r, \Delta r)}, \tag{7.24}$$

where $\Delta \Omega(r, \Delta r) \simeq 4\pi r^2 \Delta r$ is the volume element of a spherical shell with radius r and thickness Δr and $\Delta N(r, \Delta r)$ is the number of particles in the shell with the ith particle at the center of the sphere. The average is taken over the time steps as well as over the particles, if necessary.

The dynamics of the system can be measured from the displacement of the particles in the system. We can evaluate the time dependence of the mean square displacement of all the particles,

$$\Delta^2(t) = \frac{1}{N} \sum_{i=1}^{N} [\mathbf{r}_i(t) - \mathbf{r}_i(0)]^2, \tag{7.25}$$

where $\mathbf{r}_i(t)$ is the position vector of the ith particle at time t. For a solid system, $\Delta^2(t)$ is relatively small and does not grow with time, and the particles are in nondiffusive, or oscillatory, states. For a liquid system, $\Delta^2(t)$ grows linearly with time:

$$\Delta^2(t) = 6Dt + \Delta^2(0), \tag{7.26}$$

where D is the *self-diffusion coefficient* (a measure of the motion of a particle in a medium of identical particles) and $\Delta^2(0)$ is a time-independent constant. The particles are then in diffusive, or propagating, states.

The very first issue in numerical simulations for a bulk system is how to extend a finite simulation box to model the nearly infinite system. A common practice is to use a periodic boundary condition, that is, to approximate an infinite system by piling up identical simulation boxes periodically. A periodic boundary condition removes the conservation of the angular momentum of the simulated system (particles in one simulation box), but still preserves the translational symmetry of the center of mass. So the temperature is related to the average kinetic energy by

$$\langle E_k \rangle = \frac{3}{2}(N - 1)k_B T, \tag{7.27}$$

where $N - 1$ is due to the removal of the rotation around the center of mass.

The remaining issue, then, is how to include the interactions among the particles in different simulation boxes. If the interaction is a short-range interaction, one can truncate it at a length r_c. $V(r_c)$ has to be small enough so the truncation will not affect the simulation results significantly. A typical simulation box usually has much larger dimensions than r_c. For a three-dimensional cubic box with sides of length L, the total interaction potential can be evaluated with many fewer summations than $N!/2$, the number of possible pairs in the system. For example, if we have $L/2 > r_c$, and if $|x_{ij}|$, $|y_{ij}|$, and $|z_{ij}|$ are all smaller than $L/2$, we can use $V_{ij} = V(r_{ij})$; otherwise, we use the corresponding point in the neighboring box. For example, if $|x_{ij}| > L/2$, we can replace x_{ij} with $x_{ij} \pm L$ in the interaction. We can deal similarly with y and z coordinates. In order to avoid a finite jump at the truncation, one can always shift the interaction to $V(r) - V(r_c)$ to make sure that it is zero at the truncation.

The pressure of a bulk system can be evaluated from the pair-distribution function through

$$P = \rho k_B T - \frac{2\pi\rho^2}{3} \int_0^\infty \frac{\partial V(r)}{\partial r} g(r) r^3 dr, \tag{7.28}$$

which is the result of the virial theorem, which relates the average kinetic energy to the average potential energy of the system. The correction due to the truncation of the potential is then given by

$$\Delta P = -\frac{2\pi\rho^2}{3} \int_{r_c}^{\infty} \frac{\partial V(r)}{\partial r} g(r) r^3 dr, \tag{7.29}$$

which is useful for estimating the influence on the pressure from the truncation in the interaction potential. Numerically, one can also evaluate the pressure from the time average

$$\langle w \rangle = \frac{1}{3} \sum_{i>j} \langle \mathbf{r}_{ij} \cdot \mathbf{f}_{ij} \rangle, \tag{7.30}$$

since $g(r)$ can be interpreted as the probability of seeing another particle at a distance r. Then we have

$$P = \rho k_B T + \frac{\rho}{N} \langle w \rangle + \Delta P, \tag{7.31}$$

which can be evaluated quite easily, because at every time step the force $\mathbf{f}_{ij} = -\nabla V(r_{ij})$ is calculated for each particle pair.

7.3 The Verlet algorithm

Hamilton's equation given in Eq. (7.5) is equivalent to Newton's equation

$$m_i \frac{d^2 \mathbf{r}_i}{dt^2} = \mathbf{f}_i \tag{7.32}$$

for the ith particle in the system. To simplify the notation, we will use $\mathbf{R}$ to represent all the coordinates $(\mathbf{r}_1, \mathbf{r}_2, \ldots, \mathbf{r}_N)$ and use $\mathbf{G}$ to represent all the accelerations $(\mathbf{f}_1/m_1, \mathbf{f}_2/m_2, \ldots, \mathbf{f}_N/m_N)$. Thus, we can rewrite Newton's equations for all the particles as

$$\frac{d^2 \mathbf{R}}{dt^2} = \mathbf{G}. \tag{7.33}$$

If we apply the three-point formula for the second-order derivative $d^2\mathbf{R}/dt^2$, we have

$$\frac{d^2 \mathbf{R}}{dt^2} = \frac{1}{\tau^2}(\mathbf{R}_{n+1} - 2\mathbf{R}_n + \mathbf{R}_{n-1}) + O(\tau^2), \tag{7.34}$$

with $t = n\tau$. We can also apply the three-point formula for the first-order derivative to the velocity:

$$\mathbf{V} = \frac{d\mathbf{R}}{dt}$$
$$= \frac{1}{2\tau}(\mathbf{R}_{n+1} - \mathbf{R}_{n-1}) + O(\tau^2). \tag{7.35}$$

After we put all the above together, we obtain the simplest algorithm, which is called the *Verlet algorithm*, for a classical many-body system,

$$\mathbf{R}_{n+1} = 2\mathbf{R}_n - \mathbf{R}_{n-1} + \tau^2 \mathbf{G}_n + O(\tau^4), \tag{7.36}$$

$$\mathbf{V}_n = \frac{\mathbf{R}_{n+1} - \mathbf{R}_{n-1}}{2\tau} + O(\tau^2). \tag{7.37}$$

The Verlet algorithm can be started if the first two positions $\mathbf{R}_0$ and $\mathbf{R}_1$ of the particles are given. However, in practice, only the initial position $\mathbf{R}_0$ and initial velocity $\mathbf{V}_0$ are given. Therefore, we need to figure out $\mathbf{R}_1$ before we can start the recursion relations defined in Eqs. (7.36) and (7.37). A common practice is to treat the force during the first time interval $[0, \tau]$ as a constant, then apply the kinematic equation to obtain $\mathbf{R}_1$ with

$$\mathbf{R}_1 \simeq \mathbf{R}_0 + \tau \mathbf{V}_0 + \frac{\tau^2}{2} \mathbf{G}_0, \tag{7.38}$$

where $\mathbf{G}_0$ is the acceleration vector evaluated at the initial configuration $\mathbf{R}_0$. Of course, the position $\mathbf{R}_1$ can be improved if the accuracy in the first two points is critical. We can replace $\mathbf{G}_0$ in Eq. (7.38) with the average $(\mathbf{G}_0 + \mathbf{G}_1)/2$, with $\mathbf{G}_1$ evaluated with the configuration $\mathbf{R}_1$ obtained from Eq. (7.38). This procedure can be iterated several times before starting the algorithm for the velocity $\mathbf{V}_1$ and the next position $\mathbf{R}_2$.

The Verlet algorithm has advantages and disadvantages. This algorithm apparently preserves the time reversibility that is one of the important properties of Newton's equation. The rounding error may eventually destroy this time symmetry. The error in the velocity is two orders of magnitude higher than the error in the position. Another advantage of the Verlet algorithm is that in many applications, we may need information only about the positions of the particles, and the Verlet algorithm yields very high accuracy on the position. If the velocity is not needed, we can totally ignore the evaluation of the velocity, since the evaluation of the position does not depend on the velocity at each time step. The biggest disadvantage of the Verlet algorithm is that the velocity is evaluated one time step behind the position. However, this lag can be removed if the velocity is evaluated directly from the force. A two-point formula would yield

$$\mathbf{V}_{n+1} = \mathbf{V}_n + \tau \mathbf{G}_n + O(\tau^2). \tag{7.39}$$

We would get much better accuracy if we replaced $\mathbf{G}_n$ with the average $(\mathbf{G}_n + \mathbf{G}_{n+1})/2$. The new position can be obtained by treating the motion within $t \in [n\tau, (n+1)\tau]$ as the motion with a constant acceleration $\mathbf{G}_n$; that is,

$$\mathbf{R}_{n+1} = \mathbf{R}_n + \tau \mathbf{V}_n + \frac{\tau^2}{2} \mathbf{G}_n. \tag{7.40}$$

Then a variation of the Verlet algorithm with the velocity calculated at the same time step of the position is

$$\mathbf{R}_{n+1} = \mathbf{R}_n + \tau \mathbf{V}_n + \frac{\tau^2}{2}\mathbf{G}_n + O(\tau^2), \tag{7.41}$$

$$\mathbf{V}_{n+1} = \mathbf{V}_n + \frac{\tau}{2}(\mathbf{G}_{n+1} + \mathbf{G}_n) + O(\tau^2). \tag{7.42}$$

This variation has velocity and position calculated at the same time step with the same degree of accuracy.

Here we would like to demonstrate this *velocity version* of the Verlet algorithm with a very simple example, the motion of Halley's comet, which has a period of about 76 years. We were among the lucky ones who could observe it in 1986.

The potential between the comet and the Sun is given by

$$V(r) = -G\frac{Mm}{r}, \tag{7.43}$$

where r is the distance between the comet and the Sun, M and m are the respective masses of the Sun and the comet, and G is the gravitational constant. If we use the center-of-mass coordinate system as described in Chapter 2 for the two-body collision, the dynamics of the comet is governed by

$$\mu\frac{d^2\mathbf{r}}{dt^2} = \mathbf{f} = -GMm\frac{\mathbf{r}}{r^3}, \tag{7.44}$$

with the reduced mass

$$\mu = \frac{Mm}{M+m} \simeq m, \tag{7.45}$$

for the case of Halley's comet. We can take the farthest point (aphelion) as the starting point, and then we can easily obtain its whole orbit with one of the versions of the Verlet algorithm. Two quantities can be assumed as the known quantities, the total energy and the angular momentum, which are the constants of motion. We can describe the motion of the comet in the xy-plane and choose $x_0 = r_{max}$, $v_{x0} = 0$, $y_0 = 0$, and $v_{y0} = v_{min}$. From the well-known results we know that $r_{max} = 5.28 \times 10^{12}$ m and $v_{min} = 9.13 \times 10^2$ m/s. Let us apply the velocity version of the Verlet algorithm to this problem. Then we have

$$x^{(n+1)} = x^{(n)} + \tau v_x^{(n)} + \frac{\tau^2}{2}g_x^{(n)}, \tag{7.46}$$

$$v_x^{(n+1)} = v_x^{(n)} + \frac{\tau}{2}\left[g_x^{(n+1)} + g_x^{(n)}\right], \tag{7.47}$$

$$y^{(n+1)} = y^{(n)} + \tau v_y^{(n)} + \frac{\tau^2}{2}g_y^{(n)}, \tag{7.48}$$

$$v_y^{(n+1)} = v_y^{(n)} + \frac{\tau}{2} \left[g_y^{(n+1)} + g_y^{(n)} \right], \tag{7.49}$$

where the time step index is put in parentheses to distinguish it from the x-component and y-component indices. The acceleration components are given by

$$g_x = -\kappa \frac{x}{r^3}, \tag{7.50}$$

$$g_y = -\kappa \frac{y}{r^3}, \tag{7.51}$$

with $r = \sqrt{x^2 + y^2}$ and $\kappa = GM$. We can use more specific units in the numerical calculations – 76 years as the time unit and the semimajor axis of the orbital $a = 2.68 \times 10^{12}$ m as the length unit. Then we have $r_{\max} = 1.97$, $v_{\min} = 0.816$, and $\kappa = 39.5$.

The following program is the implementation of the algorithm outlined above for Halley's comet.

```
                    PROGRAM COMET
      C
      C Program for the time-dependent position of
      C Halley's comet with the velocity version of the
      C Verlet algorithm.
      C
            PARAMETER (N=20001,NP=10000,NI=200)
            DIMENSION X(N),Y(N),VX(N),VY(N),GX(N),
           *          GY(N),T(N),R(N)
            REAL KAPPA
      C
      C Initialization of the problem
      C
            H     = 1.0/NP
            KAPPA = 39.478428
            T(1)  = 0.0
            X(1)  = 1.966843
            Y(1)  = 0.0
            R(1)  = X(1)
            GX(1) = -KAPPA*X(1)/R(1)**3
            GY(1) = 0.0
            VX(1) = 0.0
            VY(1) = 0.815795
```

```
C
C Verlet algorithm for the position and velocity
C
      DO 100  I = 1, N-1
         T(I+1)   = I*H
         X(I+1)   = X(I)+H*VX(I)+H*H*GX(I)/2.0
         Y(I+1)   = Y(I)+H*VY(I)+H*H*GY(I)/2.0
         R(I+1)   = SQRT(X(I+1)*X(I+1)+Y(I+1)*Y(I+1))
         GX(I+1)  = -KAPPA*X(I+1)/R(I+1)**3
         GY(I+1)  = -KAPPA*Y(I+1)/R(I+1)**3
         VX(I+1)  = VX(I)+H*(GX(I+1)+GX(I))/2.0
         VY(I+1)  = VY(I)+H*(GY(I+1)+GY(I))/2.0
  100 CONTINUE
      WRITE (6,999) (T(I),R(I),I=1,N,NI)
  999 FORMAT (2F10.6)
      STOP
      END
```

In Fig. 7.1, we show the numerical result of the distance between the Sun and Halley's comet calculated from the above program.

One should realize that the accuracy of the velocity version of the Verlet algorithm is relatively low. In practice, however, it is usually accurate enough for simulations,

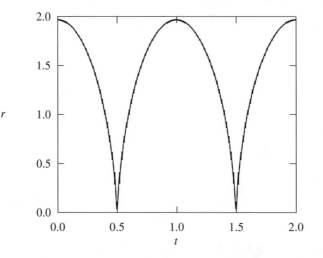

Fig. 7.1. The distance between Halley's comet and the Sun calculated with the program COMET. The period of the comet is used as the unit of time, and the semimajor axis of the orbit is used as the unit of distance.

since in most molecular dynamics simulations, physical quantities are rescaled according to, for example, the temperature of the system. The accumulated errors are thus removed. More accurate algorithms, such as the Gear predictor–corrector algorithm, will be discussed later in this chapter.

7.4 Structure of atomic clusters

One of the simplest applications of the Verlet algorithm is for the study of an isolated collection of particles. For an isolated system, there is a significant difference between a small system and a large system. For a small system, the ergodic assumption of statistical mechanics fails and the system may never reach the so-called equilibrium state. For a very large system, however, one can still apply statistical mechanics, even if it is isolated from the environment; the interactions among the particles will cause exchange of energies and drive the system to equilibrium. Very small clusters with just a few particles usually behave like molecules. What is unclear is the behavior of medium-size clusters, say, clusters with about 100 atoms.

In this section, we would like to demonstrate the application of the velocity version of the Verlet algorithm in determining the structure and dynamics of clusters of an intermediate size. We will assume that the system consists of N atoms that interact with each other through the Lennard–Jones potential

$$V(r_{ij}) = 4\epsilon \left[\left(\frac{\sigma}{r_{ij}} \right)^{12} - \left(\frac{\sigma}{r_{ij}} \right)^{6} \right], \tag{7.52}$$

with $i, j = 1, 2, \ldots, N$, and ϵ and σ being the system-dependent parameters. The force exerted on the ith particle is then given by

$$\mathbf{f}_i = \frac{48\epsilon}{\sigma^2} \sum_{j \neq i}^{N} (\mathbf{r}_i - \mathbf{r}_j) \left[\left(\frac{\sigma}{r_{ij}} \right)^{14} - \frac{1}{2} \left(\frac{\sigma}{r_{ij}} \right)^{8} \right]. \tag{7.53}$$

In order to simplify the notation, ϵ is usually used as the unit of energy, ϵ/k_B as the unit of temperature, and σ as the unit of length. Then the unit of time is given by $\sqrt{m\sigma^2/48\epsilon}$. Newton's equation for each particle then becomes dimensionless,

$$\frac{d^2\mathbf{r}_i}{dt^2} = \mathbf{g}_i = \sum_{j \neq i}^{N} (\mathbf{r}_i - \mathbf{r}_j) \left(\frac{1}{r_{ij}^{14}} - \frac{1}{2r_{ij}^{8}} \right). \tag{7.54}$$

This can be discretized easily with the Verlet algorithm,

$$\mathbf{r}_i^{(n+1)} = \mathbf{r}_i^{(n)} + \tau \mathbf{v}_i^{(n)} + \frac{\tau^2}{2} \mathbf{g}_i^{(n)}, \tag{7.55}$$

$$\mathbf{v}_i^{(n+1)} = \mathbf{v}_i^{(n)} + \frac{\tau}{2} \left[\mathbf{g}_i^{(n+1)} + \mathbf{g}_i^{(n)} \right]. \tag{7.56}$$

The update of the velocity is usually done in two steps. When $\mathbf{g}_i^{(n)}$ is evaluated, the velocity is partially updated with

$$\mathbf{v}_i^{(n+1/2)} = \mathbf{v}_i^{(n)} + \frac{\tau}{2}\mathbf{g}_i^{(n)} \tag{7.57}$$

and then updated again when $\mathbf{g}_i^{(n+1)}$ becomes available after the coordinate is updated; that is,

$$\mathbf{v}_i^{(n+1)} = \mathbf{v}_i^{(n+1/2)} + \frac{\tau}{2}\mathbf{g}_i^{(n+1)}. \tag{7.58}$$

This is equivalent to the one-step update, which requires storage of the acceleration at the last time step as well. We can then simulate the structure and dynamics of the cluster with a given initial position and velocity for each particle. However, several aspects still need special care. The initial positions of the particles are usually selected as the lattice points on a closely packed structure, for example, the face-centered cubic structure. What one wants to avoid is the breaking up of the system in the first few time steps, which happens if some particles are too close to each other. Another way to set up a relatively stable initial cluster is to cut out a piece from a bulk simulation. The corresponding bulk simulation is achieved with the application of the periodic condition. The initial velocities of the particles should also be assigned reasonably. A common practice is to assign velocities from the Maxwell distribution

$$\mathcal{W}(v_x) \propto \exp\left(\frac{-mv_x^2}{2k_BT}\right), \tag{7.59}$$

which can be achieved numerically quite easily with the availability of Gaussian random numbers. The variance in the Maxwell distribution is $\sqrt{k_BT/m}$ for each velocity component. For example, the following subroutine returns the Maxwell distribution for a given temperature.

```
      SUBROUTINE MXWL(N,M,T,V)
C
C This subroutine assigns velocities according to
C the Maxwell distribution. N is the total number of
C velocity components and M is the total number of
C degrees of freedom. T is the system temperature
C in reduced units.
C
      DIMENSION V(N)
      COMMON \CSEED\ ISEED
C
```

```
C Assign a Gaussian distribution to each
C velocity component
C
      DO 100    I = 1, N-1, 2
         CALL GRNF(V1,V2)
         V(I)    = V1
         V(I+1) = V2
  100 CONTINUE
C
C Scale the velocity to satisfy the partition theorem
C
      EK = 0.0
      DO 200    I = 1, N
         EK = EK+V(I)*V(I)
  200 CONTINUE
      VS = SQRT(EK/(M*T))
      DO 300    I = 1, N
         V(I) = V(I)/VS
  300 CONTINUE
      RETURN
      END
```

However, one would always have difficulty in defining a temperature if the system is very small, that is, where the thermodynamic limit is not applicable. In practice, we can still call a quantity associated with the kinetic energy of the system the *temperature*, which is basically a measure of the kinetic energy at each time step,

$$E_k = \frac{G}{2} k_B T, \tag{7.60}$$

with G being the total number of independent degrees of freedom in the system. Note that T now is not necessarily a constant and that G is equal to $3N - 6$, with N being the number of particles in the system, since we have to remove the center-of-mass motion and the rotation around the center of mass when we study the structure and dynamics of the isolated cluster.

The total energy of the system is given by

$$E = \sum_{i=1}^{N} \frac{m_i v_i^2}{2} + \sum_{i>j}^{N} V(r_{ij}), \tag{7.61}$$

with the kinetic energy E_k in the first term and the potential energy E_p in the second term.

One remarkable effect observed in the simulations of finite clusters is that there is a temperature region where the system would fluctuate from a liquid state to a solid state over time. A phase is identified as solid if the particles in the system only vibrate around their equilibrium positions; otherwise they are in a liquid phase. Simulations of clusters have revealed some very interesting phenomena that are unique for clusters with intermediate sizes (with $N \simeq 100$). We discussed how to analyze the structural and dynamical information of a collection of particles in Sections 7.1 and 7.2. One can evaluate the mean square of the displacement of each particle. For the solid state, $\Delta^2(t)$ is relatively small and does not grow with time, and the particles are nondiffusive but oscillatory. For the liquid state, $\Delta^2(t)$ grows with time close to the linear relation $\Delta^2(t) = 6Dt + \Delta^2(0)$, with D being the self-diffusion coefficient and $\Delta^2(0)$ being a constant. In a small system, this relation in time may not be exactly linear. One can also measure the specific structure of the cluster from the radial distribution function $g(r)$ and the orientational correlation function of the bonds. The temperature (or total kinetic energy) can be gradually changed to cool down or heat up the system.

The results from the simulations on clusters of about 100 particles show that the clusters can fluctuate from a group of solid states with lower energy levels to a group of liquid states that lie at higher energy levels in a specific temperature region, with $k_B T$ in the same order of the energy that separates the two groups. This is not expected, since the higher-energy level liquid states are not supposed to have an observable lifetime. However, the statistical mechanics may not be accurate here, since the number of particles is relatively small. More detailed simulations also reveal that the melting of the cluster usually starts from the surface. Interested readers can find more discussions in Matsuoka et al. (1992) and Kunz and Berry (1993; 1994).

7.5 The Gear predictor–corrector method

We discussed multistep predictor–corrector methods in Chapter 3 in solving initial-value problems. Another type of predictor–corrector method uses the truncated Taylor expansion of the function and its derivatives as a prediction and then evaluates the function and its derivatives at the same time step with a correction given from the restriction of the differential equation. This multivalue predictor–corrector scheme was developed by Nordsieck and Gear. Details on the derivations can be found in Gear (1971).

We will take a first-order differential equation

$$\frac{d\mathbf{r}}{dt} = \mathbf{f}(\mathbf{r}, t) \tag{7.62}$$

as an example and then generalize the method to other types of initial-value problems, such as Newton's equation. For simplicity, we will introduce the rescaled quantities

$$\mathbf{r}^{(l)} = \frac{\tau^l}{l!} \frac{d^l \mathbf{r}}{dt^l}, \tag{7.63}$$

with $l = 0, 1, 2, \ldots$. Note that $\mathbf{r}^{(0)} = \mathbf{r}$. Now if we define a vector

$$\mathbf{x} = (\mathbf{r}^{(0)}, \mathbf{r}^{(1)}, \mathbf{r}^{(2)}, \ldots), \tag{7.64}$$

we can obtain the predicted value of $\mathbf{x}$ in the next time step $\mathbf{x}_{k+1}$, with $t_{k+1} = (k+1)\tau$, from the current time step $\mathbf{x}_k$, with $t_k = k\tau$, from the Taylor expansion for each component of $\mathbf{x}_{k+1}$; that is,

$$\mathbf{x}_{k+1} = \mathbf{B}\mathbf{x}_k, \tag{7.65}$$

where $\mathbf{B}$ is the coefficient matrix from the Taylor expansion. One can show easily that $\mathbf{B}$ is an upper triangular matrix with unit diagonal and first row elements, with the other elements given by

$$B_{ij} = \binom{j-1}{i} = \frac{(j-1)!}{i!\,(j-i-1)!}. \tag{7.66}$$

Note that the dimensions of $\mathbf{B}$ are $(n+1) \times (n+1)$; that is, $\mathbf{x}$ is a vector of dimensions $n+1$, if the Taylor expansion is carried out up to the term of $d^n \mathbf{r}/dt^n$.

The correction in the Gear method is performed with the application of the differential equation under study. The difference between the predicted value of the first-order derivative $\mathbf{r}_{k+1}^{(1)}$ and the velocity field $\tau \mathbf{f}_{k+1}$ is

$$\delta \mathbf{r}_{k+1}^{(1)} = \tau \mathbf{f}_{k+1} - \mathbf{r}_{k+1}^{(1)}, \tag{7.67}$$

which would be zero if the exact solution were obtained with $\tau \to 0$. Note that $\mathbf{r}_{k+1}^{(1)}$ in the above equation is from the prediction of the Taylor expansion. So if we combine the prediction and the correction in one expression, we have

$$\mathbf{x}_{k+1} = \mathbf{B}\mathbf{x}_k + \mathbf{C}\delta \mathbf{x}_{k+1}, \tag{7.68}$$

where $\delta \mathbf{x}_{k+1}$ has only one nonzero component, $\delta \mathbf{r}_{k+1}^{(1)}$, and $\mathbf{C}$ is the correction coefficient matrix, which has nonzero elements $C_{i1} = c_i$ for $i = 0, 1, \ldots, n$. c_i are obtained by solving a matrix eigenvalue problem involving $\mathbf{B}$ and the gradient of $\delta \mathbf{x}_{k+1}$ with respect to $\mathbf{x}_{k+1}$. It is straightforward to solve for c_i if the truncation in the Taylor expansion is not very high. Interested readers can find the derivation in Gear (1971), with a detailed table listing up to the fourth-order differential equation (Gear 1971, p. 154). The most commonly used Gear scheme for the first-order differential equation is the fifth-order Gear algorithm (with the Taylor expansion carried out up to $n = 5$), with $c_0 = 95/288$, $c_1 = 1$, $c_2 = 25/24$, $c_3 = 35/72$, $c_4 = 5/48$, and $c_5 = 1/120$.

We should point out that the above procedure is not unique to first-order differential equations. The predictor part is identical for any higher-order differential equation. The only change one needs to make is the correction, which is the result of the difference between the prediction and the solution restricted by the equation. In molecular dynamics, we are interested in the solution of Newton's equation, which is a second-order differential equation with

$$\frac{d^2\mathbf{r}}{dt^2} = \frac{\mathbf{f}(\mathbf{r}, t)}{m}, \tag{7.69}$$

with $\mathbf{f}$ being the force on a specific particle of mass m. We can still formally express the algorithm as

$$\mathbf{x}_{k+1} = \mathbf{B}\mathbf{x}_k + \mathbf{C}\delta\mathbf{x}_{k+1}, \tag{7.70}$$

where $\delta\mathbf{x}_{k+1}$ also has only one nonzero element

$$\delta\mathbf{r}_{k+1}^{(2)} = \tau^2\mathbf{f}_{k+1}/2m - \mathbf{r}_{k+1}^{(2)}, \tag{7.71}$$

which is used as the correction element to all components of $\mathbf{x}$. Similarly, $\mathbf{r}_{k+1}^{(2)}$ in the above equation is from the Taylor expansion. The corrector coefficient matrix has nonzero elements $C_{i2} = c_i$ for $i = 0, 1, \ldots, n$. c_i for the second-order differential equation, which have also been worked out by Gear (1971, p. 154). The most commonly used Gear scheme for the second-order differential equation is still the fifth-order Gear algorithm (with the Taylor expansion carried out up to $n = 5$), with $c_0 = 3/20$, $c_1 = 251/360$, $c_2 = 1$, $c_3 = 11/18$, $c_4 = 1/6$, and $c_5 = 1/60$. The values of these coefficients (c_1 through c_5) are obtained with the assumption that the force in Eq. (7.69) does not have an explicit dependence on the velocity, that is, the first-order derivative of $\mathbf{r}$. If it does, $c_0 = 3/20$ needs to be changed to $c_0 = 3/16$ (Allen and Tildesley, 1987, pp. 340–2). Sometimes $c_0 = 3/20$ is replaced with $c_0 = 3/16$ for a better performance (Haile, 1992, p. 164). The readers should consult Gear (1971) for a complete discussion of the scheme.

7.6 Constant pressure, temperature, and bond length

Several issues still need to be addressed when we want to compare the simulation results with the experimental measurements. For example, environmental parameters, such as temperature or pressure, are the result of thermal or mechanical contact of the system with the environment or the result of the equilibrium of the system. For a macroscopic system, we deal with average quantities by means of statistical mechanics, so it is highly desirable to have the simulation environment as close as possible to a specific ensemble. The scheme adopted in Section 7.4 for the atomic cluster systems is closely related to the microcanonical ensemble, which has the total energy of the system conserved. It is important that we also be able to find ways to deal with constant-temperature and/or constant-pressure conditions. We should emphasize that there have been many efforts to model realistic systems with simulation boxes by introducing some specific procedures. However, all these procedures do not introduce any new concepts in physics. They are merely numerical techniques to make the simulation boxes as close as possible to the physical systems under study.

Constant pressure: The Andersen scheme

The scheme for dealing with a constant-pressure environment was devised by Andersen (1980) with the introduction of an environmental variable, the instantaneous volume of the system, to the effective Lagrangian. When the Lagrange equation is applied to the Lagrangian, the equations of motion for the coordinates of the particles and the volume result. The constant pressure is then a result of the zero average of the second-order time derivative of the volume.

The effective Lagrangian of Andersen (1980) is given by

$$\mathcal{L} = \sum_{i=1}^{N} \frac{m_i}{2} L^2 \dot{\mathbf{x}}_i^2 - \sum_{i>j} V(Lx_{ij}) + \frac{M}{2} \dot{\Omega}^2 - P_0 \Omega, \tag{7.72}$$

where the last two terms are added to deal with the constant pressure from the environment. The parameter M can be viewed here as an effective inertia associated with the expansion and contraction of the volume Ω, and P_0 is the external pressure, which introduces a potential energy $P_0 \Omega$ to the system under the assumption that the system is in contact with the constant-pressure environment. The coordinate of each particle, $\mathbf{r}_i$, is rescaled with the dimension of the simulation box, $L = \Omega^{1/3}$, since the distance between any two particles changes with L and the coordinates independent of the volume are then given by

$$\mathbf{x}_i = \mathbf{r}_i / L, \tag{7.73}$$

which are not directly related to the changing volume. Note that this effective Lagrangian is not the result of any new physical principles or concepts but is merely a method to model the effect of the environment in realistic systems. Now if we apply the Lagrange equation to the above Lagrangian, we obtain the equations of motion for the particles and the volume Ω,

$$\ddot{\mathbf{x}}_i = \frac{\mathbf{g}_i}{L} - \frac{2\dot{\Omega}}{3\Omega} \dot{\mathbf{x}}_i, \tag{7.74}$$

$$\ddot{\Omega} = \frac{P - P_0}{M}, \tag{7.75}$$

where $\mathbf{g}_i = \mathbf{f}_i/m_i$ and P is given by

$$P = \frac{1}{3\Omega}\left(\sum_i m_i L^2 \dot{\mathbf{x}}_i^2 + \sum_{i>j}^N \mathbf{r}_{ij} \cdot \mathbf{f}_{ij}\right), \tag{7.76}$$

which can be interpreted as the instantaneous pressure of the system and has a constant average P_0, since $\langle\ddot{\Omega}\rangle \equiv 0$.

After we have the effective equations of motion, the algorithm can be worked out quite easily. We will use

$$\mathbf{X} = (\mathbf{x}_1, \mathbf{x}_2, \cdots, \mathbf{x}_N), \tag{7.77}$$

$$\mathbf{G} = (\mathbf{f}_1/m_1, \mathbf{f}_2/m_2, \cdots, \mathbf{f}_N/m_N), \tag{7.78}$$

to simplify the notation. If we apply the velocity version of the Verlet algorithm, the difference equations for the volume and the rescaled coordinates are given by

$$\Omega_{k+1} = \Omega_k + \tau\dot{\Omega}_k + \frac{\tau^2(P_k - P_0)}{2M}, \tag{7.79}$$

$$\mathbf{X}_{k+1} = \left(1 - \frac{\tau^2\dot{\Omega}_k}{2\Omega_k}\right)\mathbf{X}_k + \tau\dot{\mathbf{X}}_k + \frac{\tau^2\mathbf{G}_k}{2L_k}, \tag{7.80}$$

$$\dot{\Omega}_{k+1} = \dot{\Omega}_{k+1/2} + \frac{\tau(P_{k+1} - P_0)}{2M}, \tag{7.81}$$

$$\dot{\mathbf{X}}_{k+1} = \left(1 - \frac{\tau\dot{\Omega}_{k+1}}{2\Omega_{k+1}}\right)\dot{\mathbf{X}}_{k+1/2} + \frac{\tau\mathbf{G}_{k+1}}{2L_{k+1}}, \tag{7.82}$$

where the values with index $k + 1/2$ are intermediate values before the pressure and force are updated,

$$\dot{\Omega}_{k+1/2} = \dot{\Omega}_k + \frac{\tau(P_k - P_0)}{2M},$$

$$\dot{\mathbf{X}}_{k+1/2} = \left(1 - \frac{\tau\dot{\Omega}_k}{2\Omega_k}\right)\dot{\mathbf{X}}_k + \frac{\tau\mathbf{G}_k}{2L_{k+1}}, \tag{7.83}$$

which are usually evaluated right after the volume and the coordinates are updated.

In practice, one first needs to set up the initial positions and velocities of the particles and the initial volume and its time derivative. The initial volume is determined from the given particle number and density, and its initial time derivative is usually set to be zero. The initial coordinates of the particles are usually arranged on a densely packed lattice, for example, a face-centered cubic lattice, and the initial velocities are usually drawn from the Maxwell distribution. One should test the program with different M in order to find the value of M that minimizes the fluctuation.

A generalization of the Andersen constant-pressure scheme was introduced by Parrinello and Rahman (1980; 1981) to allow the shape of the simulation box to change as well. The importance of this generalization is in the study of the structural phase transition. With the shape of the simulation box allowed to vary, the particles can easily move to the lattice points of the structure with the lowest free energy. The idea of Parrinello and Rahman can be summarized in the Lagrangian

$$\mathcal{L} = \frac{1}{2} \sum_{i=1}^{N} m_i \dot{\mathbf{x}}_i^T \mathbf{B} \dot{\mathbf{x}}_i - \sum_{i>j}^{N} V(x_{ij}) + \frac{M}{2} \sum_{i,j=1}^{3} \dot{A}_{ij}^2 - P_0 \Omega, \qquad (7.84)$$

where $\mathbf{y}_i$ is the coordinate of the ith particle in the vector representation of the simulation box, $\Omega = \mathbf{a} \cdot (\mathbf{b} \times \mathbf{c})$, with

$$\mathbf{r}_i = x_i^{(1)} \mathbf{a} + x_i^{(2)} \mathbf{b} + x_i^{(3)} \mathbf{c}. \qquad (7.85)$$

$\mathbf{A}$ is the matrix representation of $(\mathbf{a}, \mathbf{b}, \mathbf{c})$ in the Cartesian coordinates, and $\mathbf{B} = \mathbf{A}^T \mathbf{A}$. Instead of a single variable Ω, there are nine variables A_{ij}, with $i, j = 1, 2, 3$; this allows both the volume size and the shape of the simulation box to change. The equations of motion for $\mathbf{x}_i$ and A_{ij} can be derived by applying the Lagrange equation to the above Lagrangian. An external stress can also be included in such a procedure (Parrinello and Rahman, 1981).

Constant temperature: The Nosé scheme

The constant-pressure scheme discussed above is usually performed with an ad hoc constant-temperature constraint, which is done by rescaling the velocities during the simulation to ensure the relation between the total kinetic energy and the desired temperature in the canonical ensemble.

This rescaling can be shown to be equivalent to a force constraint up to the first order in the time step τ. The constraint method for the constant-temperature simulation is achieved by introducing an artificial force $\mathbf{f}_i^c = -\eta \mathbf{p}_i$, which is similar to a frictional force if η is greater than zero or to a heating process if η is less than zero. The equations of motion are modified under this force to

$$\dot{\mathbf{r}}_i = \frac{\mathbf{p}_i}{m_i}, \qquad (7.86)$$

$$\dot{\mathbf{v}}_i = \frac{\mathbf{f}_i}{m_i} - \eta \mathbf{v}_i, \qquad (7.87)$$

where $\mathbf{p}_i$ is the momentum of the ith particle and η is the constraint parameter, which can be obtained from the relevant Lagrange multiplier (Evans *et al.*, 1983) in the Lagrange equations with

$$\eta = \frac{2 \sum \mathbf{f}_i \cdot \mathbf{v}_i}{\sum m_i \mathbf{v}_i^2} = -\frac{dE_p/dt}{Gk_BT},$$

(7.88)

which can be evaluated at every time step. G here is the total number of degrees of freedom of the system, and E_p is the total potential energy. So one can simulate the canonical ensemble averages from the equations for $\mathbf{r}_i$ and $\mathbf{v}_i$ given in Eqs. (7.86) and (7.87).

The most popular constant-temperature scheme is that of Nosé (1983; 1984), who introduced a fictitious dynamical variable to take the constant-temperature environment into account. The idea is very similar to that of Andersen for the constant-pressure case. In fact, one can put both fictitious variables together to have simulations for constant pressure and constant temperature together. Here we will briefly discuss the Nosé scheme.

We can introduce a rescaled effective Lagrangian

$$\mathcal{L} = \sum_{i=1}^{N} \frac{m_i}{2} s^2 \dot{\mathbf{x}}_i^2 - \sum_{i>j} V(x_{ij}) + \frac{m_s}{2} v_s^2 - Gk_BT \ln s,$$

(7.89)

where s and v_s are the coordinate and velocity of an introduced fictitious variable that rescales the time and the kinetic energy in order to have the constraint of the canonical ensemble satisfied. The rescaling is achieved by replacing the time element dt with dt/s and holding the coordinates unchanged, that is, $\mathbf{x}_i = \mathbf{r}_i$. The velocity is rescaled with time: $\dot{\mathbf{r}}_i = s\dot{\mathbf{x}}_i$. One can then obtain the equation of motion for the coordinate $\mathbf{x}_i$ and the variable s by applying the Lagrange equation. Hoover (1985) showed later that the Nosé Lagrangian leads to a set of equations very similar to the result of the constraint force scheme discussed at the beginning of this subsection. The Nosé equations of motion are given in the Hoover version by

$$\dot{\mathbf{r}}_i = \mathbf{v}_i,$$

$$\dot{\mathbf{v}}_i = \frac{\mathbf{f}_i}{m_i} - \eta \mathbf{v}_i,$$

(7.90)

where η is given in a differential form,

$$\dot{\eta} = \frac{1}{m_s} \left(\sum_{i=1}^{N} \frac{\mathbf{p}_i^2}{m_i} - Gk_BT \right),$$

(7.91)

and the original variable s, introduced by Nosé, is related to η by

$$s = s_0 e^{\eta(t-t_0)},$$

(7.92)

with s_0 being the initial value of s at $t = t_0$. One can discretize the above equation set easily with either the Verlet algorithm or one of the Gear schemes. Note that the behavior of the parameter s is no longer directly related to the simulation; it is merely a parameter Nosé introduced to accomplish the microscopic processes happening in the constant-temperature environment. We can also combine the Andersen constant-pressure scheme with the Nosé constant-temperature scheme in a single effective Lagrangian

$$\mathcal{L} = \sum_{i=1}^{N} \frac{m_i}{2} s^2 L^2 \dot{\mathbf{x}}_i^2 - \sum_{i>j} V(Lx_{ij}) + \frac{m_s}{2} \dot{s}^2 - Gk_B T \ln s + \frac{M}{2} \dot{\Omega}^2 - P_0 \Omega,$$

$$(7.93)$$

which is worked out in detail in the original work of Nosé (1984). Another constant-temperature scheme was introduced by Berendsen et al. (1984) with the parameter η given by

$$\eta = \frac{1}{m_s} \left(\sum_{i=1}^{N} m_i \mathbf{v}_i^2 - Gk_B T \right), \qquad (7.94)$$

which can be interpreted as a similar form of the constraint that differs from the Hoover–Nosé form in the choice of η. For a recent review on the subject, see Nóse (1991).

Constant bond length

Another issue we have to deal with in practice is that for large molecular systems, such as biopolymers, the bond length of a pair of nearest neighbors does not change very much even though the angle between a pair of nearest bonds does. If we want to obtain accurate simulation results, we have to choose the time step much smaller than the period of the vibration of each pair of atoms. This is going to cost a lot of computing time and might limit the applicability of the simulation to more complicated systems, such as biopolymers.

A procedure commonly known as the SHAKE algorithm (Ryckaert, Ciccotti, and Berendsen, 1977; van Gunsteren and Berendsen, 1977) was introduced to deal with the constraint on the distance between a pair of particles in the system. The idea of this procedure is to adjust each pair of particles iteratively to have

$$(\mathbf{r}_{ij}^2 - d_{ij}^2)/d_{ij}^2 \leq \Delta \qquad (7.95)$$

in each time step. Here d_{ij} is the distance constraint between the ith and jth particles and Δ is the tolerance in the simulation. The adjustment of the position of each particle is performed after each time step of the molecular dynamics simulation.

Assume that we are working on a specific pair of particles and for the lth constraint we would like to have

$$(\mathbf{r}_{ij} + \delta\mathbf{r}_{ij})^2 - d_{ij}^2 = 0, \tag{7.96}$$

where $\mathbf{r}_{ij} = \mathbf{r}_j - \mathbf{r}_i$ is the new position vector difference after a time step starting from $\mathbf{r}_{ij}^{(0)}$ and the adjustments for the first $l - 1$ constraints have been completed. Here $\delta\mathbf{r}_{ij} = \delta\mathbf{r}_j - \delta\mathbf{r}_i$ is the total amount of adjustment needed for both particles.

One can show, in conjunction with the Verlet algorithm, that the adjustments needed are given by

$$m_i\delta\mathbf{r}_i = -g_{ij}\mathbf{r}_{ij}^{(0)} = -m_j\delta\mathbf{r}_j, \tag{7.97}$$

with g_{ij} being a parameter to be determined. The center of mass of these two particles is unmodified by the adjustment.

If we substitute $\delta\mathbf{r}_i$ and $\delta\mathbf{r}_j$ given in Eq. (7.97) into Eq. (7.96), we obtain the equation for g_{ij}:

$$\left(r_{ij}^{(0)}\right)^2 g_{ij}^2 + 2\mu_{ij}\mathbf{r}_{ij}^{(0)} \cdot \mathbf{r}_{ij}g_{ij} + \mu_{ij}^2(r_{ij}^2 - d^2) = 0, \tag{7.98}$$

where $\mu_{ij} = m_i m_j/(m_i + m_j)$ is the reduced mass from the two particles. If we keep only the linear term in g_{ij}, we have

$$g_{ij} = \frac{\mu_{ij}}{2\mathbf{r}_{ij}^{(0)} \cdot \mathbf{r}_{ij}}(d_{ij}^2 - r_{ij}^2), \tag{7.99}$$

which is reasonable, since g_{ij} is a small number during the simulation. More importantly, by the end of the iteration, all the constraints will be satisfied as well, since all g_{ij} go to zero at the convergence. Eq. (7.99) is used to estimate each g_{ij} for each constraint in each iteration. After one has the estimate of g_{ij} for each constraint, the positions of the relevant particles are all adjusted. The adjustments have to be performed several times until the convergence is reached. For more details on the algorithm, see Ryckaert et al. (1977).

This procedure has been used in the simulation of chainlike systems as well as of proteins and nucleic acids. Interested readers can find some detailed discussions on the dynamics of proteins and nucleic acids in McCammon and Harvey (1987).

7.7 Structure and dynamics of real materials

In this section, we will discuss some typical methods used to extract information about the structure and dynamics of real materials in recent molecular dynamics simulations.

A numerical simulation of a specific material starts with a determination of the interaction potential in the system. In most cases, the interaction potential is formulated in a parameterized form, which is usually determined separately from the available experimental data, first principles calculations, and condition of the system under study. The accuracy of the interaction potential will determine the validity of the simulation results. Accurate model potentials have been developed in the last few years for many realistic materials, for example, Au(100) surface (Ercolessi, Tosatti, and Parrinello, 1986) and Si_3N_4 ceramics (Vashishta et al., 1995). In the next section, we will discuss an ab initio molecular dynamics scheme in which the interaction potential is obtained by calculating the electronic structure of the system at each particle configuration.

One then needs to set up a simulation box with a periodic boundary condition. Since most experiments are performed in a constant-pressure environment, one typically uses the scheme developed by Andersen (1980) or its generalization (Parrinello and Rahman, 1980; 1981). The size of the simulation box has to be decided together with the available computing resources and the accuracy required for the quantities to be evaluated. The initial positions of the particles are usually assigned at the lattice points of a closely packed structure, for example, a face-centered cubic structure. The initial velocities of the particles are drawn from the Maxwell distribution for a given temperature. The temperature can be changed by rescaling the velocities. This is extremely useful in the study of phase transitions with varying temperature, such as the transition between different lattice structures, glass transition under quenching, or liquid–solid transition when the system is cooled down slowly. The advantage of the simulation over the actual experiment also shows up when one wants to observe some behavior that is not achievable experimentally due to the limitations of the technique or equipment. For example, the glass transition in the Lennard–Jones system is observed in molecular dynamics simulations but not in the experiments for liquid Ar, since the necessary quenching rate is so high that it is impossible to achieve it experimentally.

Studying the dynamics of different materials requires a more general time-dependent density–density correlation function

$$C(\mathbf{r}, \mathbf{r}'; t) = \langle \hat{\rho}(\mathbf{r} + \mathbf{r}', t)\hat{\rho}(\mathbf{r}', 0)\rangle, \tag{7.100}$$

with the time-dependent density operator given by

$$\hat{\rho}(\mathbf{r}, t) = \sum_{i=1}^{N} \delta[\mathbf{r} - \mathbf{r}_i(t)]. \tag{7.101}$$

If the system is homogeneous, we can integrate out $\mathbf{r}'$ in the time-dependent density–density correlation function to reach the van Hove time-dependent distribution function (van Hove, 1954)

$$G(\mathbf{r}, t) = \frac{1}{\rho N} \left\langle \sum_{i,j}^{N} \delta\{\mathbf{r} - [\mathbf{r}_i(t) - \mathbf{r}_j(0)]\} \right\rangle. \tag{7.102}$$

The *dynamical structure factor* measured in an experiment, for example, neutron scattering, is given by the Fourier transform of $G(\mathbf{r}, t)$,

$$S(\mathbf{k}, \omega) = \frac{\rho}{2\pi} \int e^{i(\omega t - \mathbf{k} \cdot \mathbf{r})} G(\mathbf{r}, t) \, d\mathbf{r} \, dt. \tag{7.103}$$

The above equation reduces to the static case with

$$S(k) - 1 = 4\pi\rho \int \frac{\sin kr}{kr} [g(r) - 1] r^2 \, dr, \tag{7.104}$$

if one realizes that

$$G(\mathbf{r}, 0) = g(\mathbf{r}) + \frac{\delta(\mathbf{r})}{\rho} \tag{7.105}$$

and

$$S(\mathbf{k}) = \int_{-\infty}^{\infty} S(\mathbf{k}, \omega) \, d\omega, \tag{7.106}$$

where $g(\mathbf{r})$ is the pair distribution discussed earlier in this chapter and $S(k)$ is the angular average of $S(\mathbf{k})$. $G(\mathbf{r}, t)$ can be interpreted as the probability of observing one of the particles at $\mathbf{r}$ at time t if a particle was observed at $\mathbf{r} = 0$ at $t = 0$. This leads to the numerical evaluation of $G(r, t)$, which is the angular average of $G(\mathbf{r}, t)$. If we write $G(r, t)$ into two parts,

$$G(r, t) = G_s(r, t) + G_d(r, t), \tag{7.107}$$

with $G_s(r, t)$ being the probability of observing the same particle that was at $r = 0$ at $t = 0$, and $G_d(r, t)$ being the probability of observing other particles,

$$G_d(r, t) \simeq \frac{1}{\rho} \frac{\langle \Delta N(r, \Delta r; t) \rangle}{\Delta \Omega(r, \Delta r)}, \tag{7.108}$$

where $\Delta\Omega(r, \Delta r) \simeq 4\pi r^2 \Delta r$ is the volume of a spherical shell with radius r and thickness Δr, and $\Delta N(r, \Delta r; t)$ is the number of particles in the spherical shell at time t. The position of each particle at $t = 0$ is chosen as the origin in the evaluation of $G_d(r, t)$ and the average is taken from all the particles in the system. Note that this is different from the evaluation of $g(r)$, in which we always select a particle position as the origin and take the average over time. One can also take the average over all the particles in the evaluation of $g(r)$. $G_s(r, t)$ can be evaluated in a similar fashion. Since $G_s(r, t)$ represents the probability that a particle is at a distance r at the time t away from its original position at $t = 0$, we have

$$\Delta^{2n}(t) = \frac{1}{N}\left\langle \sum_{i=1}^{N}[\mathbf{r}_i(t) - \mathbf{r}_i(0)]^{2n} \right\rangle = \int r^{2n} G_s(r, t)\, d\mathbf{r}, \tag{7.109}$$

which can be used in the evaluation of the diffusion coefficient with $n = 1$. The diffusion coefficient can also be evaluated from the autocorrelation function of the velocity,

$$c(t) = \langle \mathbf{v}(t) \cdot \mathbf{v}(0) \rangle = \frac{1}{N} \sum_{i=1}^{N}[\mathbf{v}_i(t) \cdot \mathbf{v}_i(0)], \tag{7.110}$$

with

$$D = \frac{1}{3c(0)} \int_0^\infty c(t)\, dt, \tag{7.111}$$

since the velocity of each particle at each time step is known from the simulation. The velocity correlation function can also be used to obtain the power spectrum of the velocity by the Fourier transform

$$P(\omega) = \frac{6}{\pi c(0)} \int_0^\infty c(t) \cos \omega t\, dt, \tag{7.112}$$

which has many features similar to those of the phonon spectrum of the system, for example, a broad peak for the glassy state and more structures for a crystalline state.

Thermodynamical quantities can also be evaluated from molecular dynamics simulations. For example, if a simulation is performed under the constant-pressure condition, one can obtain physical quantities such as the particle density, pair-distribution function, and so on, at different temperatures. The inverse of the particle density is called the specific volume, notated as $V_P(T)$. The thermal expansion coefficient under the constant-pressure condition is then given by

$$\alpha_P = \frac{\partial V_P(T)}{\partial T}, \tag{7.113}$$

which is quite different when the system is in a liquid phase or a solid phase. Furthermore, one can calculate the temperature-dependent enthalpy

$$H = E + P\Omega, \tag{7.114}$$

with E being the internal energy given by

$$E = \left\langle \sum_{i=1}^{N} \frac{m_i}{2} v_i^2 + \sum_{i \neq j}^{N} V(r_{ij}) \right\rangle, \tag{7.115}$$

P being the pressure, and Ω being the volume of the system. The specific heat under the constant-pressure condition is then obtained from

$$C_P = \frac{1}{N}\frac{\partial H}{\partial T}. \tag{7.116}$$

The specific heat under the constant-volume condition can be derived from the fluctuation of the internal energy, $\langle(\delta E)^2\rangle = \langle[E - \langle E\rangle]^2\rangle$, with

$$C_V = \frac{\langle(\delta E)^2\rangle}{k_B T^2}. \tag{7.117}$$

The isothermal compressibility κ_T is then obtained from the identity

$$\kappa_T = \frac{T\Omega\alpha_P^2}{C_P - C_V}, \tag{7.118}$$

which is also quite different for the liquid phase and for the solid phase. For more discussions on the molecular dynamics simulation of glass transition, see Yonezawa (1991).

More aspects related to the structure and dynamics of a system can be studied in molecular dynamics simulations. The advantage of molecular dynamics over a typical stochastic simulation is that molecular dynamics can give all the information on the time dependence of the system, which is necessary for analyzing the structural and dynamical properties of the system. Molecular dynamics is therefore the method of choice in computer simulations of many-particle systems. However, stochastic simulations, such as Monte Carlo simulations, are sometimes easier to perform for some systems and are closely related to the simulations of quantum systems.

7.8 Ab initio molecular dynamics

In this section, we would like to briefly outline a very interesting simulation scheme that combines the calculation of the electronic structure and the molecular dynamics simulation of the system. This is known as *ab initio molecular dynamics*, which was devised and put into practice by Car and Parrinello (1985).

The maturity of molecular dynamics simulation schemes and the great increase in computing capacity have made it possible to perform molecular dynamics simulations in studies of amorphous materials, biopolymers, and other complex systems. However, in order to obtain an accurate description of a specific system, one has to know the precise behavior of the interactions among the particles, that is, the ions in the system. Electrons move much faster than ions because the electron mass is much smaller than that of an ion. The position dependence of the interactions among the ions in a given system is therefore determined by the distribution of the electrons (electronic structure) at the specific moment. Thus, a good approximation of the electronic structure in a calculation can be obtained with all the nuclei fixed in space for that moment. This is the essence of the Born–Oppenheimer approximation, which allows the degrees of freedom of the electrons to be treated separately from those of the ions.

In the past, the interactions among the ions were given in a parameterized form based on experimental data, quantum chemistry calculations, or specific conditions of the system under study. All these procedures are limited due to the complexity of the electronic structure of the actual materials. One can easily obtain accurate parameterized interactions for the inertial gases, such as Ar, but would have a lot of difficulties in obtaining an accurate parameterized interaction for H_2O – one that can correctly produce the structures of ice in molecular dynamics simulations, for example.

It seems that a combined scheme is highly desirable. One can calculate the many-body interactions among the ions in the system from the electronic structure calculated at every molecular dynamics time step and then determine the next configuration from such ab initio interactions. This can be achieved in principle, but in practice the scheme will be required to work with existing computing capacity. The scheme devised by Car and Parrinello (1985) was the first in its class and also the first to be applied to real materials.

Density functional theory

Hohenberg, Kohn, and Sham (Hohenberg and Kohn, 1964; Kohn and Sham, 1965) introduced a scheme about thirty years ago to cope with the many-electron system in a very simple formalism. The density functional theorem proved by Hohenberg and Kohn (1964) states that the ground-state energy of an interacting system is the optimized value of an energy functional $E[\rho(\mathbf{r})]$ of the electron density $\rho(\mathbf{r})$ and that the corresponding density distribution of the optimization is the unique ground-state density distribution. Symbolically, one can write

$$E[\rho(\mathbf{r})] = E_{ext}[\rho(\mathbf{r})] + E_h[\rho(\mathbf{r})] + E_k[\rho(\mathbf{r})] + E_{xc}[\rho(\mathbf{r})], \qquad (7.119)$$

where $E_{ext}[\rho(\mathbf{r})]$ is the contribution from the external potential $U_{ext}(\mathbf{r})$ with

$$E_{ext}[\rho(\mathbf{r})] = \int U_{ext}(\mathbf{r})\rho(\mathbf{r})\,d\mathbf{r}, \qquad (7.120)$$

$E_h[\rho(\mathbf{r})]$ is the Hartree type of contribution due to the electron–electron interaction,

$$E_h[\rho(\mathbf{r})] = \frac{e^2}{2} \int \frac{\rho(\mathbf{r}')\rho(\mathbf{r})}{|\mathbf{r}' - \mathbf{r}|}\,d\mathbf{r}\,d\mathbf{r}', \qquad (7.121)$$

E_k is the contribution of the kinetic energy, and E_{xc} is the rest of the contributions termed as the exchange-correlation energy functional.

In general, we can express the electron density in a spectral representation,

$$\rho(\mathbf{r}) = \sum_i \psi_i^+(\mathbf{r})\psi_i(\mathbf{r}), \tag{7.122}$$

where $\psi_i^+(\mathbf{r})$ is the complex conjugate of the wavefunction $\psi_i(\mathbf{r})$ and the summation is over all the degrees of freedom, that is, all the occupied states with different spin orientations. Then the kinetic energy functional can be written as

$$E_k[\rho(\mathbf{r})] = -\frac{\hbar^2}{2m} \int \sum_i \psi_i^+(\mathbf{r})\nabla^2\psi_i(\mathbf{r})\,d\mathbf{r}. \tag{7.123}$$

There is a constraint that the number of electrons in the system be a constant, namely,

$$\int \rho(\mathbf{r})\,d\mathbf{r} = N, \tag{7.124}$$

which introduces the Lagrange multipliers into the variation. If we use the spectral representation of the density in the energy functional and apply the Euler equation with the Lagrange multipliers,

$$\frac{\delta E[\rho(\mathbf{r})]}{\delta \psi_i^+(\mathbf{r})} - \epsilon_i \psi_i(\mathbf{r}) = 0, \tag{7.125}$$

we arrive at the Kohn–Sham equation

$$\left[-\frac{\hbar^2}{2m}\nabla^2 + V_e(\mathbf{r}) \right]\psi_i(\mathbf{r}) = \epsilon_i \psi_i(\mathbf{r}), \tag{7.126}$$

where $V_e(\mathbf{r})$ is an effective potential given by

$$V_e(\mathbf{r}) = U_{\text{ext}}(\mathbf{r}) + V_h(\mathbf{r}) + V_{xc}(\mathbf{r}), \tag{7.127}$$

with

$$V_{xc}(\mathbf{r}) = \frac{\delta E_{xc}[\rho(\mathbf{r})]}{\delta\rho(\mathbf{r})}, \tag{7.128}$$

which cannot be obtained exactly. A common practice is to approximate it by its homogeneous density equivalence, the so-called local approximation, in which one assumes that $V_{xc}(\mathbf{r})$ is given by the same quantity of a uniform electron gas with density equal to $\rho(\mathbf{r})$. This is termed as *local density approximation*. Local density approximation has been successfully applied to many physical systems, including atomic, molecular, and condensed-matter systems. The unexpected success of local density approximation in materials research has made it a standard technique for calculating electronic properties of new materials and systems. The procedure of calculating the electronic structure with local density approximation can be described in several steps. One first constructs the local approximation of $V_{xc}(\mathbf{r})$ with a guessed density distribution. Then the Kohn–Sham equation is solved, and a new density distribution is constructed from the solution. With the new density distribution, one can improve $V_{xc}(\mathbf{r})$ and then solve the Kohn–Sham equation again. This procedure is repeated until a convergence is reached. Interested readers can find detailed discussions on the density functional theory in many monographs or review articles, for example, Kohn and Vashishta (1983), and Jones and Gunnarsson (1989).

The Car–Parrinello simulation scheme

The Hohenberg–Kohn energy functional forms the Born–Oppenheimer potential surface for the ions in the system. The idea of ab initio molecular dynamics is similar to the relaxation scheme we discussed in Chapter 6. We introduced a functional

$$U = \int \left\{ \frac{1}{2}\epsilon(x)\left[\frac{d\psi(x)}{dx}\right]^2 - 4\pi\rho(x)\psi(x) \right\} dx \qquad (7.129)$$

for the one-dimensional Poisson equation. Note that $\rho(x)$ here is the charge density instead of the particle density. The physical meaning of this functional is the electrostatic energy of the system. After applying the trapezoid rule to the integral and taking a partial derivative of U with respect to ϕ_i, we obtain the corresponding difference equation

$$(H_i + 4\pi\rho_i)\psi_i = 0 \qquad (7.130)$$

for the one-dimensional Poisson equation. Here $H_i\phi_i$ is a brief notation for $\epsilon_{i+1/2}\phi_{i+1} + \epsilon_{i-1/2}\phi_{i-1} - (\epsilon_{i+1/2} + \epsilon_{i-1/2})\phi_i$. If we combine the above equation with the relaxation scheme discussed in Section 6.5, we have

$$\psi_i^{(k+1)} = (1-p)\psi_i^{(k)} + p(H_i + 4\pi\rho_i + 1)\psi_i^{(k)}, \qquad (7.131)$$

which would optimize (minimize) the functional (the electrostatic energy) as $k \to \infty$. The indices k and $k+1$ are for iteration steps, and the index i is for the spatial points. The iteration can be interpreted as a fictitious time step, since we can rewrite the above equation into

$$\frac{\psi_i^{(k+1)} - \psi_i^{(k)}}{p} = (H_i + 4\pi\rho_i)\psi_i^{(k)}, \qquad (7.132)$$

with p acting like a fictitious time step. The solution converges to the true solution of the Poisson equation as k goes to infinity if the functional U decreases during the iterations.

The ab initio molecular dynamics is devised by introducing a fictitious time-dependent equation for the electron degrees of freedom,

$$\mu \frac{d^2 \psi_i(\mathbf{r}, t)}{dt^2} = -\frac{1}{2} \frac{\delta E[\rho(\mathbf{r}, t); \mathbf{R}_n]}{\delta \psi_i^+(\mathbf{r}, t)} + \sum_j \Lambda_{ij} \psi_j(\mathbf{r}), \qquad (7.133)$$

where μ is an adjustable parameter introduced for convenience; Λ_{ij} is the Lagrange multiplier, introduced to ensure the orthonormal condition of the wavefunctions $\psi_i(\mathbf{r}, t)$; and the summation is for all the occupied states. Note that the potential energy surface E is a functional of the electron density as well as a function of the ionic coordinates $\mathbf{R}_n$ for $n = 1, 2, \ldots, N_c$ with N_c ions in the system. In practice, one can also consider the first-order time derivative equation, with $d^2 \psi_i(\mathbf{r}, t)/dt^2$ replaced by the first-order derivative $d\psi_i(\mathbf{r}, t)/dt$, since either the first- or second-order derivative will approach zero at the limit of convergence. Second-order derivatives were used in the original work of Car and Parrinello and were later shown to yield a fast convergence if a special damping term is introduced (Tassone, Mauri, and Car, 1994). The ionic degrees of freedom are then simulated from Newton's equation

$$M_n \frac{d^2 \mathbf{R}_n}{dt^2} = -\frac{\partial E[\rho(\mathbf{r}, t); \mathbf{R}_n]}{\partial \mathbf{R}_n}, \qquad (7.134)$$

where M_n and $\mathbf{R}_n$ are the mass and the position vector of the nth ion. The advantage of the ab initio molecular dynamics is that the electron degrees of freedom and the ionic degrees of freedom are simulated simultaneously by the above equations. Since its introduction by Car and Parrinello in 1985, the method has been applied to many systems, especially to those without a crystalline structure, namely, liquids and amorphous materials. We will not go into more detail on the method or its applications; interested readers can find them in the recent review by Tassone et al. (1994). Recent progress in the ab initio molecular dynamics also includes mapping the Hamiltonian onto a tight-binding model in which the evaluation of the electron degrees of freedom is drastically simplified (Wang, Chan, and Ho, 1989). This approach has also been applied to many systems, for example, amorphous carbon and carbon clusters. One can find more discussions on the method in several recent review articles, for example, Oguchi and Sasaki (1991).

Exercises

7.1 Show that the Verlet algorithm preserves the time reversal of Newton's equation.

7.2 Write a program that uses the velocity version of the Verlet algorithm to simulate the small charge clusters discussed in Chapter 4. Use the parameters given in Section 4.4 and set up the initial configuration as the equilibrium configuration. Assign initial velocities from the Maxwell distribution. Discuss the time dependence of the kinetic energy with different total energies.

7.3 Develop a program with the fifth-order Gear predictor–corrector scheme and apply it to the damped pendulum under a sinusoidal driving force. Study the properties of the pendulum with different values of the parameters.

7.4 Derive the Hoover equations from the Nosé Lagrangian. Show that the generalized Lagrangian given in Section 7.6 can provide correct equations of motion to ensure the constraint of constant temperature as well as that of constant pressure.

7.5 Show that the Parrinello–Rahman Lagrangian allows the shape of the simulation box to change, and derive equations of motion from it. Find the Lagrangian that combines the Parrinello–Rahman Lagrangian and Nosé Lagrangian and derive the equations of motion from this generalized Lagrangian.

7.6 For quantum many-body systems, the n-body density function is defined by

$$\rho_n(\mathbf{r}_1, \mathbf{r}_2, \cdots, \mathbf{r}_n) = \frac{1}{\mathcal{Z}} \frac{N!}{(N-n)!} \int |\Psi(\mathbf{R})|^2 \, d\mathbf{r}_{n+1} \, d\mathbf{r}_{n+2} \cdots d\mathbf{r}_N,$$

where $\Psi(\mathbf{R})$ is the ground-state wavefunction of the many-body system and $\mathcal{Z}$ is the normalization constant given by

$$\mathcal{Z} = \int |\Psi(\mathbf{R})|^2 \, d\mathbf{r}_1 \, d\mathbf{r}_2 \cdots d\mathbf{r}_N.$$

The pair-distribution function is related to the two-body density function through

$$\rho(\mathbf{r})g(\mathbf{r}, \mathbf{r}')\rho(\mathbf{r}') = \rho_2(\mathbf{r}, \mathbf{r}').$$

Show that

$$\int \rho(\mathbf{r})[\tilde{g}(\mathbf{r}, \mathbf{r}') - 1] \, d\mathbf{r} = -1,$$

where

$$\tilde{g}(\mathbf{r}, \mathbf{r}') = \int_0^1 g(\mathbf{r}, \mathbf{r}'; \lambda) \, d\lambda,$$

with $g(\mathbf{r}, \mathbf{r}'; \lambda)$ being the pair-distribution function under the scaled interaction given by

$$V(\mathbf{r}, \mathbf{r}'; \lambda) = \lambda V(\mathbf{r}, \mathbf{r}').$$

7.7 Show that the exchange-correlation energy functional is given by

$$E_{xc}[\rho(\mathbf{r})] = \frac{1}{2} \int \rho(\mathbf{r}) V(\mathbf{r}, \mathbf{r}') [\tilde{g}(\mathbf{r}, \mathbf{r}') - 1] \rho(\mathbf{r}') \, d\mathbf{r} \, d\mathbf{r}',$$

with $\tilde{g}(\mathbf{r}, \mathbf{r}')$ given from Exercise 7.6.

8

Modeling continuous systems

It is usually more difficult to simulate continuous than discrete systems, especially when the systems are described by equations with nonlinear terms. The systems can be so complex that the length scale at the atomic level can be as important as the length scale at the macroscopic level. The basic strategy in dealing with complex systems is similar to a divide-and-conquer concept, that is, dividing the systems with an understanding of the length scales involved in the physical phenomena and then solving the problem with an appropriate method. Length scale is typically associated with an involved energy scale, such as the average temperature of the system or the average interaction of each pair of particles. The divide-and-conquer schemes are quite powerful in a wide range of applications. However, each method has its advantages and disadvantages, depending on the specific system.

8.1 Hydrodynamic equations

In this chapter, we will discuss several methods used in simulating continuous systems. First we will discuss a quite mature method, the *finite element method*, which sets up the idea of partitioning the system according to physical condition. Then we will discuss another method, the *particle-in-cell method*, which adopts a mean-field concept to deal with a large system involving many, many atoms, for example, 10^{23} atoms. This method has been very successful in the simulation of plasma, galactic, hydrodynamic, and magnetohydrodynamic systems. Then we will briefly highlight a relatively new method, the *Boltzmann lattice-gas method*, which is closely related to the lattice-gas method of *cellular automata* and has recently been applied to the simulation of several continuous systems.

Before going into the details of each method, we would like to introduce the hydrodynamic equations. The equations are obtained by analyzing the dynamics of a small element of fluid in the system and then taking the limit of continuity. We can obtain three basic equations by examining the changes in the mass, momentum, and energy of a small element in the system. For simplicity, let us first assume that the fluid is neutral and not under any external field. The three fundamental hydrodynamic equations are given by

$$\frac{\partial \rho}{\partial t} + \nabla \cdot (\rho \mathbf{v}) = 0, \tag{8.1}$$

$$\frac{\partial}{\partial t}(\rho \mathbf{v}) + \nabla \cdot \mathbf{\Pi} = 0, \tag{8.2}$$

$$\frac{\partial}{\partial t}\left(\rho \epsilon + \frac{1}{2}\rho v^2\right) + \nabla \cdot \mathbf{j}_e = 0, \tag{8.3}$$

where the first equation is commonly known as the *continuity equation*, which is the result of mass conservation. The second equation is known as the Navier–Stokes equation, which is Newton's equation for the element. The last equation is the result of the *work-energy theorem*. Here ρ is the mass density, $\mathbf{v}$ is the velocity, ϵ is the internal energy per unit mass, $\mathbf{\Pi}$ is the momentum flux tensor, and $\mathbf{j}_e$ is the energy flux density. The momentum flux tensor can be written as

$$\mathbf{\Pi} = \rho \mathbf{v}\mathbf{v} - \mathbf{\Gamma}, \tag{8.4}$$

with $\mathbf{\Gamma}$ being the stress tensor of the fluid,

$$\mathbf{\Gamma} = \eta[\nabla \mathbf{v} + (\nabla \mathbf{v})^T] + \left[\left(\zeta - \frac{2\eta}{3}\right)\nabla \cdot \mathbf{v} - P\right]\mathbf{I}, \tag{8.5}$$

where η and ζ are coefficients of bulk and shear viscosity, P is the pressure in the fluid, and $\mathbf{I}$ is the unit tensor. The energy flux density is given by

$$\mathbf{j}_e = \mathbf{v}\left(\rho \epsilon + \frac{1}{2}\rho v^2\right) - \mathbf{v} \cdot \mathbf{\Gamma} - \kappa \nabla T, \tag{8.6}$$

where the last term is the thermal energy flow due to the gradient of the temperature T and κ is the thermal conductivity.

The above set of equations can be solved together with the equation of state,

$$f(\rho, P, T) = 0, \tag{8.7}$$

which relates several thermodynamic quantities of the system.

Now we can include the effects of external fields. For example, if gravity is important, we can modify the second equation to

$$\frac{\partial}{\partial t}(\rho \mathbf{v}) + \nabla \cdot \mathbf{\Pi} = \rho \mathbf{g}, \tag{8.8}$$

with $\mathbf{g}$ being the gravitational field given by

$$\mathbf{g} = -\nabla \Phi, \tag{8.9}$$

where Φ is the gravitational potential from the Poisson equation

$$\nabla^2 \Phi = 4\pi G\rho, \tag{8.10}$$

with G being the gravitational constant.

In cases where the system is charged and electromagnetic fields become important, we need to modify the momentum flux tensor, energy density, and energy flux density. A term $\mathbf{BB} - \mathbf{I}B^2/2$ should be added to the momentum flux tensor. A magnetic field energy density term $B^2/2$ should be added to the energy density, and an extra energy flux term $\mathbf{E} \times \mathbf{B}$ should be added to the energy flux density. We have used the Gaussian units for these terms. The electric current density $\mathbf{j}$, electric field $\mathbf{E}$, and magnetic field $\mathbf{B}$ are related through the Maxwell equations and the charge transport equation. We will come back to this point in the later sections of this chapter.

These hydrodynamic or magnetohydrodynamic equations can be solved in principle with the finite difference methods discussed in Chapter 6. But in this chapter, we would like to introduce some other methods, mainly based on the divide-and-conquer schemes. For more detailed discussions on the hydrodynamic and magnetohydrodynamic equations, see Landau and Lifshitz (1987) and Lifshitz and Pitaevskii (1981).

8.2 The basic finite element method

In order to show how a typical finite element method works for a specific problem, let us take the one-dimensional Poisson equation as an example. Assume that the charge distribution is $\rho(x)$ and the equation for the electrostatic potential is

$$\frac{d^2\phi(x)}{dx^2} = -4\pi\rho(x). \tag{8.11}$$

In order to simplify our discussion here, let us assume that the boundary condition is given as $\phi(0) = \phi(1) = 0$. As discussed in Chapter 5, we can always express our solutions in terms of a complete set of orthogonal basis functions $u_i(x)$ for $i = 1, 2, \ldots, \infty$:

$$\phi(x) = \sum_i a_i u_i(x), \tag{8.12}$$

and the basis functions satisfy

$$\int_0^1 u_j(x)u_i(x)\,dx = \delta_{ij}, \tag{8.13}$$

and $u_i(0) = u_i(1) = 0$. We have assumed that $u_i(x)$ is a real function and that the summation is over all the basis functions. An example in this case is that

$u_j(x) = \sqrt{2} \sin j\pi x$. In order to obtain the actual value of $\phi(x)$, we need to solve all the coefficients a_i by applying the differential equation and the orthogonal condition in Eq. (8.13). This becomes quite a difficult task when the boundary of the system does not have a regular shape.

The finite element method is designed to find a good approximation of the solution for irregular boundary conditions or in situations of rapid change of the solution. Typically, the approximation is still written in a series

$$\phi_n(x) = \sum_{i=1}^{n} a_i u_i(x), \tag{8.14}$$

with a finite number of terms. $u_i(x)$ is a set of linearly independent local functions, each defined in a small region around a *node* x_i. If we use this approximation in the differential equation, we have a nonzero value

$$r_n(x) = \phi_n''(x) + 4\pi \rho(x), \tag{8.15}$$

which would be zero if $\phi_n(x)$ were the exact solution. The goal now is to set up a scheme that would make $r_n(x)$ small over the whole region during the process of determining all a_i. The selection of $u_i(x)$ and the procedure for optimizing $r_n(x)$ determine how accurate the solution would be for a given n. One can always improve the accuracy with a higher n. One scheme for performing optimization is to introduce a weighted integral

$$g_i = \int_0^1 r_n(x) w_i(x) \, dx, \tag{8.16}$$

which is then forced to be zero for the determination of a_i. Here $w_i(x)$ is a selected weight.

The procedure just described is the essence of the finite element method. The region of interest is divided into many small pieces, usually of the same topology but not the same size, for example, different triangles for a two-dimensional domain. Then a set of linearly independent local functions $u_i(x)$ are selected, with each defined around a node and its neighborhood. One then selects the weight $w_i(x)$, which is usually chosen as a local function as well. For the one-dimensional Poisson equation, the weighted integral becomes

$$g_i = \int_0^1 \left[\sum_{j=1}^{n} a_j u_j''(x) + 4\pi \rho(x) \right] w_i(x) \, dx = 0, \tag{8.17}$$

which is equivalent to a linear equation problem

$$\mathbf{Aa} = \mathbf{b}, \tag{8.18}$$

with

$$A_{ij} = -\int_0^1 u_i''(x) w_j(x) \, dx, \tag{8.19}$$

and

$$b_i = 4\pi \int_0^1 \rho(x) w_i(x) \, dx. \tag{8.20}$$

The advantage of choosing $u_i(x)$ and $w_i(x)$ as local functions is in the solution of the linear equation set given in Eq. (8.18). Basically, one needs to solve only a tridiagonal or a band matrix problem, which can be solved quite quickly. The idea is to make the problem computationally simple but numerically accurate. That means the choice of the weight $w_i(x)$ is rather an art. A common choice is that $w_i(x) = u_i(x)$, which is called the *Galerkin method*.

Now let us go into some detail for the one-dimensional Poisson equation with the Galerkin method. We can divide the region $[0, 1]$ into $n + 1$ equal intervals with $x_0 = 0$ and $x_{n+1} = 1$ and take

$$u_i(x) = \begin{cases} (x - x_{i-1})/h & \text{for} \quad x \in [x_{i-1}, x_i], \\ (x_{i+1} - x)/h & \text{for} \quad x \in [x_i, x_{i+1}], \\ 0 & \text{otherwise.} \end{cases} \tag{8.21}$$

Here $h = x_i - x_{i-1} = 1/(n + 1)$ is the spatial interval. Note that this choice of $u_i(x)$ satisfies the boundary condition. Let us take a very simple charge distribution

$$\rho(x) = \frac{\pi}{4} \sin \pi x \tag{8.22}$$

for illustrative purposes. We can easily carry out the weighted integrals to obtain the matrix elements

$$A_{ij} = \int_0^1 u_i'(x) u_j'(x) \, dx = \begin{cases} 2/h & \text{for} \quad i = j, \\ -1/h & \text{for} \quad i = j \pm 1, \\ 0 & \text{otherwise} \end{cases} \tag{8.23}$$

and the vector

$$\begin{aligned} b_i &= 4\pi \int_0^1 \rho(x) u_i(x) \, dx \\ &= \frac{\pi}{h} (x_{i-1} + x_{i+1} - 2x_i) \cos \pi x_i \\ &\quad + \frac{1}{h} (2 \sin \pi x_i - \sin \pi x_{i-1} - \sin \pi x_{i+1}). \end{aligned} \tag{8.24}$$

Now we are ready to put it into a program. One may have noticed that since the local basis $u_i(x)$ is confined in the region $[x_{i-1}, x_{i+1}]$, the matrix $\mathbf{A}$ is automatically a symmetric tridiagonal matrix. As we discussed in Chapter 6, an LU decomposition scheme can easily be devised to solve this tridiagonal matrix problem. Here is the implementation of the algorithm.

```
        PROGRAM GALERKIN
C
C This program solves the one-dimensional Poisson
C equation with the Galerkin method.
C
        PARAMETER (N=99)
        DIMENSION B(N),A(N),Y(N),W(N),U(N)
C
        PI  =  4.0*ATAN(1.0)
        XL  =  1.0
        H   =  XL/(N+1)
        D   =  2.0/H
        E   =  -1.0/H
        B0  =  PI/H
        B1  =  1.0/H
C
C Find the elements in L and U
C
        W(1) =  D
        U(1) =  E/D
        DO      100 I = 2, N
          W(I) = D-E*U(I-1)
          U(I) = E/W(I)
  100 CONTINUE
C
C Assign the array B
C
        DO      200 I = 1, N
          XIM  = H*(I-1)
          XI   = H*I
          XIP  = H*(I+1)
          B(I) = B0*COS(PI*XD)*(XIM+XIP-2.0*XI)
     *             +B1*(2.0*SIN(PI*XI)-SIN(PI*XIM)
     *             -SIN(PI*XIP))
  200 CONTINUE
C
```

```
C Find the solution
C
      Y(1) = B(1)/W(1)
      DO      300 I = 2, N
        Y(I) = (B(I)-E*Y(I-1))/W(I)
  300 CONTINUE
C
      A(N) = Y(N)
      DO      400 I = N-1,1,-1
        A(I) = Y(I)-U(I)*A(I+1)
  400 CONTINUE
C
      WRITE (6,999) (I*H,A(I),I=1,N)
      STOP
  999 FORMAT(2F16.8)
      END
```

The result from the above program is plotted in Fig. 8.1. As one can easily show, the above problem has an analytic solution, $\phi(x) = \sin \pi x$. There is one more related issue involved in the Galerkin method: how to deal with different boundary conditions. We have selected a very simple boundary condition in the above example. In general, we may have nonzero boundary conditions, for example,

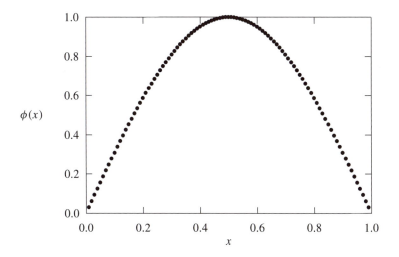

Fig. 8.1. Numerical solution of the one-dimensional Poisson equation with the program GALERKIN.

$\phi(0) = \phi_0$ and $\phi(1) = \phi_1$. We can write the solution as

$$\phi_n(x) = (1 - x)\phi_0 + x\phi_1 + \sum_{i=1}^{n} a_i u_i(x), \tag{8.25}$$

which has the first two terms to satisfy the boundary conditions; the summation part is zero at the boundaries. Similarly, if the boundary conditions are of the Neumann type or a mixed type, special care can be taken by adding a couple of terms and the local basis $u_i(x)$ can be made to satisfy homogeneous boundary conditions only. We will discuss a few of the other situations in the next section.

8.3 The Ritz variational method

The basic finite element method discussed above resembles the idea of the discretization scheme of the finite difference method discussed in Chapter 6. The idea developed there was to construct a functional that would lead to the original differential equation when optimized. For example, the functional for the one-dimensional Helmholtz equation

$$\frac{d^2\phi(x)}{dx^2} + k^2\phi(x) = -s(x), \tag{8.26}$$

with k^2 being a parameter and $s(x)$ being a function of x defined in the region $[0, 1]$, is given by

$$E[\phi(x)] = \int_0^1 \left\{ \frac{1}{2}[\phi'^2(x) - k^2\phi^2(x)] - s(x)\phi(x) \right\} dx, \tag{8.27}$$

which leads to the original differential equation from the Euler–Lagrange equation

$$\frac{\delta E[\phi(x)]}{\delta \phi(x)} = 0, \tag{8.28}$$

with the condition that the variations at the boundaries are zero. What we did in Chapter 6 was to discretize the integrand of the functional and then consider the solution at each lattice point as an independent variable when the optimization was performed. A difference equation

$$\frac{\partial E[\phi_i(x)]}{\partial \phi_i(x)} = 0 \tag{8.29}$$

is obtained from the Euler–Lagrange equation and will act as a finite difference approximation of the original differential equation.

The variational principle used in the above Euler–Lagrange equation is called the *Ritz variational principle*. The true solution of the differential equation has the lowest value of $E[\phi(x)]$. An approximate solution is better the lower its value of $E[\phi(x)]$. Thus, there are two basic steps in the Ritz variational scheme. A proper functional for a specific differential equation has to be constructed first. Then one can approximate the solution of the equation piece by piece in each finite element and apply the Ritz variational principle to obtain the linear equation set for the coefficients of the approximate solution.

Let us here use the one-dimensional Helmholtz equation to demonstrate some details of the Ritz variational method. The approximate solution is still expressed as a linear combination of a set of local functions

$$\phi_n(x) \simeq \sum_{i=1}^{n} a_i u_i(x), \tag{8.30}$$

which are linearly independent. $u_i(x)$ is usually defined around the ith node and its neighborhood. The essence of the Ritz variational method here is to treat each coefficient a_i as an independent variational parameter. The optimization is then done with respect to each a_i:

$$\frac{\partial E[(\phi_n(x)]}{\partial a_i} = 0 \tag{8.31}$$

for $i = 1, 2, \ldots, n$, which produces a linear equation set

$$\mathbf{Aa} = \mathbf{b}, \tag{8.32}$$

with

$$A_{ij} = \frac{\partial^2 E[\phi_n(x)]}{\partial a_i \partial a_j} = \int_0^1 \left[u_i'(x) u_j'(x) - k^2 u_i(x) u_j(x) \right] dx, \tag{8.33}$$

and

$$b_i = \int_0^1 s(x) u_i(x) \, dx. \tag{8.34}$$

As one can see, the Ritz variational scheme actually produces the exact same set of linear equations as the Galerkin method in this specific problem, since

$$\int_0^1 u_i''(x) u_j(x) \, dx = -\int_0^1 u_i'(x) u_j'(x) \, dx, \tag{8.35}$$

if $u_i(x)$ or $u_i'(x)$ is zero at the boundaries. In fact, if the first two terms in $E[\phi(x)]$ together are positive definite, the Ritz method is equivalent to the Galerkin method for problems with homogeneous boundary conditions in general.

In most cases, it is much easier to deal with the so-called weak form of the original differential equation. Here we would like to demonstrate how to obtain the weak form of a differential equation. If we take the Helmholtz equation discussed above as an example, we can multiply the equation by a function $\psi(x)$ and then integrate it:

$$\int_0^1 [\phi''(x) + k^2\phi(x) + s(x)]\psi(x)\,dx = 0, \tag{8.36}$$

where $\psi(x)$ is assumed to be squarely integrable, with $\psi(0) = \psi(1) = 0$. The reverse is also true from the fundamental variational theorem, which states that if we have

$$\int_a^b r(x)\psi(x)\,dx = 0 \tag{8.37}$$

for any given function $\psi(x)$, with $\psi(a) = \psi(b) = 0$, then the equation

$$r(x) = 0 \tag{8.38}$$

will result. The only restriction on $\psi(x)$ is that it has to be squarely integrable. Now if we integrate the first term of Eq. (8.36) by parts, we have

$$\int_0^1 \{\phi'(x)\psi'(x) - [k^2\phi(x) + s(x)]\psi(x)\}\,dx = 0, \tag{8.39}$$

which is the so-called *weak form* of the original differential equation, since now only the first-order derivatives are involved in the above integral. One needs to realize that the solution of the weak form is not necessarily the same as the original differential equation, especially when the second-order derivative of $\phi(x)$ is not well-behaved. The weak form can be solved approximately by taking the linear combination for the solution

$$\phi_n(x) = \sum_{i=0}^n a_i u_i(x), \tag{8.40}$$

and the basis itself for the auxiliary function

$$\psi(x) = u_j(x). \tag{8.41}$$

Then the weak form of the equation becomes a set of linear equations,

$$\sum_{i=1}^n a_i \int_0^1 [u_i'(x)u_j'(x) - k^2 u_i(x)u_j(x) - s(x)u_i(x)]\,dx = 0, \tag{8.42}$$

with $j = 1, 2, \ldots, n$. This is exactly the same as the equation derived from the Ritz variational scheme earlier in this section.

In most cases, the first part and the second part of the matrix **A** are written separately, with

$$K_{ij} = \int_0^1 u_i'(x)u_j'(x)\,dx \tag{8.43}$$

referred to as the *stiffness matrix* and

$$M_{ij} = -k^2 \int_0^1 u_i(x)u_j(x)\,dx \tag{8.44}$$

referred to as the *mass matrix*.

For the case where $s(x) = 0$, we have an eigenvalue problem with $\mathbf{A}$ being still an $n \times n$ matrix and $\mathbf{b} = \mathbf{0}$. The actual form of the matrix elements are given by the choice of $u_i(x)$. A good example is the choice of $u_i(x)$ discussed in the preceding section. The eigenvalues can be obtained from

$$|\mathbf{A}| = 0. \tag{8.45}$$

When we have the exact forms of $\mathbf{K}$ and $\mathbf{M}$, we can use the schemes developed in Chapter 4 to find all the quantities associated with them. Note that if we choose $u_i(x)$ to have $\mathbf{K}$ and $\mathbf{M}$ as tridiagonal matrices, the numerical difficulty is reduced drastically.

As we mentioned in Section 8.2, the boundary condition always needs special attention in the choice of $u_i(x)$ and the construction of $\phi_n(x)$. Here we would like to illustrate the modification under the different boundary conditions from the solution of the Sturm–Liouville equation

$$[p(x)\phi'(x)]' - r(x)\phi(x) + \lambda w(x)\phi(x) = s(x), \tag{8.46}$$

where λ is the eigenvalue and $p(x)$, $r(x)$, and $w(x)$ are functions of x. The Sturm–Liouville equation is a general form of some very important equations in physics and engineering, for example, the Legendre equation and the spherical Bessel equation. We can rewrite the Sturm–Liouville equation as

$$-[p(x)\phi'(x)]' + q(x)\phi(x) + s(x) = 0, \tag{8.47}$$

for the convenience of the variational procedure and assume that $p(x)$, $q(x)$, and $s(x)$ are well-behaved functions; that is, they are continuous and squarely integrable functions in most cases. One can easily show that, for homogeneous boundary conditions, the functional

$$E[\phi(x)] = \frac{1}{2}\int_0^1 [p(x)\phi'^2(x) + q(x)\phi^2(x) + 2s(x)\phi(x)]\,dx \tag{8.48}$$

will produce the differential equation under optimization. Now if the boundary condition is the so-called natural boundary condition,

$$\phi'(0) + \alpha\phi(0) = 0, \tag{8.49}$$

$$\phi'(1) + \beta\phi'(1) = 0, \tag{8.50}$$

with α and β being the given parameters, one can also show that the functional

$$E[\phi(x)] = \frac{1}{2} \int_0^1 [p(x)\phi'^2(x) + q(x)\phi^2(x) + 2s(x)\phi(x)] \, dx$$

$$- \frac{\alpha}{2} p(0)\phi^2(0) + \frac{\beta}{2} p(1)\phi^2(1) \tag{8.51}$$

will produce the Sturm–Liouville equation with the correct boundary condition. One can convince oneself by taking the functional variation on $E[\phi(x)]$ with the given boundary condition; the Sturm–Liouville equation should result. When we set up the finite elements, we can include the boundary points in the expression of the approximate solution, and the modified functional will take care of the necessary adjustment. Of course, the local functions $u_0(x)$ and $u_{n+1}(x)$ are set to be zero outside of the region $[0, 1]$.

However, when the boundary values are not zero, as we discussed in Section 8.2, special care is needed in the construction of $u_i(x)$ and $\phi_n(x)$. For example, for the inhomogeneous Dirichlet boundary condition, we have discussed that one can introduce two more terms that explicitly take care of the boundary condition. Another way of doing this is to include the boundary points in the approximation, for example,

$$\phi_n(x) = \sum_{i=0}^{n+1} a_i u_i(x). \tag{8.52}$$

If we substitute the above approximate solution into the inhomogeneous Dirichlet boundary condition, we have

$$a_0 = \phi(0)/u_0(0), \tag{8.53}$$

$$a_{n+1} = \phi(1)/u_{n+1}(1). \tag{8.54}$$

We have assumed that $u_i(0) = u_i(1) = 0$ for $i \neq 0, n + 1$. The coefficients a_0 and a_{n+1} are obtained immediately if the form of $u_i(x)$ is given. The approximation $\phi_n(x)$ can be substituted into the functional before taking the Ritz variation. The resulting linear equation set is still in an $n \times n$ matrix form. We will come back to this point again in the next section when we discuss higher-dimensional systems.

8.4 Higher-dimensional systems

The Galerkin or Ritz scheme can be generalized for higher-dimensional systems. For example, if we want to study a two-dimensional system, the nodes can be assigned at the vertices of the polygons that cover the specified domain. For convenience in solving the linear equation set, one needs to construct all the elements with a similar shape, for example, all triangular. The finite element coefficient matrix is then a band matrix, which is much easier to solve.

The simplest way to construct the approximate solution is to choose $u_i(x, y)$ as a function around the ith node, for example, a pyramidal function with $u_i(x_i, y_i) = 1$ that linearly goes to zero at the nearest neighboring nodes. Then the approximate solution of a differential equation is represented by a linear combination of the local functions $u_i(x, y)$ as

$$\phi_n(x, y) = \sum_{i=1}^{n} a_i u_i(x, y), \tag{8.55}$$

where i runs through all the internal nodes if the boundary satisfies the homogeneous Dirichlet condition. For illustrative purposes, let us still take the Helmholtz equation

$$\nabla^2 \phi(x, y) + k^2 \phi(x, y) = -s(x, y) \tag{8.56}$$

as an example. The corresponding functional with the approximate solution is given by

$$E[\phi_n(x, y)] = \frac{1}{2} \int \{ [\nabla \phi_n(x, y)]^2 - k^2 \phi_n^2(x, y)$$
$$- 2s(x, y)\phi_n(x, y) \} \, dx \, dy, \tag{8.57}$$

which is optimized with the Ritz variational procedure:

$$\frac{\partial E[\phi_n(x, y)]}{\partial a_i} = 0. \tag{8.58}$$

This variation produces a linear equation set

$$\mathbf{Aa} = \mathbf{b} \tag{8.59}$$

with

$$A_{ij} = \int [\nabla u_i(x, y) \cdot \nabla u_j(x, y) - k^2 u_i(x, y) u_j(x, y)] \, dx \, dy, \tag{8.60}$$

and

$$b_i = \int s(x, y) u_i(x, y) \, dx \, dy. \tag{8.61}$$

Note that the node index runs through all the vertices of the finite elements in the system except the ones on the boundary. The linear equation set can be solved with any of the numerical schemes developed in Chapter 4.

In reality, the homogeneous Dirichlet boundary condition is not always the case. A more general situation is given by the boundary condition

$$\nabla_n \phi(x, y) + \alpha(x, y)\phi(x, y) + \beta(x, y) = 0, \tag{8.62}$$

which includes most cases in practice. $\alpha(x, y)$ and $\beta(x, y)$ are assumed to be the given functions, and $\nabla_n \phi(x, y)$ is the gradient projected outward perpendicular to the boundary line. It is straightforward to show that the functional

$$E[\phi(x, y)] = \frac{1}{2} \int \{[\nabla \phi(x, y)]^2 - k^2 \phi^2(x, y) - 2s(x, y)\phi(x, y)\} \, dx \, dy$$

$$+ \int_B [\alpha(x, y)\phi^2(x, y) + \beta(x, y)\phi(x, y)] \, d\ell \tag{8.63}$$

can produce the differential equation with the given general boundary condition if the Euler–Lagrange equation is applied. Here the line integral is performed along the boundary B with the domain on the left side. As we discussed for one-dimensional equations, the boundary points can be included in the expansion of the approximate solution for the inhomogeneous boundary conditions. The local functions $u_i(x, y)$ at the boundaries are the same as other local functions except that they are set to be zero outside the domain of the problem specified. When the local functions for the internal nodes are selected so as to be zero at the boundary, the coefficients for the terms from the nodes at the boundary points are given by the values of the local function at those points.

A very efficient way of constructing the stiffness matrix and the mass matrix is to view the system as a collection of elements and the solution of the equation is given in each individual element separately. For example, we can divide the domain into many triangles and label all the triangles sequentially. The solution of the equation is then given in each triangle separately. For simplicity, we will use the pyramidal functions $u_i(x, y)$ as the local functions for the expansion of the approximate solution. $u_i(x, y)$ is 1 at the ith node and goes to zero linearly at the nearest neighboring nodes. Let us select a triangle element labeled by σ. Assume that the three nodes at the three corners of the selected element are labeled i, j, and k. The approximate solution in the element can be expressed as

$$\phi_{n\sigma}(x, y) = a_i u_i(x, y) + a_j u_j(x, y) + a_k u_k(x, y), \tag{8.64}$$

since the local basis $u_l(x)$ is taken to be a function that is zero beyond the nearest neighbors of the lth node. As one can show, for the case of a local pyramidal function, the approximation is given by a linear function

$$\phi_{n\sigma}(x, y) = \alpha + \beta x + \gamma y, \tag{8.65}$$

with the constant α given by a determinant,

$$\alpha = \frac{1}{\Delta} \begin{vmatrix} a_i & a_j & a_k \\ x_i & x_j & x_k \\ y_i & y_j & y_k \end{vmatrix}, \tag{8.66}$$

and the coefficients β and γ given by two other determinants,

$$\beta = \frac{1}{\Delta} \begin{vmatrix} 1 & 1 & 1 \\ a_i & a_j & a_k \\ y_i & y_j & y_k \end{vmatrix} \tag{8.67}$$

and

$$\gamma = \frac{1}{\Delta} \begin{vmatrix} 1 & 1 & 1 \\ x_i & x_j & x_k \\ a_i & a_j & a_k \end{vmatrix}. \tag{8.68}$$

The constant Δ can also be expressed in a determinant form,

$$\Delta = \begin{vmatrix} 1 & 1 & 1 \\ x_i & x_j & x_k \\ y_i & y_j & y_k \end{vmatrix}. \tag{8.69}$$

The functional can now be expressed as the summation of integrals performed on each triangular element. The elements of the coefficient matrix are then obtained from

$$A_{ij} = \sum_{\sigma} A_{ij}^{\sigma}, \tag{8.70}$$

where the summation on σ is for the integrals over two adjacent elements that share the nodes i and j if $i \neq j$, and six elements that share the node i if $i = j$. For the Helmholtz equation, the integral

$$A_{ij}^{\sigma} = \int_{\sigma} [\nabla u_i(x) \cdot \nabla u_j(x) - k^2 u_i(x) u_j(x)] \, dx \, dy \tag{8.71}$$

is performed over the element σ. Similarly, the constant vector element can also be obtained with

$$b_i = \sum_{\sigma} b_i^{\sigma}, \tag{8.72}$$

where the summation is over all the six elements that share the node i. As we discussed in earlier sections in this chapter, there are other choices of $u_i(x, y)$ than just a linear pyramidal function. Interested readers can find these choices in Strang and Fix (1973). A Fortran program that solves the extended Helmholtz equation in two dimensions (a two-dimensional version of the Sturm–Liouville equation) is given in Vichnevetsky (1981, pp. 269–78). One can also find extensive discussions on the method and its applications in Burnett (1987).

An extremely important issue is the analysis of errors in the finite element methods, which would require more space than is available here. Similar procedures can also be developed for three-dimensional systems. Since most applications of the finite element method are in one-dimensional or two-dimensional problems, we will not go any further here. Interested readers can find examples in the current literature.

8.5 The finite element method for nonlinear equations

The examples discussed in this chapter so far are all confined to linear equations. However, the finite element method can also be applied to solve nonlinear equations. In this section, we would like to use the two-dimensional Navier–Stokes equation as an illustrative example to demonstrate the method for nonlinear equations.

For simplicity, we will assume that the system is stationary; that is, it has no time dependence, and under constant temperature and density; that is, it is a stationary isothermal and incompressible fluid. The Navier–Stokes equation under such conditions is given by

$$\rho \mathbf{v} \cdot \nabla \mathbf{v} + \nabla P - \eta \nabla^2 \mathbf{v} = \rho \mathbf{g}, \tag{8.73}$$

where ρ is the density, $\mathbf{v}(x, y)$ is the velocity, η is the viscosity coefficient, $P(x, y)$ is the pressure, and $\mathbf{g}(x, y)$ is the external force field. The continuity equation under the incompressible condition is simply given by

$$\nabla \cdot \mathbf{v} = 0. \tag{8.74}$$

Since the system is two-dimensional, we have three coupled equations for $v_x(x, y)$, $v_y(x, y)$, and $P(x, y)$. Two of them are from the vector form of the Navier–Stokes equation. Note that the first term in the Navier–Stokes equation is nonlinear.

Before we introduce the numerical scheme to solve this equation set, we still need to introduce the appropriate boundary condition. Let us denote the domain of interest as D and the boundary of the domain as B. Since the solution and the existence of the solution are very sensitive to the boundary condition, one has to be very careful to select a valid one (Bristeau et al., 1985). A common choice in numerical solution of the Navier–Stokes equation is to have an extended domain $D_e > D$ with a boundary B_e. Either the velocity

$$\mathbf{v}(x, y) = \mathbf{v}_e(x, y) \tag{8.75}$$

for $\mathbf{n} \cdot \mathbf{v} < 0$ or the combination

$$P(x, y)\mathbf{n} - \eta \mathbf{n} \cdot \nabla \mathbf{v}(x, y) = \mathbf{q}(x, y) \tag{8.76}$$

for $\mathbf{n} \cdot \mathbf{v} \geq 0$ is given at the boundary B_e. Here $\mathbf{n}$ is the normal unit vector at the boundary pointing outward. The weak form of the Navier–Stokes equation is obtained if we take a dot product of the equation with a vector function $\mathbf{u}(x, y)$ and then integrate both sides:

$$\int_{D_e} (\rho \mathbf{v} \cdot \nabla \mathbf{v} + \nabla P - \eta \nabla^2 \mathbf{v}) \cdot \mathbf{u} \, dx \, dy = \int_{D_e} \rho \mathbf{g} \cdot \mathbf{u} \, dx \, dy, \tag{8.77}$$

which can be transformed into the weak form with integration by parts to the Laplace term and the pressure gradient term. Then we have

$$\int_{D_e} [\rho(\mathbf{v} \cdot \nabla \mathbf{v}) \cdot \mathbf{u} - P \nabla \cdot \mathbf{u} - \eta \nabla \mathbf{v} : \nabla \mathbf{u}] \, dx \, dy$$

$$= \int \rho \mathbf{g} \cdot \mathbf{u} \, dx \, dy - \int_{B_e} (P \mathbf{n} - \eta \mathbf{n} \cdot \nabla \mathbf{v}) \cdot \mathbf{u} \, d\ell, \tag{8.78}$$

where : means a double dot product, a combination of the dot product between ∇ and ∇ and the dot product between $\mathbf{v}$ and $\mathbf{u}$. The line integral is along the boundary with the domain on the left.

The weak form for the continuity equation can be obtained by multiplying the equation by a scalar function $f(x, y)$ and then having it integrated in the domain D_e,

$$\int_{D_e} [\nabla \cdot \mathbf{v}(x, y)] f(x, y) \, dx \, dy = 0, \tag{8.79}$$

which can be solved together with the weak form of the Navier–Stokes equation. The line integral along the boundary can be replaced with the term with the given $\mathbf{v}(x, y)$ and $P(x, y)$ at the boundary or with $\int_{B_e} \mathbf{q} \cdot \mathbf{u} \, d\ell$ after applying Eq. (8.76). Either way, this integral is given after we select $\mathbf{u}(x, y)$.

Let us represent the approximate solutions by linear combinations of the local functions,

$$\mathbf{v}_n(x, y) = \sum_{i=1}^{n} \mathbf{e}_i u_i(x, y), \tag{8.80}$$

$$P_n(x, y) = \sum_{i=1}^{n} c_i w_i(x, y), \tag{8.81}$$

where $\mathbf{e}_i = (a_i, b_i)$ is a two-component vector corresponding to the two components of $\mathbf{v}$. $u_i(x, y)$ and $w_i(x, y)$ are chosen to be different because in the weak forms, the pressure appears only in its own form but the velocity field also appears in the form of its first-order derivative. This means that we can choose a simpler function $w_i(x, y)$ (e.g., differentiable once) than $u_i(x, y)$ (e.g., differentiable twice) in order to have the same level of smoothness in $P(x, y)$ and $\mathbf{v}(x, y)$.

In order to convert the weak form into a matrix form, we will use $\mathbf{u} = (u_i, u_j)$ and $f(x, y) = w_i(x, y)$. More important, we have to come up with a scheme for the nonlinear term. An iterative scheme can be devised as follows:

(1) We first make a guess of the solution, $a_i^{(0)}$, $b_i^{(0)}$, and $c_i^{(0)}$, for $i = 1, 2, \ldots, n$. This can be achieved in most cases by solving the problem with a mean-field type of approximation to the nonlinear term, for example,

$$\mathbf{v} \cdot \nabla \mathbf{v} \simeq \mathbf{v}_0 \cdot \nabla \mathbf{v}, \tag{8.82}$$

where $\mathbf{v}_0$ is a constant vector that is more or less the average of $\mathbf{v}$ over the whole domain.

(2) Then we can improve the solution iteratively. For the $(k + 1)$th iteration, we can use the results of the kth iteration for part of the nonlinear term; for example, the first term in the weak form of the Navier–Stokes equation can be written as $\mathbf{v}^{(k)} \cdot \nabla \mathbf{v}^{(k+1)}$. Then we can solve $a_i^{(k+1)}$, $b_i^{(k+1)}$, and $c_i^{(k+1)}$ for $i = 1, 2, \ldots, n$. The weak form at each iteration is a linear equation set, since $a_i^{(k)}$, $b_i^{(k)}$, and $c_i^{(k)}$ have already been solved in the previous step.

This iterative scheme is rather general. One can devise similar schemes to solve any other nonlinear equations with the nonlinear terms rewritten into linear terms with part of each term represented by the results of the last step. Note that the scheme outlined here is similar to the scheme discussed in Chapter 4 for solving multivariable nonlinear problems. The scheme outlined can also be interpreted as the special choice of a relaxation scheme (Fortin and Thomasset, 1983)

$$\mathbf{v}^{(k+1)} = (1 - p)\mathbf{v}^{(k)} + p\mathbf{v}^{(k+1/2)}, \tag{8.83}$$

where $\mathbf{v}^{(k+1/2)}$ is the solution of

$$\begin{cases} \rho \mathbf{v}^{(k)} \cdot \nabla \mathbf{v}^{(k+1/2)} + \nabla P^{(k+1)} - \eta \nabla^2 \mathbf{v}^{(k+1/2)} = \rho \mathbf{g}, \\ \nabla \cdot \mathbf{v}^{(k+1/2)} = 0. \end{cases} \tag{8.84}$$

Here p is a parameter that can be adjusted to achieve the best convergence.

An important aspect of the solution of the Navier–Stokes equation is the solution of the time-dependent equation, which requires us to combine the spatial solution just discussed and the initial-value problem discussed in Chapter 6. Interested readers can find the discussions on the numerical algorithms for solving the time-dependent Navier–Stokes equation in Bristeau et al. (1985).

8.6 The particle-in-cell method

In order to simulate a large continuous system such as a hydrodynamic fluid, one has to devise a method that can deal with the macroscopic phenomena observed.

Finite element methods and finite difference methods can be used in the solution of macroscopic equations that describe the dynamics of continuous systems. However, the behavior of the systems may involve different length scales, which will make a finite difference or finite element method quite difficult.

One can, in principle, treat each atom or molecule as an individual particle and then solve Newton's equations for all the particles in the many-body system. This can be a good method if the structural and dynamical properties under study are at the length scale of the average interparticle distance, which is clearly the case when we study the behavior of salt melting, glass formation, or structure factors of specific liquids. This was the subject of the preceding chapter, on molecular dynamics simulation.

Schemes dealing with the simulation of the dynamics of each individual particle in the system will run into difficulty when applied to a typical hydrodynamic system, since such a system has more than 10^{23} particles and the phenomena take place at macroscopic length scales. The common practice is to solve the set of macroscopic equations discussed in Section 8.1 with either the finite difference method discussed in Chapter 6 or the finite element method discussed in the last few sections. However, we cannot always do that, since some dynamical phenomena, such as the nonlinear properties observed in plasma or galactic systems, may root back to the fundamental interaction in the system – for example, the Coulomb interaction between electrons and ions, or gravity between stars. More importantly, density fluctuations around a phase transition or the onset of a chaotic behavior might involve several orders of magnitude in the length scale. So a scheme that can bridge the microscopic dynamics of each particle to the macroscopic phenomena in some systems is highly desirable. The *particle-in-cell method* was first introduced by Evans and Harlow (1957) for this purpose. For a review on the early development of the method, see Harlow (1964).

Let us assume that the system we are interested in is a large continuous system. We can divide the system into many cells, which are similar to the finite elements discussed earlier in this chapter. In each cell there are still a large number of particles. For example, if we are interested in a two-dimensional system with 10^{14} particles and divide it into one million cells, we still have 100 million particles in each cell. A common practice is to construct pseudoparticles, which are still collections of many true particles. For example, if we take one million particles as one pseudoparticle for the two-dimensional system with 10^{14} particles, we can easily divide the system into 10,000 cells, of which each has 10,000 pseudo particles. This becomes a manageable problem with currently available computing capacity. Note that since the pseudoparticles are still collections of particles, simulations with such pseudoparticles are not completely described by microscopic dynamics. However, since the division of the system into cells allows us to cover several orders of magnitude in length scales, we can simulate many phenomena that involve different length scales in macroscopic systems.

The size of the cells is usually determined by the relevant length scales in the system. If we are interested in a collection of charged particles, we must have each cell small enough that the detailed distribution of the charges in each cell will not affect the behavior of each particle or pseudoparticle very much. This means the contribution to the energy of each particle due to the other charges in the same cell can be ignored in comparison with the dominant energy scale, for example, the thermal energy of each particle in this case. The electrostatic potential energy of a particle due to the uniform distribution of the charges around it inside a sphere with a radius a is given by

$$U_c = 4\pi n e^2 a^2, \tag{8.85}$$

which should be much smaller than the thermal energy of one degree of freedom, $k_B T/2$. The dimensions of the cell then must satisfy

$$a \ll \lambda_D = \sqrt{\frac{k_B T}{4\pi n e^2}}, \tag{8.86}$$

where λ_D is called the Debye length, e is the charge of each particle, and n is the number density. However, since we are interested in the hydrodynamics of the system, we cannot choose the cell too small. It has to be much larger than the average volume occupied by each particle, that is,

$$a \gg a_0 = \frac{1}{n^{1/3}}, \tag{8.87}$$

for a three-dimensional system. Similar relations can be established for other systems, such as galaxies. After the size of the cells is determined, one can decide whether or not a pseudoparticle picture is needed based on how many particles are in the cell. Basically, one has to make sure that the finite size effect of each cell will not dominate the behavior of the system under study.

Let us first sketch the idea of the method in a specific example: a three-dimensional charged particle system. If the cell size is much larger than the average distance between two nearest neighbors, the system is nearly collisionless. The dynamics of each particle or pseudoparticle is then governed by the instantaneous velocity ($\mathbf{v}$) of the particle and the average electrostatic field ($\mathbf{E}$) at the position of the particle. For example, at time $t = n\tau$, where τ is the time step, the ith particle has a velocity $\mathbf{v}_i^{(n)}$ and an acceleration

$$\mathbf{g}_i^{(n)} = e\mathbf{E}_i^{(n)}/m, \tag{8.88}$$

with e being the charge and m being the mass of the particle. Then the next position $(\mathbf{r}_i^{(n+1)})$ and velocity $(\mathbf{v}_i^{(n+1)})$ of the same particle are given by

$$\mathbf{r}_i^{(n+1)} = \mathbf{r}_i^{(n)} + \tau \mathbf{v}_i^{(n)} + \frac{\tau^2}{2} \mathbf{g}_i^{(n)}, \tag{8.89}$$

$$\mathbf{v}_i^{(n+1)} = \mathbf{v}_i^{(n)} + \frac{\tau}{2} \left(\mathbf{g}_i^{(n)} + \mathbf{g}_i^{(n+1)} \right), \tag{8.90}$$

with the Verlet velocity algorithm. We have assumed that the charge and the mass of each particle or pseudoparticle are e and m, respectively. Note that there is no difference between the current situation and the molecular dynamics case except that we are dealing only with an average acceleration $\mathbf{g}_i$, which is an interpolated value from the fields at the grid points around the particle, instead of a calculation from the particle–particle interactions. The "particles" here are pseudoparticles, each equivalent to many, many particles. It is worth pointing out that other numerical schemes developed in molecular dynamics can be applied here as well. As we already discussed in molecular dynamics, the most time-consuming part of the simulation is the evaluation of the interactions among the particles. The basic idea of the particle-in-cell method is to circumvent such an effort. Instead, we try to solve the field on grid points and then find the average field at the particle from the interpolation of the values at the grid points around it. For a charged particle system, this can be achieved by solving the Poisson equation for the electrostatic potential at the grid points, which are typically set at the centers of the cells. The accuracy of the interpolation kernel is the key to a stable and accurate algorithm. We will here incorporate the idea of the improved interpolation kernels, introduced by Monaghan (1985), into our discussions. For convenience of notation, we will use Greek letters in the indices for the grid points and Latin letters for the particles. The charge density on a specific grid point can be determined from the distribution of the particles in the system at the same moment, that is,

$$\rho^{(n+1)}(\mathbf{r}_\sigma) = e \sum_{i=1}^{N} \mathcal{W}\left(\mathbf{r}_\sigma - \mathbf{r}_i^{(n+1)}, h\right), \tag{8.91}$$

where $\mathcal{W}(\mathbf{r}, h)$ is the interpolation kernel, which can be interpreted as a distribution function and satisfies

$$\int \mathcal{W}(\mathbf{r}, h) \, d\mathbf{r} = 1, \tag{8.92}$$

and N is the total number of pseudoparticles in the system. Note that $\mathcal{W}(\mathbf{r}, h)$ is usually a function ranged with the grid spacing h. This is due to the fact that $\mathcal{W}(\mathbf{r}, h)$ has to be very small beyond $|\mathbf{r}| = h$. Monaghan (1985) has shown that one can improve interpolation accuracy by choosing

$$\mathcal{W}(\mathbf{r}, h) = \frac{1}{2}\left[(d+2)\mathcal{G}(\mathbf{r}, h) - h\frac{\partial \mathcal{G}(\mathbf{r}, h)}{\partial h}\right], \tag{8.93}$$

where $\mathcal{G}(\mathbf{r}, h)$ is a smooth interpolation function and d is the number of the spatial dimensions of the system. For example, if $\mathcal{G}(\mathbf{r}, h)$ is a Gaussian function,

$$\mathcal{G}(\mathbf{r}, h) = \frac{1}{\pi^{3/2}}e^{-r^2/h^2}, \tag{8.94}$$

$\mathcal{W}(\mathbf{r}, h)$ is given by

$$\mathcal{W}(\mathbf{r}, h) = \frac{1}{\pi^{3/2}}\left(\frac{d+2}{2} - \frac{r^2}{h^2}\right)e^{-r^2/h^2}. \tag{8.95}$$

From the interpolated grid values of the density, we can solve the Poisson equation

$$\nabla^2\Phi(\mathbf{r}) = -4\pi\rho(\mathbf{r}) \tag{8.96}$$

with any finite difference method discussed in Chapter 6. Then the electric field $\mathbf{E} = -\nabla\Phi(\mathbf{r})$ at each grid point can be evaluated through, for example, a three-point or two-point formula. In order to have a more accurate value of the field at the particle site, one can also interpolate the potential at the particle site and then evaluate the field

$$\mathbf{E}_i^{(n+1)} = -\sum_{\sigma=1}^{z}\Phi_\sigma^{(n+1)}\nabla_\sigma\mathcal{U}(\mathbf{r}_i - \mathbf{r}_\sigma), \tag{8.97}$$

where $\mathcal{U}(\mathbf{r})$ is another kernel that can be either the same as or different from the kernel $\mathcal{W}(\mathbf{r}, h)$. The gradient is taken on the continuous function of $\mathbf{r}_\sigma$, and z is the number of grid points nearby that one would like to include in the interpolation scheme. For example, if only the nearest neighbors are included, $z = 4$ for a square lattice. The new velocity of the particle in the Verlet algorithm above is then given by

$$\mathbf{v}_i^{(n+1)} = \mathbf{v}_i^{(n)} + \frac{e\tau}{2m}\left(\mathbf{E}_i^{(n)} + \mathbf{E}_i^{(n+1)}\right), \tag{8.98}$$

The number of points to be included in the summation is determined from the spatial dimensions of the system and the geometry of the cells. Like $\mathcal{W}(\mathbf{r}, h), \mathcal{U}(\mathbf{r})$ also has to be a smooth even function of $\mathbf{r}$ and it must satisfy

$$\sum_{\sigma=1}^{z}\mathcal{U}(\mathbf{r}_\sigma - \mathbf{r}_i) = 1, \tag{8.99}$$

for any given $\mathbf{r}_i$. This can be done by normalizing $\mathcal{U}(\mathbf{r}_\sigma - \mathbf{r}_i)$ numerically at every time step. Note that the order of the velocity update and the position update in Eq. (8.90) can be interchanged. In practice, this will make no difference in the numerical difficulty, but will provide some flexibility to improve the accuracy for a specific problem.

So a typical particle-in-cell scheme has four major steps: solving the microscopic equations, interpolating microscopic quantities to the grid points, solving the macroscopic equations on the grid points, and then interpolating the macroscopic quantities back to the particles. The order of the steps may vary depending on the specific problem in question, but the goal is always the same: to avoid direct simulation of the system from the microscopic equations while still maintaining the input from the microscopic scales to the macroscopic phenomena. The particle-in-cell method is very powerful in many practical applications, ranging from plasma simulations to galactic dynamics. In the next section, we will discuss a typical application of the method in hydrodynamics and magnetohydrodynamics. For more details of the method, see Potter (1977), Hockney and Eastwood (1988), and Monaghan (1985).

8.7 Hydrodynamics and magnetohydrodynamics

Now let us turn to hydrodynamic systems. We already discussed the equations of hydrodynamics at the beginning of this chapter. To simplify our discussion here, we will first consider the case of an ideal fluid, that is, one for which it can be assumed that the viscosity, temperature gradient, and external field in the system are all very small and can be ignored. The Navier–Stokes equation then becomes

$$\rho \frac{\partial \mathbf{v}}{\partial t} + \rho \mathbf{v} \cdot \nabla \mathbf{v} + \nabla P = 0, \tag{8.100}$$

where the first term on the left-hand side is the convective (nonlinear) term and the second term results from the change of the pressure. The continuity equation is still the same,

$$\frac{\partial \rho}{\partial t} + \nabla \cdot (\rho \mathbf{v}) = 0. \tag{8.101}$$

Since there is no dissipation, the equation for the internal energy per unit mass is given by

$$\rho \frac{\partial \epsilon}{\partial t} + \rho \mathbf{v} \cdot \nabla \epsilon + P \nabla \cdot \mathbf{v} = 0. \tag{8.102}$$

The pressure P, the density ρ, and the internal energy ϵ are all related by the equation of state,

$$f(P, \rho, \epsilon) = 0. \tag{8.103}$$

Sometimes entropy is a more convenient choice than the internal energy.

Once again we will incorporate the improved interpolation scheme of Monaghan (1985) into our discussion. We divide the hydrodynamic system into many cells, and in each cell we have many particles or pseudoparticles. Assume that the equation of state and the initial conditions are given, so we can set up the initial density, velocity, energy per unit mass, and pressure associated with the lattice points as well as the particles.

If we use the two-point formula for the partial time derivative and an interpolated pressure from an interpolation kernel $\mathcal{V}(\mathbf{r})$, the velocity at the lattice points can be partially updated as

$$\mathbf{v}_\sigma^{(n+1/2)} = \mathbf{v}_\sigma^{(n)} - \frac{\tau}{\rho_\sigma^{(n)}} \sum_\mu P_\mu \nabla_\sigma \mathcal{V}(\mathbf{r}_\sigma - \mathbf{r}_\mu), \tag{8.104}$$

which does not include the effect due to the convective term $\rho\mathbf{v} \cdot \nabla\mathbf{v}$. Here index μ runs through all the nearest neighbors of the lattice point $\mathbf{r}_\sigma$ and $\mathcal{V}(\mathbf{r}_\sigma - \mathbf{r}_\mu)$ is normalized as

$$\sum_\mu \mathcal{V}(\mathbf{r}_\sigma - \mathbf{r}_\mu) = 1. \tag{8.105}$$

We will explain later that the velocity at the lattice point is modified by convection to

$$\mathbf{v}_\sigma^{(n+1)} = \frac{1}{2}(\mathbf{v}_\sigma^{(n)} + \mathbf{v}_\sigma^{(n+1/2)}), \tag{8.106}$$

if the particle positions are updated as discussed below. The velocity $\mathbf{v}_\sigma^{(n+1)}$ can be used to update the internal energy partially at the lattice point,

$$\epsilon_\sigma^{(n+1/2)} = \epsilon_\sigma^{(n)} - \frac{\tau p_\sigma^{(n)}}{\rho_\sigma^{(n)}} \sum_\mu \mathbf{v}_\mu^{(n+1)} \cdot \nabla_\sigma \mathcal{V}(\mathbf{r}_\sigma - \mathbf{r}_\mu), \tag{8.107}$$

which does not include the effect of the convective term $\rho\mathbf{v} \cdot \nabla\epsilon$ in Eq. (8.102). This is also done with the application of the two-point formula to the partial time derivative.

The total thermal energy per unit mass at the lattice point with the partial update of the velocity and the internal energy is then given by

$$e_\sigma^{(n+1/2)} = \epsilon_\sigma^{(n+1/2)} + \frac{1}{2}(v_\sigma^{(n+1/2)})^2, \tag{8.108}$$

which in turn can be used to interpolate the energy per unit mass at the particle site,

$$e_j^{(n+1/2)} = \sum_{\sigma=1}^{z} e_\sigma^{(n+1/2)} \mathcal{U}(\mathbf{r}_i - \mathbf{r}_\sigma), \tag{8.109}$$

with $\mathcal{U}(\mathbf{r})$ being the interpolation kernel from lattice points to particle positions used in the preceding section. The new particle velocity is obtained from the interpolation of the velocity $\mathbf{v}_\sigma$,

$$\mathbf{v}_i^{(n+1)} = \sum_{\sigma=1}^{z} \mathbf{v}_\sigma^{(n+1)} \mathcal{U}(\mathbf{r}_i - \mathbf{r}_\sigma). \tag{8.110}$$

which can then be used to update the particle positions with a simple two-point formula:

$$\mathbf{r}_i^{(n+1)} = \mathbf{r}_i^{(n)} + \tau \mathbf{v}_i^{(n+1)}, \tag{8.111}$$

which completes the convective motion. One can easily show that the updating formula for the particle position is exactly the same as in the Verlet algorithm

$$\mathbf{r}_i^{(n+1)} = \mathbf{r}_i^{(n)} + \tau \mathbf{v}_i^{(n)} + \frac{\tau^2}{2}\mathbf{g}_i^{(n)}, \tag{8.112}$$

with

$$\mathbf{g}(\mathbf{r}) = -\frac{1}{\rho(\mathbf{r})}\nabla P(\mathbf{r}). \tag{8.113}$$

This has exactly the same mathematical structure as the electrostatic field discussed in the preceding section. The gradient of the pressure at the particle position can therefore be obtained from the pressure at the lattice points,

$$\nabla P_i = \sum_{\sigma=1}^{z} P_\sigma \nabla_\sigma \mathcal{U}(\mathbf{r}_\sigma - \mathbf{r}_i). \tag{8.114}$$

One can also show that the updated particle velocity is also equivalent to the Verlet algorithm

$$\mathbf{v}_i^{(n+1)} = \mathbf{v}_i^{(n)} + \frac{\tau}{4}\left(\mathbf{g}_i^{(n)} + \mathbf{g}_i^{(n+1)}\right) \tag{8.115}$$

to the same order of accuracy with the convective term included. Note that the second term is only half of the corresponding term in the nonconvective case.

Now we can update the quantities associated with the lattice points. Following the discussions in the preceding section, we have

$$\rho_\sigma^{(n+1)} = m \sum_{i=1}^{N} \mathcal{W}(\mathbf{r}_\sigma - \mathbf{r}_i^{(n+1)}), \tag{8.116}$$

$$\rho_\sigma^{(n+1)}\mathbf{v}_\sigma^{(n+1)} = m \sum_{i=1}^{N} \mathbf{v}_i \mathcal{W}(\mathbf{r}_\sigma - \mathbf{r}_i^{(n+1)}), \tag{8.117}$$

$$\rho_\sigma^{(n+1)}e_\sigma^{(n+1)} = m \sum_{i=1}^{N} e_i^{(n+1)}\mathcal{W}(\mathbf{r}_\sigma - \mathbf{r}_i^{(n+1)}), \tag{8.118}$$

$$\epsilon_\sigma^{(n+1)} = e_\sigma^{(n+1)} - \frac{1}{2}(v_\sigma^{(n+1)})^2. \tag{8.119}$$

The values of ρ_σ and e_σ can then be used in the equation of state to obtain the new values for the pressure at the lattice points. Then the steps outlined above can be repeated for the next time step.

The above scheme can easily be extended to include viscosity and magnetic field effects. The Navier–Stokes equation is then given by

$$\frac{\partial}{\partial t}(\rho \mathbf{v}) = -\nabla \cdot \mathbf{\Pi}, \tag{8.120}$$

with

$$\mathbf{\Pi} = \rho \mathbf{vv} - \eta[\nabla \mathbf{v} + (\nabla \mathbf{v})^T] - \left[\left(\zeta - \frac{2\eta}{3}\right)\nabla \cdot \mathbf{v} - P - \frac{B^2}{2}\right]\mathbf{I} - \mathbf{BB}, \tag{8.121}$$

and the magnetic field satisfies

$$\frac{\partial \mathbf{B}}{\partial t} = \nabla \times [\mathbf{v} \times \mathbf{B} - \rho \cdot (\nabla \times \mathbf{B})], \tag{8.122}$$

where ρ is the resistivity tensor of the system. The energy equation is then given by

$$\frac{\partial}{\partial t}\left(\epsilon + \rho\frac{v^2}{2} + \frac{B^2}{2}\right) = \nabla \cdot \mathbf{j}_e, \tag{8.123}$$

with

$$\mathbf{j}_e = \mathbf{v}\left(\rho\epsilon + \frac{1}{2}\rho v^2 + \frac{B^2}{2}\right) - \mathbf{v} \cdot \mathbf{G} - \kappa \nabla T \tag{8.124}$$

as the energy current. All these extra terms can be incorporated into the particle-in-cell scheme outlined earlier in this section. We will not go into more details here, but interested readers can find those discussions in Monaghan (1985).

8.8 The Boltzmann lattice-gas method

As discussed in the last few sections, continuous systems can be properly represented by discrete models if the choice of discretization can still account for the basic physical process involved in the systems. This basic idea was put forward by Frisch, Haalacher, and Pomeau (1986), Frisch et al. (1987), and Wolfram (1986) into a model now known as lattice-gas cellular automata. The model treats the system as a lattice gas and the occupancy at each site of the lattice as a Boolean number, 0 or 1, corresponding to an unoccupied or occupied state of a few possible velocities. One can recover the macroscopic equations of fluid dynamics from the microscopic Boltzmann equation under the constraint of mass and momentum conservation. This lattice-gas model has the advantage of using Boolean numbers in simulations and therefore is free of rounding errors. Another advantage of the lattice-gas model is that it is completely local and can easily be implemented in parallel or distributed computing environments. However, the lattice-gas model also has some disadvantages in simulations, such as large fluctuations in the density distribution and other physical quantities due to the discrete occupancy of each site and an exponential increase with the number of nearest vertices in the collision rules. These two disadvantages were overcome by the introduction of the Boltzmann lattice-gas model (McNamara and Zanetti, 1988; Higuera and Jiménez, 1989), which preserves most advantages of the lattice-gas model in simulations but allows a smooth occupancy, which reduces the fluctuations. The introduction of the BGK (Bhatnagar–Gross–Krook) form of the collision term in the Boltzmann equation also reduces the exponential complexity in the collision rules with the number of the nearest vertices.

Before introducing the lattice-gas model and the Boltzmann lattice-gas model, let us first have a brief outline of the Boltzmann kinetic theory. All the ideas behind the classic kinetic theory are based on the Boltzmann transport equation. Assume that $f(\mathbf{r}, \mathbf{v}, t)$ is the distribution function at the point $(\mathbf{r}, \mathbf{v})$ in the phase space at time t. Then the number of particles in the phase space volume element $d\mathbf{r}\, d\mathbf{v}$ is $f(\mathbf{r}, \mathbf{v}, t)\, d\mathbf{r}\, d\mathbf{v}$. Since the number of particles and the volume element in the phase space are conserved in equilibrium based on Liouville's theorem, we have

$$\frac{df(\mathbf{r}, \mathbf{v}, t)}{dt} = 0 \tag{8.125}$$

if the particles in the system do not interact with each other. Since both $\mathbf{r}$ and $\mathbf{v}$ are functions of time, we can then rewrite the above derivative as

$$\frac{df}{dt} = \frac{\partial f}{\partial t} + \mathbf{v} \cdot \nabla f + \mathbf{g} \cdot \nabla_{\mathbf{v}} f = 0, \tag{8.126}$$

where we have used the definition of the velocity $\mathbf{v} = d\mathbf{r}/dt$, the acceleration $\mathbf{g} = d\mathbf{v}/dt$, and the gradient in velocity $\nabla_{\mathbf{v}} = \partial/\partial \mathbf{v}$ for convenience. We will consider only the zero external field case, that is, $\mathbf{g} = 0$, for simplicity. The above equation will not hold if the collisions in the system cannot be ignored. Instead, the equation is modified to

$$\frac{\partial f}{\partial t} + \mathbf{v} \cdot \nabla f = \Omega(\mathbf{r}, \mathbf{v}, t), \tag{8.127}$$

where $\Omega(\mathbf{r}, \mathbf{v}, t)$ is a symbolic representation of the effects of all the collisions among the particles in the system. $\Omega(\mathbf{r}, \mathbf{v}, t)$ can be expressed in terms of an integral that involves the product of $f(\mathbf{r}, \mathbf{v}, t)$ at different velocities as well as the scattering cross section of many-body collisions in the integrand. We will not derive this expression here, since we will need only the sum rules of $\Omega(\mathbf{r}, \mathbf{v}, t)$. The simplest approximation is the BGK form

$$\Omega(\mathbf{r}, \mathbf{v}, t) = -\frac{f(\mathbf{r}, \mathbf{v}, t) - f_0(\mathbf{r}, \mathbf{v}, t)}{\tau_c}, \tag{8.128}$$

where f_0 is the equilibrium distribution and τ_c is the relaxation time due to the collisions. The conservation laws are reflected in the sum rules of $\Omega(\mathbf{r}, \mathbf{v}, t)$. For example, the conservation of the total mass ensures

$$\int m\Omega(\mathbf{r}, \mathbf{v}, t)\, d\mathbf{r}\, d\mathbf{v} = 0, \tag{8.129}$$

and the conservation of the linear momentum would require

$$\int m\mathbf{v}\Omega(\mathbf{r}, \mathbf{v}, t)\, d\mathbf{r}\, d\mathbf{v} = 0. \tag{8.130}$$

One can also obtain the hydrodynamic equations from the Boltzmann equation (Lifshitz and Pitaevskii, 1981) after taking the statistical averages of the relevant physical quantities. The point we want to make is that the macroscopic behavior of a dynamic system is the result of the average effects of all the particles involved.

As we discussed in the last two sections, it is impossible to simulate 10^{23} particles from their equations of motion with current computing capacity; it would be nice if we could do so. Other methods have to be devised in the simulation of hydrodynamics. As we discussed earlier in this chapter, one can use the particle-in-cell method to simulate hydrodynamic systems.

The idea of the lattice-gas model resembles the particle-in-cell concept in several respects: The particles in the lattice are pseudoparticles, and the lattice sites are similar to the grid points in the particle-in-cell schemes. The difference is in the assignment of velocities to the particles. In the lattice-gas model, only a few discrete velocities are allowed. Let us take a triangular lattice as an example. At each site, there cannot be more than seven particles, each of which has to have a unique velocity

$$v_\sigma = \frac{\mathbf{e}_\sigma}{\tau_c}, \tag{8.131}$$

where $\mathbf{e}_\sigma$ is one of the vectors pointing to the nearest neighboring vertices or 0, that is,

$$|\mathbf{v}_\sigma| = \begin{cases} a/\tau_c & \text{for} \quad \sigma = 1, \ldots, 6, \\ 0 & \text{for} \quad \sigma = 0, \end{cases} \tag{8.132}$$

where a is the distance to a nearest neighbor. Here the index σ runs from 0 to z, with z being the number of nearest neighbors of a lattice point. Now if we take τ_c as the unit of time and $f_\sigma(\mathbf{r}, t)$ as the occupancy of the site $\mathbf{r}$ in the state with the velocity $\mathbf{v}_\sigma$, the Boltzmann equation for each lattice point becomes

$$f_\sigma(\mathbf{r} + \mathbf{e}_\sigma, t + 1) - f_\sigma(\mathbf{r}, t) = \Omega_\sigma[f_\mu(\mathbf{r}, t)], \tag{8.133}$$

with $\sigma, \mu = 0, 1, \ldots, z$. Note that f_σ here is a Boolean number that can be only 0 for the unoccupied state and 1 for the occupied state. Now if we require the mass and momentum at each site to be constant as $\tau \to 0$, we have

$$\frac{\partial \rho}{\partial t} = -\nabla \cdot (\rho \mathbf{v}),$$

$$\frac{\partial}{\partial t}(\rho \mathbf{v}) = -\nabla \cdot \mathbf{\Pi}, \tag{8.134}$$

where ρ is the local density,

$$\rho(\mathbf{r}, t) = m \sum_{\sigma=0}^{z} f_\sigma(\mathbf{r}, t), \tag{8.135}$$

$\rho \mathbf{v}$ is the local momentum density,

$$\rho(\mathbf{r}, t)\mathbf{v}(\mathbf{r}, t) = m \sum_{\sigma=0}^{z} \mathbf{v}_\sigma f_\sigma(\mathbf{r}, t), \tag{8.136}$$

and $\mathbf{\Pi}$ is the momentum flux density,

$$\mathbf{\Pi} = m \sum_{\sigma=0}^{z} \mathbf{v}_\sigma \mathbf{v}_\sigma f_\sigma(\mathbf{r}, t). \tag{8.137}$$

One can show that the above equations will lead to a macroscopic equation that resembles the Navier–Stokes equation if one performs an expansion of $f_\sigma(\mathbf{r}, t)$ in terms of the velocities $\mathbf{v}_\sigma$ and $\mathbf{v}$ up to the second order, the so-called Champman–Enskog expansion (Frisch et al., 1987; Wolfram, 1986).

Two very important aspects of lattice-gas automaton simulations are the geometry of the lattice and the collision rules at each vertex. For example, one has to use the triangular lattice in two-dimensional systems, since other simple lattice structures do not have isotropic macroscopic behavior. For three-dimensional systems, there is no such simple lattice structure to ensure isotropic macroscopic behavior; one way to ensure macroscopic isotropy is to simulate a three-dimensional projection of a four-dimensional face-centered cubic lattice. For more details on lattice-gas automata, see the articles in Doolen (1991).

Now we would like to turn to the Boltzmann lattice-gas model. As we have stated, the distribution at each site for a specific state is assumed to be a smooth function of the position instead of a Boolean number. We will take

$$n_\sigma(\mathbf{r}, t) = \langle f_\sigma(\mathbf{r}, t) \rangle \tag{8.138}$$

as the distribution function. The Boltzmann lattice equation then becomes

$$n_\sigma(\mathbf{r} + \mathbf{e}_\sigma, t + 1) - n_\sigma(\mathbf{r}, t) = \Omega_\sigma[n_\mu(\mathbf{r}, t)] \tag{8.139}$$

for $\sigma, \mu = 0, 1, \ldots, z$. The collision term can be further approximated by keeping only the linear term of the deviation of the distribution from its equilibrium value

(Higuera and Jiménez, 1989),

$$n_\sigma(\mathbf{r} + \mathbf{e}_\sigma, t + 1) - n_\sigma(\mathbf{r}, t) = \sum_{\mu=0}^{z} \Delta_{\sigma\mu} \left[n_\mu(\mathbf{r}, t) - n_\mu^0(\mathbf{r}, t) \right], \qquad (8.140)$$

where the matrix Δ is determined from the symmetry of the lattice and conservation of mass and momentum. One can expand the equilibrium distribution function $n_\sigma^0(\mathbf{r}, t)$ into a power series in $\mathbf{v}$, and then one can show that the macroscopic equations resemble the hydrodynamic equations. The selection of the parameters in the expansion can also lead to the equation of state for a specific system under study (Chen, Chen, and Matthaeus, 1992). Interested readers can find more detailed discussions in Benzi, Succi, and Vergassola (1992), and Succi, Amati, and Benzi (1995).

Exercises

8.1 Solve the one-dimensional Poisson equation

$$\phi''(x) = -4\pi\rho(x)$$

by the Galerkin method. Assume that the boundary condition is $u(0) = u(1) = 0$ and the density distribution is given by

$$\rho(x) = \begin{cases} x^2 & \text{for} \quad x \in [0, 0.5], \\ (1-x)^2 & \text{for} \quad x \in [0.5, 1]. \end{cases}$$

Use the basis

$$u_i(x) = \begin{cases} (x - x_{i-1})/h & \text{for} \quad x \in [x_{i-1}, x_i], \\ (x_{i+1} - x)/h & \text{for} \quad x \in [x_i, x_{i+1}], \\ 0 & \text{otherwise}, \end{cases}$$

to construct the matrix $\mathbf{A}$ and the vector $\mathbf{b}$. Write a program that solves the linear equation set through the LU decomposition.

8.2 Solve the Helmholtz equation

$$\phi''(x) + k^2\phi(x) = -s(x)$$

through its weak form. Use the boundary condition $\phi(0) = \phi(1) = 0$ and assume that $k = \pi$ and $s(x) = \delta(x - 0.5)$. Is the solution of the weak form the same as the solution of the equation?

8.3 Show that the optimization of the generalized functional introduced in Section 8.3 for the Sturm–Liouville equation with the natural boundary condition can produce the equation with the correct boundary condition. Similarly, show that the optimization of the functional introduced in Section 8.4

for the two-dimensional Helmholtz equation with the general boundary condition can produce the equation with the correct boundary condition.

8.4 Show that the finite element solution for a specific element in two dimensions with triangular elements and the linear local pyramidal functions is given by Eq. (8.65). Work out the coefficient matrix and the constant vector if the two-dimensional space is divided into a triangular lattice with a lattice constant h.

8.5 An infinite cylindrical conducting shell of radius a is cut along its axis into four equal parts, which are then insulated from each other. If the potential on the first and third quarters is V and the potential on the second and fourth quarters is $-V$, find the potential in the plan through a finite element method. Compare the numerical result with the exact solution.

8.6 Obtain the Debye length for a galaxy. Estimate the size of the cell needed if it is simulated by the particle-in-cell method.

9

Monte Carlo simulations

One of the major numerical techniques developed in the last half century for evaluating multidimensional integrals or solving integral equations is the Monte Carlo method. The basic idea of the method is to select the points in the region enclosed by the boundary and then take the weighted data as the estimated value of the integral. Early Monte Carlo simulations go back to the 1950s, when the first computers became available. In this chapter, we will introduce the basic idea of the Monte Carlo method with applications in statistical physics, quantum mechanics, and related fields and highlight some recent developments.

9.1 Sampling and integration

Let us first use a simple example to illustrate how a basic Monte Carlo scheme works. If we want to find the numerical value of the integral

$$S = \int_0^1 f(x)\, dx, \tag{9.1}$$

we can simply divide the region $[0, 1]$ evenly into M slices with $x_0 = 0$ and $x_M = 1$, and then the integral can be approximated as

$$S = \frac{1}{M} \sum_{n=1}^{M} f(x_n) + O(h^2), \tag{9.2}$$

which is equivalent to sampling from a set of points $x_1, x_2, \ldots, x_M$ in the region $[0, 1]$ with an equal weight, in this case, 1, at each point. We can, on the other hand, select x_n with $n = 1, 2, \ldots, M$ from a uniform random number generator in the region $[0, 1]$ to accomplish the same goal. If M is very large, we would expect that x_n is a set of numbers uniformly distributed in the region $[0, 1]$ with fluctuations proportional to $1/\sqrt{M}$. Then the integral can be approximated by the average of the sampled integrand values,

$$S \simeq \frac{1}{M} \sum_{n=1}^{M} f(x_n), \tag{9.3}$$

where x_n is a set of M points generated from a uniform random number generator in the region [0, 1]. Note that the error in the integral is now given by the fluctuation of the distribution of x_n. If we use the standard deviation of the statistics of random sampling, the possible error in the sampling, ΔS, is given by

$$(\Delta S)^2 = \frac{1}{M} (\langle f_n^2 \rangle - \langle f_n \rangle^2). \tag{9.4}$$

Here the average of a quantity is defined as

$$\langle A_n \rangle = \frac{1}{M} \sum_{n=1}^{M} A_n, \tag{9.5}$$

with A_n being the sampled data. We have used $f_n = f(x_n)$ for convenience.

Now we would like to illustrate how the scheme works in an actual numerical example. In order to demonstrate the algorithm clearly, let us take a very simple integrand $f(x) = x^2$. The exact result of the integral is $1/3$ for this simple example. The following program implements the direct sampling Monte Carlo algorithm just discussed.

```
      PROGRAM MCDS
C
C Integration with the direct sampling Monte Carlo
C scheme. The integrand is f(x) = x*x.
C
      PARAMETER (M=1000000)
      COMMON /CSEED/ ISEED
      INTEGER*4 time,STIME,T(9)
C
C Initial seed from the system time and forced to be odd
C
      STIME = time(%REF(0))
      CALL gmtime_(STIME,T)
      ISEED = T(6)+70*(T(5)+12*(T(4)
     *                 +31*(T(3)+23*(T(2)+59*T(1)))))
      IF (MOD(ISEED,2).EQ.0) ISEED = ISEED-1
```

```
C
      SUM1 = 0.0
      SUM2 = 0.0
      DO 100 J = 1, M
        X = RANF()
        SUM1 = SUM1+F(X)
        SUM2 = SUM2+F(X)**2
100 CONTINUE
      S  = SUM1/M
      DS = SQRT(ABS(SUM2/M-(SUM1/M)**2)/M)
      WRITE(6,999) S,DS
      STOP
999 FORMAT (2F14.8)
      END
C

      FUNCTION F(X)
        F = X*X
      RETURN
      END
```

We have used the uniform random number generator introduced in Chapter 2. The numerical results for the integral and its estimated error from the program MCDS are 0.3330 ± 0.0003. More reliable results can be obtained from the average of several independent runs. For example, with four independent runs, we have obtained an average of 0.3332 with an estimated error of 0.0002. This error is obtained from the variance of the four runs.

As one can see, this simple Monte Carlo quadrature does not show any advantage in this numerical example. The result from the trapezoid rule yields much higher accuracy with the same number of points. The reason is that the Monte Carlo quadrature yields an estimated error given by Eq. (9.4), which means that

$$\Delta S \propto \frac{1}{M^{1/2}}, \tag{9.6}$$

while the trapezoid rule yields an estimated error of

$$\Delta S \propto \frac{1}{M^2}, \tag{9.7}$$

which is much better than that of the Monte Carlo evaluation in the program MCDS. The real advantage of Monte Carlo simulations is in the estimate of multidimensional integrals. For example, if we are interested in a many-body system, such as the neon atom, we have ten electrons, and the integrals for the expectation value can be as high as thirty dimensions. For a d-dimensional space, the Monte Carlo quadrature will still yield the same error behavior; that is, the error will be proportional to $1/\sqrt{M}$, with M being the number of points sampled. However, the trapezoid rule will yield an estimated error

$$\Delta S \propto \frac{1}{M^{2/d}}, \tag{9.8}$$

which becomes greater than the Monte Carlo error estimate with the same number of points if d is greater than four. The point is that the Monte Carlo quadrature can still produce a reasonable estimate of a d-dimensional integral when d is very large. There are some specially designed numerical quadratures that would still work better than the Monte Carlo quadrature when d is slightly larger than four. But for a real many-body system, the typical number of dimensions in an integral is $3N$, where N is the number of particles in the system. When N is on the order of ten or larger, any other workable quadrature would perform worse than the Monte Carlo quadrature.

9.2 The Metropolis algorithm

From the discussion in the last section, one should have noticed that for an arbitrary integrand such as $f(x) = x^2$ in the region [0,1], the accuracy of the integral evaluated from the Monte Carlo quadrature is very low. We used one million points in the numerical example and obtained an accuracy of only 0.1%. Is there any way to increase the accuracy for some specific types of integrands?

If $f(x)$ were a constant, one would need only one point to have the exact result from the Monte Carlo quadrature. Or if $f(x)$ is smooth and close to a constant, the accuracy from the Monte Carlo and other quadratures would be much higher. Here we would like to discuss a more general case in which the number of dimensions in the integral is $3N$, that is, $\mathbf{R} = (\mathbf{r}_1, \mathbf{r}_2, \ldots, \mathbf{r}_N)$, with each $\mathbf{r}_i$ for $i = 1, 2, \ldots, N$ being a three-dimensional vector. The $3N$-dimensional integral is then written as

$$S = \int_D F(\mathbf{R}) \, d\mathbf{R}, \tag{9.9}$$

where D is the domain of the integral.

In many cases, the function $F(\mathbf{R})$ is not a smooth function. The idea of importance sampling introduced by Metropolis et al. (1953) is to sample the points from a nonuniform distribution. If a distribution function $\mathcal{W}(\mathbf{R})$ can mimic the drastic changes in $F(\mathbf{R})$, one should expect a much faster convergence with

$$S \simeq \frac{1}{M} \sum_{i=1}^{M} \frac{F(\mathbf{R}_i)}{\mathcal{W}(\mathbf{R}_i)}, \tag{9.10}$$

where M is the total number of points of the configurations $\mathbf{R}_i$ sampled according to the distribution function $\mathcal{W}(\mathbf{R})$.

Now we would like to show some detail of how this sampling scheme was reached. We can, of course, rewrite the integral into

$$S = \int \mathcal{W}(\mathbf{R})G(\mathbf{R})\,d\mathbf{R}, \tag{9.11}$$

where $\mathcal{W}(\mathbf{R})$ is positive definite and satisfies the normalization condition

$$\int \mathcal{W}(\mathbf{R})\,d\mathbf{R} = 1, \tag{9.12}$$

which ensures that $\mathcal{W}(\mathbf{R})$ can be viewed as a probability function. From the two expressions of the integral one can easily see that $G(\mathbf{R}) = F(\mathbf{R})/\mathcal{W}(\mathbf{R})$. The problem is solved if $G(\mathbf{R})$ is a smooth function of $\mathbf{R}$, that is, nearly a constant. One can now imagine a statistical process that leads to an equilibrium distribution $\mathcal{W}(\mathbf{R})$ and the integral S is merely a statistical average of $G(\mathbf{R})$. This can be compared with the canonical ensemble average of a specific physical quantity A, which is a function of coordinate $\mathbf{R}$ only,

$$\langle A \rangle = \int A(\mathbf{R})\mathcal{W}(\mathbf{R})\,d\mathbf{R}, \tag{9.13}$$

where the probability or distribution function $\mathcal{W}(\mathbf{R})$ is given by

$$\mathcal{W}(\mathbf{R}) = \frac{e^{-U(\mathbf{R})/k_B T}}{\int e^{-U(\mathbf{R}')/k_B T}\,d\mathbf{R}'}, \tag{9.14}$$

with $U(\mathbf{R})$ being the potential energy of the system for a given configuration $\mathbf{R}$. k_B is the Boltzmann constant, and T is the temperature of the system.

A procedure introduced by Metropolis et al. (1953) is extremely powerful in the evaluation of the multidimensional integral defined in Eq. (9.11). Here we give a brief outline of the procedure and refer to it as the *Metropolis algorithm*. The selection of the sampling points is viewed as a Markov process in an equilibrium system with a distribution function $\mathcal{W}(\mathbf{R})$. In equilibrium, the values of the distribution function at different points of the phase space are related by

$$\mathcal{W}(\mathbf{R})T(\mathbf{R} \to \mathbf{R}') = \mathcal{W}(\mathbf{R}')T(\mathbf{R}' \to \mathbf{R}), \tag{9.15}$$

with $T(\mathbf{R} \to \mathbf{R}')$ being the transition rate from the state characterized by $\mathbf{R}$ to the state characterized by $\mathbf{R}'$. This is usually referred to as *detailed balance* in statistical mechanics. Now the points are no longer sampled randomly but rather by following the Markov chain. The transition from one point $\mathbf{R}$ to another point $\mathbf{R}'$ is accepted if the ratio of the transition rates satisfies

$$\frac{T(\mathbf{R} \to \mathbf{R}')}{T(\mathbf{R}' \to \mathbf{R})} = \frac{\mathcal{W}(\mathbf{R}')}{\mathcal{W}(\mathbf{R})} \geq w_i, \tag{9.16}$$

where w_i is a uniform random number in the region $[0, 1]$. Let us here outline the steps used for the evaluation of the integral defined in Eq. (9.11) or Eq. (9.13). One first randomly selects a configuration $\mathbf{R}_0$ inside the specified domain D. Then $\mathcal{W}(\mathbf{R}_0)$ is evaluated. A new configuration $\mathbf{R}_1$ is attempted with

$$\mathbf{R}_1 = \mathbf{R}_0 + \Delta\mathbf{R}, \tag{9.17}$$

where $\Delta\mathbf{R}$ is a $3N$-dimensional vector with each component from a uniform distribution between $[-\Delta, \Delta]$, which is achieved, for example, with

$$\Delta x_i = \Delta(2\eta_i - 1), \tag{9.18}$$

for the x-component of $\mathbf{r}_i$. Here η_i is a uniform random number generated in the region $[0, 1]$. The actual value of the step size is determined from the desired accepting rate (the ratio of the accepted to the attempted steps). A large Δ will result in a small accepting rate. In practice, Δ is commonly chosen such that the accepting rate is around 50%. After all the components of $\mathbf{R} = \mathbf{r}_1, \mathbf{r}_2, \ldots, \mathbf{r}_N$ are attempted, one can evaluate the distribution function at the new configuration, $\mathcal{W}(\mathbf{R}_1)$. The new configuration $\mathbf{R}_1$ is accepted with the probability

$$p = \frac{\mathcal{W}(\mathbf{R}_1)}{\mathcal{W}(\mathbf{R}_0)}, \tag{9.19}$$

that is, comparing p with a uniform random number w_i in the region $[0, 1]$. If $p \geq w_i$, the new configuration is accepted; otherwise, the old configuration is taken as the new configuration. This procedure is repeated, and at each step one can evaluate the physical quantity $A(\mathbf{R}_k)$ for $k = n_1, n_1 + n_0, \ldots, n_1 + (M-1)n_0$. The numerical result of the integral is then given by

$$\langle A \rangle \simeq \frac{1}{M} \sum_{k=k_1}^{k_M} A(\mathbf{R}_k), \tag{9.20}$$

where $\mathbf{R}_k$ is a set of configurations from $(M-1)n_0 + n_1$ Metropolis steps. n_1 steps are used to remove the influence of the initial selection of the configuration, that is, $\mathbf{R}_0$. The data are taken n_0 steps apart to avoid high correlation among the data points, since they are generated consecutively and one has to skip several points before the next data point can be used.

One can, however, minimize the correlation and at the same time minimize the number of steps needed to reach a required accuracy by analyzing the autocorrelation function of $A(\mathbf{R}_n)$ for $n = 1, 2, \ldots$. The autocorrelation function is defined as

$$C(l) = \frac{\langle A_{n+l} A_n \rangle - \langle A_n \rangle^2}{\langle A_n^2 \rangle - \langle A_n \rangle^2}, \tag{9.21}$$

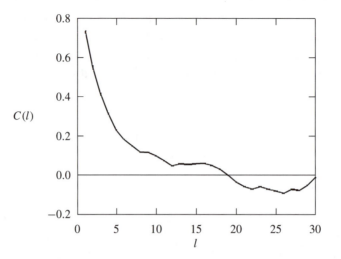

Fig. 9.1. Autocorrelation function obtained from the numerical example in the program MCMA in this section.

where the average is taken from consecutive steps, that is,

$$\langle A_{n+l} A_n \rangle = \frac{1}{M} \sum_{n=1}^{M} A_{n+l} A_n. \tag{9.22}$$

We have used A_n for $A(\mathbf{R}_n)$ for simplicity of notation. A typical $C(l)$ in the numerical example given later in this section is shown in Fig. (9.1). As one can see, the autocorrelation function starts to decrease at the beginning and then saturates around $l = l_c \simeq 12$, which is a good choice of the number of steps to be skipped, $n_0 = l_c$, for the two adjacent data points used in the average in Eq. (9.20).

We would like to remark that in most cases, such as the average done in statistical mechanics, the distribution function $\mathcal{W}(\mathbf{R})$ varies by several orders of magnitude, whereas $A(\mathbf{R})$ stays a smooth or near-constant function. The sampling scheme outlined above will accumulate points according to the distribution $\mathcal{W}(\mathbf{R})$. In other words, there will be more points in the sampling from the configurations with higher $\mathcal{W}(\mathbf{R})$ values. This sampling by *importance* is much more efficient than direct random sampling in statistical mechanics calculations and similar situations when the distribution function has sharp features.

Now we would like to compare this procedure numerically with the direct random sampling discussed in the preceding section with exactly the same integral

$$S = \int_0^1 f(x)\,dx, \tag{9.23}$$

with $f(x) = x^2$. We can choose the distribution function as

$$W(x) = \frac{1}{\mathcal{Z}}(e^{x^2} - 1),\qquad(9.24)$$

which is positive definite. The normalization constant $\mathcal{Z}$ is given by

$$\mathcal{Z} = \int_0^1 (e^{x^2} - 1)\,dx = 0.46265167,\qquad(9.25)$$

which is calculated from another numerical scheme for convenience. Then the corresponding function $g(x) = f(x)/W(x)$ is given by

$$g(x) = \mathcal{Z}\frac{x^2}{e^{x^2} - 1}.\qquad(9.26)$$

Now we are ready to put all of these into a program that is a realization of the Metropolis algorithm for the integral specified.

```
      PROGRAM MCMA
C
C Integral evaluated with the Metropolis Monte Carlo
C scheme. The distribution function is W(x) = (exp(x*x)
C -1)/Z and the function sampled is g(x) = Z*x*x/
C (exp(x*x)-1).
C
      PARAMETER (ITHM=10000,ISIZ=10000,L=15)
      INTEGER*4 time,STIME,T(9)
      COMMON /CSEED/ ISEED
      COMMON /CPSTN/ X,B,W,ICPT
      COMMON /CARRY/ ECRR(ISIZ)
C
      B = 0.4
      Z = 0.46265167
C
C Initial seed from the system time and forced to
C be odd
C
      STIME = time(%REF(0))
      CALL gmtime_(STIME,T)
      ISEED = T(6)+70*(T(5)+12*(T(4)
     *             +31*(T(3)+23*(T(2)+59*T(1)))))
      IF (MOD(ISEED,2).EQ.0) ISEED = ISEED-1
```

```
C
C Runs to remove the initial configuration dependence
C
      X = RANF()
      W = WFNT(X)
      DO 100 J = 1, ITHM
        CALL MTSP
  100 CONTINUE
C
C Collect data every L steps
C
      SUM1 = 0.0
      SUM2 = 0.0
      ICPT = 0
      DO 200 J = 1, ISIZ*L
        CALL MTSP
        IF(MOD(J,L).EQ.0) THEN
          SUM1 = SUM1+G(X)
          SUM2 = SUM2+G(X)*G(X)
        ELSE
        ENDIF
        IF (ICRL.EQ.1) ECRR(J) = G(X)
  200 CONTINUE
C
      S  = Z*SUM1/ISIZ
      DS = Z*SQRT(ABS(SUM2/ISIZ-(SUM1/ISIZ)**2)/ISIZ)
      ACPT = 100.0*ICPT/(ISIZ*L)
C
      WRITE(6,999) S,DS,ACPT
  999 FORMAT (3F14.8)
      STOP
      END
C
      SUBROUTINE MTSP
C
C  Subroutine for one Metropolis step
C
      PARAMETER (ITHM=10000,ISIZ=10000,L=15)
      COMMON /CSEED/ ISEED
      COMMON /CPSTN/ X,B,W,ICPT
```

```
C
      XSAV = X
C
      RNDX = RANF()
      X    = X+2.0*B*(RNDX-0.5)
      IF ((X.LT.0).OR.(X.GT.1)) THEN
        X = XSAV
        RETURN
      ELSE
      ENDIF
C
      WTRY = WFNT(X)
      RNDW = RANF()
      IF (WTRY.GT.(W*RNDW)) THEN
        W    = WTRY
        ICPT = ICPT+1
        RETURN
      ELSE
      ENDIF
      X    = XSAV
      RETURN
      END
C
      FUNCTION G(X)
        G = X*X/(EXP(X*X)-1.0)
      RETURN
      END
C
      FUNCTION WFNT(X)
        WFNT = EXP(X*X)-1.0
      RETURN
      END
```

The numerical result obtained with the above program is 0.3330 ± 0.0004. The step size is adjustable, and one should try to keep it such that the accepting rate of the new configurations is less than or around 50%. In practice, it seems that such a choice of accepting rate is compatible with considerations of speed and accuracy. The above program is structured so that one can easily modify it to study other problems.

9.3 Applications in statistical physics

The Metropolis algorithm was first applied in the simulation of the structure of classical liquids and is still used in current research in the study of glass transitions and polymer systems. In this section, we would like to discuss some applications of the Metropolis algorithm in statistical physics. We will use the classical liquid system and the Ising model as the two illustrative examples. The classical liquid system is continuous in spatial coordinates of atoms or molecules in the system, whereas the Ising model is a discrete lattice system.

Structure of classical liquids

First we will consider the classical liquid system. Let us assume that the system is in good contact with a thermal bath and has a fixed number of particles and volume size; that is, that the system is described by the canonical ensemble. Then the average of a physical quantity A is given by

$$\langle A \rangle = \frac{1}{\mathcal{Z}} \int A(\mathbf{R}) e^{-U(\mathbf{R})/k_B T} d\mathbf{R}, \tag{9.27}$$

with $\mathcal{Z}$ being the partition function of the system, which is given by

$$\mathcal{Z} = \int e^{-U(\mathbf{R})/k_B T} d\mathbf{R}. \tag{9.28}$$

Here $\mathbf{R} = (\mathbf{r}_1, \mathbf{r}_2, \ldots, \mathbf{r}_N)$ is a $3N$-dimensional vector for the coordinates of all the particles in the system. We have suppressed the velocity dependence in the expression with the understanding that the distribution of the velocity is given by the Maxwell distribution, which can be used to calculate the averages of any velocity-related physical quantities. For example, the average kinetic energy component of a particle is given by

$$\left\langle \frac{m v_{ik}^2}{2} \right\rangle = \frac{1}{2} k_B T, \tag{9.29}$$

where $i = 1, 2, \ldots, N$ is the index for the ith particle and $k = 1, 2, 3$ is the index for the directions x, y, and z. As a matter of fact, the average of any physical quantity associated with velocity can be obtained through the partition theorem. For example, a quantity $B(\mathbf{V})$ with $\mathbf{V} = (\mathbf{v}_1, \mathbf{v}_2, \ldots, \mathbf{v}_N)$ can always be expanded in the Taylor series of the velocities, and then each term can be evaluated with

$$\langle v_{ik}^n \rangle = \begin{cases} (n-1)!! \, (k_B T/m)^{n/2} & \text{for even } n, \\ 0 & \text{otherwise.} \end{cases} \tag{9.30}$$

Now one can easily apply the Metropolis algorithm to evaluate a specific average of a physical quantity that depends on the spatial coordinates of all the particles in the system, such as the total potential energy, the pair distribution, and the pressure. The physical quantity under evaluation is then given by the average of a set of discrete values,

$$\langle A \rangle = \frac{1}{\mathcal{Z}} \int A(\mathbf{R}) e^{-U(\mathbf{R})/k_B T} d\mathbf{R} = \frac{1}{M} \sum_{i=1}^{M} A(\mathbf{R}_i), \qquad (9.31)$$

with $\mathbf{R}_i$ being a set of points in the phase space sampled according to the distribution function

$$\mathcal{W}(\mathbf{R}) = \frac{e^{-U(\mathbf{R})/k_B T}}{\mathcal{Z}}. \qquad (9.32)$$

Assume that we have a pairwise interaction between any two particles. The total potential energy for a specific configuration is then given by

$$U(\mathbf{R}) = \sum_{j>i}^{N} V(r_{ij}). \qquad (9.33)$$

In the preceding section, we discussed how to update all the components in the old configuration in order to reach a new configuration. However, sometimes it is more efficient to update the coordinates of just one particle at a time in the old configuration to reach the new configuration, especially when the system is very close to equilibrium or near a phase transition. Here we would like to show how to update the coordinates of one particle to obtain the new configuration. The coordinates for the ith particle are updated with

$$x_i^{(n+1)} = x_i^{(n)} + \Delta_x (2\alpha_i - 1), \qquad (9.34)$$

$$y_i^{(n+1)} = y_i^{(n)} + \Delta_y (2\beta_i - 1), \qquad (9.35)$$

$$z_i^{(n+1)} = z_i^{(n)} + \Delta_z (2\gamma_i - 1), \qquad (9.36)$$

with Δ_k as the step size along the kth direction and α_i, β_i, and γ_i as uniform random numbers in the region $[0, 1]$. The acceptance of the move is determined by importance sampling, that is, by comparing the ratio of the distribution functions,

$$p = \frac{\mathcal{W}(\mathbf{R}_{n+1})}{\mathcal{W}(\mathbf{R}_n)}, \qquad (9.37)$$

with a uniform random number w_i. The attempted move is accepted if $p \geq w_i$ and rejected if $p < w_i$. Note that in the new configuration only the coordinates of the ith particle are moved, so we do not need to evaluate the whole $U(\mathbf{R}_{n+1})$ again in order to obtain the ratio p for the pairwise interactions. We can express the ratio in terms of the potential energy difference between the old configuration and the new configuration,

$$p = e^{-\Delta U / k_B T}, \tag{9.38}$$

which is given by the summation of all the other coordinates in $\mathbf{R}$,

$$\Delta U = \sum_{j \neq i}^{N} [V(|\mathbf{r}_i^{(n+1)} - \mathbf{r}_j^{(n)}|) - V(|\mathbf{r}_i^{(n)} - \mathbf{r}_j^{(n)}|)], \tag{9.39}$$

which is usually truncated for particles within a distance $r_{ij} = r_c$. r_c is a typical distance at which the effect of the interaction $V(r_c)$ is negligible. For example, the interaction in a simple liquid is typically the Lennard–Jones potential, which decreases drastically with the separation of the two particles. The typical distance for the truncation in the Lennard–Jones potential is about $r_c = 3\sigma$, where σ is the separation of the zero potential. As we discussed earlier, one should not take *all* the discrete data points in the sampling for the average of the physical quantity, since the autocorrelation of the data is very strong if they are not far apart. Typically, one needs to skip about ten to fifteen data points before taking another value for the average. The evaluations for physical quantities such as total potential energy, structural factor, and pressure are performed almost exactly the same way as in molecular dynamics if one thinks of the Metropolis Monte Carlo steps as the time steps in molecular dynamics simulations.

Another issue for the simulation of infinite systems is the extension of the finite box with a periodic boundary condition. Long-range interactions, such as the Coulomb interaction, cannot be truncated; a summation of the interaction between particles in all the periodic boxes is needed. The Ewald sum is used to include the interactions between a charged particle and the images in other boxes constructed with the periodic boundary condition. One can find discussions of the Ewald summation in standard textbooks of solid-state physics, for example, Madelung (1978).

The Monte Carlo step size Δ is determined from the desired rejection rate. For example, if one wants 70% of the moves to be rejected, one can adjust Δ to satisfy such a rejection rate. A larger Δ produces a higher rejection rate. In practice, it is clear that a higher rejection rate will produce data points with fewer fluctuations. However, one also has to consider the computing time needed. The decision is made according to the optimization of the computing time and the accuracy of the data desired. Typically, people use 50% to 75% as the rejection rate in most Monte Carlo simulations. Δ along each different direction, that is, Δ_x, Δ_y, or Δ_z, is determined according to the length scale along that direction. An isotropic system would have $\Delta_x = \Delta_y = \Delta_z$. A system confined along the z-direction, for example, would have a smaller Δ_z.

Properties of lattice systems

Now let us turn to discrete model systems. The scheme is more or less the same. The difference now comes mainly from the way that the new configuration is attempted. Since the variables are discrete, instead of moving the particles in the system, we need to update the configuration by changing the local state of each lattice site. Here we will use the classical three-dimensional Ising model as an illustrative example. The Hamiltonian of the Ising model is

$$\mathcal{H} = -J \sum_{\langle ij \rangle}^{N} s_i s_j - B \sum_{i=1}^{N} s_i, \qquad (9.40)$$

where J is the exchange interaction strength, B is the external field, $\langle ij \rangle$ means the summation over all nearest neighbors, and N is the total number of sites in the system. The spin s_i for $i = 1, 2, \ldots, N$ can take only values of either 1 or -1, so the summation for the interactions is for energies carried by all the bonds between nearest neighboring sites.

The Ising model was historically used to study magnetic phase transitions. The magnetization is defined as

$$m = \frac{1}{N} \sum_{i=1}^{N} \langle s_i \rangle, \qquad (9.41)$$

which is a function of the temperature and external magnetic field. For the $B = 0$ case, there is a critical temperature T_c that separates different phases of the system. For example, the system is ferromagnetic if $T < T_c$, paramagnetic if $T > T_c$, and unstable if $T = T_c$. The complete plot of T, m, and B forms the so-called phase diagram. Another interesting application of the Ising model is that it is also a generic model for binary lattices; that is, two types of particles can occupy the lattice sites with two different on-site energies whose difference is $2B$. So the results obtained from the study of the Ising model apply also other systems in its class, such as the binary lattices. Readers who are interested in these aspects can find good discussions in Parisi (1988). Other quantities of interest include but are not limited to the total energy E of the system, which is the average of the Hamiltonian,

$$E = \langle \mathcal{H} \rangle, \qquad (9.42)$$

and the specific heat C per site, which is obtained from the fluctuation of the total energy,

$$C = \frac{\langle \mathcal{H}^2 \rangle - \langle \mathcal{H} \rangle^2}{N k_B T^2}. \qquad (9.43)$$

In order to simulate the Ising model, for example, in the calculation of m, we can apply more or less the same idea for the continuous system. The statistical average of the spin at each site is given by

$$m = \frac{1}{\mathcal{Z}} \sum_{\sigma} s_{\sigma} e^{-\mathcal{H}_{\sigma}/k_B T}, \tag{9.44}$$

where $s_{\sigma} = S_{\sigma}/N$, with $S_{\sigma} = \sum s_i$ as the total spin of a specific configuration labeled by σ, and $\mathcal{H}_{\sigma}$ is the corresponding Hamiltonian (energy). The summation in Eq. (9.44) is over all the possible configurations. Here $\mathcal{Z}$ is the partition function given by

$$\mathcal{Z} = \sum_{\sigma} e^{-\mathcal{H}_{\sigma}/k_B T}. \tag{9.45}$$

The average of a physical quantity, such as the magnetization, can be obtained from

$$m \simeq \frac{1}{M} \sum_{\sigma=1}^{M} s_{\sigma}, \tag{9.46}$$

with $\sigma = 1, 2, \ldots, M$ indicating the configurations sampled according to the distribution function

$$\mathcal{W}(S_{\sigma}) = \frac{\exp(-\mathcal{H}_{\sigma}/k_B T)}{\mathcal{Z}}. \tag{9.47}$$

Let us go into some more details for the actual simulations. First we randomly assign 1 and -1 to the spins on all the lattice sites. Then we select one site, which can be picked up either randomly or sequentially. Assume that the ith site is selected to be updated. The update is attempted with the spin at the site reversed; that is,

$$s_i^{(n+1)} = -s_i^{(n)}, \tag{9.48}$$

which is accepted into the new configuration if

$$p = e^{-\Delta\mathcal{H}/k_B T} \geq w_i, \tag{9.49}$$

where w_i is a uniform random number in the region $[0, 1]$ and $\Delta\mathcal{H}$ is the energy difference caused by the reversal of the spin,

$$\Delta\mathcal{H} = -2J s_i^{(n+1)} \sum_{j=1}^{z} s_j^{(n)}, \tag{9.50}$$

where j runs through all the nearest neighboring sites of i. Note that the quantity $\sum s_j^{(n)}$ associated with a specific site may be stored and updated during the simulation. Every time, if a spin is reversed and its new value is accepted, one can update the summations accordingly. If the reversal is rejected in the new configuration, no update is needed. The detailed balance condition discussed in the preceding section,

$$W(S_\sigma)T(S_\sigma \to S_\mu) = W(S_\mu)T(S_\mu \to S_\sigma), \qquad (9.51)$$

does not uniquely determine the transition rate $T(S_\sigma \to S_\mu)$; therefore we can also use the other properly normalized probability for the Metropolis steps, for example,

$$q = \frac{1}{1 + e^{\Delta \mathcal{H}/k_B T}}, \qquad (9.52)$$

instead of p in the simulation without changing the equilibrium results. The use of q speeds up the convergence at higher temperatures. There have been considerable discussions on Monte Carlo simulations for the Ising model or related lattice models, such as percolation models. Interested readers can find these discussions in Binder and Heermann (1988), and Binder (1986; 1987; 1992).

9.4 Critical slowing down and block algorithms

There is a practical problem with the Metropolis algorithm in statistical physics when the system under study is approaching a critical point. Imagine that the system is a three-dimensional Ising system. At very high temperatures, almost all the spins are uncorrelated, and therefore flipping of a spin has a very high chance of being accepted into the new configuration. So the different configurations can be accessed rather quickly and provide an average close to ergodic behavior. However, when the system is moved toward the critical temperature from above, domains with lower energy start to form. If the interaction is ferromagnetic, one will start to see large clusters with all the spins pointing in the same direction. Now, if only one spin is flipped, the configuration becomes much less favorable because of the increase in energy. The favorable configurations are the ones with all the spins in the domain reversed, a point that it will take a very long time to reach. It means that one needs to have every attempted spin flipping in the domain accepted in a very long sequence. This phenomenon is known as the *critical slowing down*.

Another way to view it is from the autocorrelation function of the total energy of the system. Since now most steps are not accepted in generating new configurations, the total energy evaluated at each Monte Carlo step is highly correlated. In fact, the time needed to have the autocorrelation function decreased to near zero, t_c, is proportional to the power of the correlation length ξ, which diverges at the critical point in a bulk system; that is,

$$t_c = c\xi^z, \qquad (9.53)$$

where c is a constant and the exponent z is called the *dynamical critical exponent*, which is about 2, estimated from standard Monte Carlo simulations. Note that the correlation length is bounded by the size of the simulation box L with

$$t_c = cL^z, \tag{9.54}$$

when the system is very close to the critical point.

The first solution of this problem was devised by Swendsen and Wang (1987). Their block update scheme is based on the nearest-neighbor pair picture of the Ising model, in which the partition function is a result of the contributions of all the nearest pairs of sites in the system. Let us examine closely the effect of a specific pair between sites i and j on the total partition function. We will write the partition function contributed from the rest pairs in the system, that is, the partition function of the system without the interaction term between sites i and j in the Hamiltonian, as Z_p or Z_f, where Z_p is the partition function of the rest pairs with the spins at sites i and j parallel and Z_f is the corresponding partition function without any restriction on the spins at sites i and j. The total partition function of the system can then be expressed as

$$Z = qZ_p + (1 - q)Z_f, \tag{9.55}$$

with

$$q = e^{-4\beta J}, \tag{9.56}$$

which can be interpreted as the probability of having a pair of correlated nearest neighbors decoupled. This is the basis on which Swendsen and Wang devised the block algorithm to remove the critical slowing down.

Here is a summary of their algorithm. One first clusters the sites next to each other that have the same spin orientation. A bond is introduced for any pair of nearest neighbors in a cluster. Each bond is then removed with the probability q. After all the bonds are attempted with some removed and the rest kept, there are more, but smaller, clusters still connected through the remaining bonds. The spins in each small cluster are then flipped all together with a probability of 50%. The new configuration is then used for the next Monte Carlo step. This procedure in fact updates the configuration by flipping blocks of parallel spins, which is similar to the physical process that one would expect when the system is close to the critical point. The speedup in this block algorithm is extremely significant. Numerical simulations show that the dynamical critical exponent is now significantly reduced, with $z \simeq 0.35$ for the two-dimensional Ising model and $z \simeq 0.53$ for the three-dimensional Ising model, instead of $z \simeq 2$ in the spin-by-spin update scheme.

The algorithm of Swendsen and Wang is important but still does not eliminate the critical slowing down completely, since the time needed to reach an uncorrelated energy still increases with the correlation length. An algorithm devised later by Wolff (1989) provides about the same improvement for the two-dimensional Ising model as the Swendsen–Wang algorithm but greater improvement for the three-dimensional Ising model. More interestingly, the Wolff algorithm eliminates the critical slowing down completely, that is, $z \simeq 0$, for the four-dimensional Ising model, whereas the Swendsen–Wang algorithm has $z \simeq 1$.

The idea of Wolff is very similar to the idea of Swendsen and Wang. Instead of removing bonds, Wolff proposed constructing a cluster with nearest sites having the same spin orientation. One first selects a site randomly from the system and adds its nearest neighbors with the same spin orientation to the cluster with the probability $p = 1 - q$. This is continued, with sites added to the cluster, until all the sites in the system are attempted. All the spins in the constructed cluster are then flipped together to reach a new spin configuration of the system. One can easily show that the Wolff algorithm ensures detailed balance.

The Swendsen–Wang and Wolff algorithms can also be generalized to other spin models, such as the xy model and the Heisenberg model (Wolff, 1989). The idea comes from the construction of the probability q in both of the above-mentioned algorithms. At every step, one first generates a random unit vector $\mathbf{e}$ with each component being a uniform random number x_i,

$$\mathbf{e} = \frac{1}{r}(x_1, x_2, \ldots, x_n), \tag{9.57}$$

with

$$r^2 = \sum_{i=1}^{n} x_i^2, \tag{9.58}$$

where n is the dimension of the vector space of the spin. One can then define the probability of breaking a bond as

$$q = \min \left\{ 1, e^{4\beta J (\mathbf{e} \cdot \mathbf{s}_i)(\mathbf{e} \cdot \mathbf{s}_j)} \right\}, \tag{9.59}$$

which reduces to Eq. (9.56) for the Ising model. The Swendsen–Wang algorithm and the Wolff algorithm have also shown significant improvement in the xy model, the Heisenberg model, and the Potts model. For more details, see Swendsen, Wang, and Ferrenberg (1992).

9.5 Variational quantum Monte Carlo simulations

The Monte Carlo method used in the simulations of classical many-body systems can be generalized to study quantum systems as well. In this section, we will introduce the most direct generalization of the Metropolis algorithm in the framework of the variational principle. We will start our discussion with a general quantum many-body problem, that is, the approximate solution of the Hamiltonian

$$
\mathcal{H} = \sum_{i=1}^{N} \left[-\frac{\hbar^2}{2m_i} \nabla_i^2 + U_{\text{ext}}(\mathbf{r}_i) \right] + \sum_{i>j} V(\mathbf{r}_i, \mathbf{r}_j), \tag{9.60}
$$

where $-\hbar^2 \nabla_i^2 / 2m_i$ is the kinetic energy and $U_{\text{ext}}(\mathbf{r}_i)$ is the external potential of the ith particle and $V(\mathbf{r}_i, \mathbf{r}_j)$ is the interaction potential between the ith and jth particles. In most cases, the interaction is spherically symmetric, $V(\mathbf{r}_i, \mathbf{r}_j) = V(r_{ij})$. The masses of all particles are the same for an identical-particle system, of course.

We can symbolically write the time-independent many-body Schrödinger equation as

$$
\mathcal{H} \Psi_n(\mathbf{R}) = E_n \Psi_n(\mathbf{R}), \tag{9.61}
$$

where $\mathbf{R} = (\mathbf{r}_1, \mathbf{r}_2, \ldots, \mathbf{r}_N)$ and $\Psi_n(\mathbf{R})$ and E_n are the nth eigenstate and its corresponding eigenvalue of $\mathcal{H}$. One usually cannot obtain analytic or exact solutions for a system with more than two particles due to the complication of the interaction potential. Approximate schemes are thus important tools for studying the different aspects of the many-body problem. It is easier to study the ground state of a many-body system. Properties associated with the ground state include the order of the state, collective excitations, and structural information.

Based on the variational principle that any other state would have a higher expectation value of the total energy than the ground state of the system, we can always introduce a trial state $\Phi(\mathbf{R})$ to approximate the ground state. $\Phi(\mathbf{R})$ can be a parameterized function or some specific function form. The parameters or the variational function in $\Phi(\mathbf{R})$ can be optimized through the variational principle

$$
E[\alpha_i] = \frac{\langle \Phi | \mathcal{H} | \Phi \rangle}{\langle \Phi | \Phi \rangle} \geq E_0, \tag{9.62}
$$

where α_i can be a set of variational parameters or functions of $\mathbf{R}$, which are linearly independent. This inequality can easily be shown by expanding $\Phi(\mathbf{R})$ in terms of the eigenstates of the Hamiltonian $\Psi_n(\mathbf{R})$ as

$$
\Phi(\mathbf{R}) = \sum_{n=0}^{\infty} a_n \Psi_n(\mathbf{R}), \tag{9.63}
$$

provided that $\Phi(\mathbf{R})$ satisfies the same boundary condition as all $\Psi_n(\mathbf{R})$. The expansion above is a generalization of the Fourier theorem, since $\Psi_n(\mathbf{R})$ form a complete basis set. The variational principle results if one substitutes the expansion for $\Phi(\mathbf{R})$ into the left side of Eq. (9.62), with the understanding that

$$
E_n \geq E_0 \tag{9.64}
$$

for $n > 0$ and that the eigenstates of a Hermitian operator are orthogonal to each other:

$$\langle \Psi_n \mid \Psi_m \rangle = \delta_{nm}. \tag{9.65}$$

In order to obtain the optimized wavefunction, one can treat the variational parameters or specific functions in the variational wavefunction as independent variables in the Euler–Lagrange equation

$$\frac{\delta E[\alpha_i]}{\delta \alpha_j} = 0, \tag{9.66}$$

with $j = 1, 2, 3, \ldots$. To simplify our discussion, let us assume that the system is a continuum and that α_i is a set of variational parameters. So we can write the expectation value in the form of an integral

$$E[\alpha_i] = \frac{\int \Phi^+(\mathbf{R}) \mathcal{H} \Phi(\mathbf{R}) \, d\mathbf{R}}{\int |\Phi(\mathbf{R}')|^2 \, d\mathbf{R}'} = \int \mathcal{W}(\mathbf{R}) \epsilon(\mathbf{R}) \, d\mathbf{R}, \tag{9.67}$$

where

$$\mathcal{W}(\mathbf{R}) = \frac{|\Phi(\mathbf{R})|^2}{\int |\Phi(\mathbf{R}')|^2 \, d\mathbf{R}'} \tag{9.68}$$

can be interpreted as a distribution function and

$$\epsilon(\mathbf{R}) = \frac{1}{\Phi(\mathbf{R})} \mathcal{H} \Phi(\mathbf{R}), \tag{9.69}$$

can be viewed as a local energy with a specific configuration $\mathbf{R}$. The expectation value can then be evaluated by the Monte Carlo quadrature if the expressions of $\epsilon(\mathbf{R})$ and $\mathcal{W}(\mathbf{R})$ for a specific set of parameters α_i are given. In practice, $\Phi(\mathbf{R})$ is parameterized to have the important physics built in. The variational parameters α_i in the trial wavefunction $\Phi(\mathbf{R})$ are then optimized with the minimization of the expectation value $E[\alpha_i]$. It needs to be emphasized that the variational process usually cannot lead to any new physics and that the physics is in the variational wavefunction, which is constructed intuitively. However, in order to observe the physics of a nontrivial wavefunction, one usually needs to carry out the calculations.

A common choice of the trial wavefunction for quantum liquids has the general form

$$\Phi(\mathbf{R}) = D(\mathbf{R}) e^{-U(\mathbf{R})}, \tag{9.70}$$

where $D(\mathbf{R})$ is a constant for boson systems and a Slater determinant of single-particle orbitals for fermion systems, so the Pauli principle is implemented in the wavefunction. $U(\mathbf{R})$ is the Jastrow correlation factor, which can be written in terms of one-body terms, two-body terms, and so on:

$$U(\mathbf{R}) = \sum_{i=1}^{N} u_1(\mathbf{r}_i) + \sum_{i>j}^{N} u_2(\mathbf{r}_i, \mathbf{r}_j) + \sum_{i>j>k}^{N} u_3(\mathbf{r}_i, \mathbf{r}_j, \mathbf{r}_k) + \cdots, \qquad (9.71)$$

which is usually truncated at the two-body terms in most Monte Carlo simulations. Both $u_1(\mathbf{r})$ and $u_2(\mathbf{r}, \mathbf{r}')$ are parameterized according to the physical understanding of the source of these terms. One can show that when the external potential becomes a dominant term, $u_1(\mathbf{r})$ is uniquely determined by the form of the external potential, and when the interaction between the ith and jth particles dominates, the form of $u_2(\mathbf{r}_i, \mathbf{r}_j)$ is uniquely determined by $V(\mathbf{r}_i, \mathbf{r}_j)$. These are the important aspects used in determining the form of u_1 and u_2 in practice.

First let us consider the case of bulk helium liquids. Since the external potential is zero, we can choose $u_1(\mathbf{r}) = 0$. The two-body term is translationally invariant, $u_2(\mathbf{r}_i, \mathbf{r}_j) = u_2(r_{ij})$. The expression of $u_2(r)$ at the limit of $r \to 0$ can be obtained by solving a two-body problem. In the center-of-mass coordinate system, we have

$$\left[-\frac{\hbar^2}{2\mu} \nabla^2 + V(r) \right] e^{-u_2(r)} = E e^{-u_2(r)}, \qquad (9.72)$$

where μ is the reduced mass $m/2$, $V(r)$ is the interaction potential that is given by $\epsilon(\sigma/r)^{12}$ at the limit of $r \to 0$, and the Laplacian can be decoupled in the angular momentum eigenstates as

$$\nabla^2 = \frac{d^2}{dr^2} + \frac{2}{r}\frac{d}{dr} - \frac{l(l+1)}{r^2}. \qquad (9.73)$$

One can show that in order to have the divergence in the potential energy canceled by the kinetic energy part at the limit of $r \to 0$, one has to have

$$u_2(r) = \left(\frac{a}{r} \right)^5. \qquad (9.74)$$

The condition needed to remove the divergence of the potential energy with the kinetic energy term constructed from the wavefunction is called the *cusp condition*. This condition is extremely important in all quantum Monte Carlo simulations, since it is the major means of stabilizing the algorithms. The behavior of $u_2(r)$ at longer range is usually dominated by the density fluctuation or zero-point motion of phonons, which is proportional to $1/r^2$. One can also show that the three-body term and the effects due to the backflow are also important and can be incorporated into the variational wavefunction. The Slater determinant for liquid ^{3}He can be constructed from plane waves. The key, as we stressed earlier, is to find a variational wavefunction that contains the essential physics of the system, which is always nontrivial. A recent work by Bouchaud and Lhuillier (1989) replaced the Slater determinant with a BCS (Bardeen–Cooper–Schrieffer) ansatz for liquid ^{3}He. This changes the structure of the ground state of the system from a Fermi liquid, which is characterized by a Fermi surface with a finite jump in the particle distribution, to a superconducting glass, which does not have any Fermi surface. For more details on the variational Monte Carlo simulations of helium systems, see Ceperley and Kalos (1986).

Electronic systems are another class of systems studied with the variational quantum Monte Carlo method. Significant results are obtained for atomic systems, molecules, and even solids. Typical approaches treat electrons and nuclei separately with the so-called Born–Oppenheimer approximation; that is, the electronic state is adjusted quickly for a given nucleus configuration so that the potential due to the nuclei can be treated as an external potential of the electronic system. Assume that we can obtain the potential of the nuclei or ions with some other methods. $u_1(\mathbf{r}_i)$ can be parameterized to ensure the cusp condition between an electron and a nucleus or ion. For the ion with an effective charge Ze,

$$u_1(r) = Zr/a_0 \qquad (9.75)$$

as $r \to 0$. Here a_0 is the Bohr radius. $u_2(r_{ij})$ is obtained from the electron–electron interaction, and the cusp condition requires

$$u_2(r_{ij}) = -\frac{\sigma_{ij} r_{ij}}{2a_0} \qquad (9.76)$$

as $r_{ij} \to 0$. Here $\sigma_{ij} = 1$ if the ith and jth electrons have different spin orientations; otherwise $\sigma_{ij} = 1/2$. The Slater determinant can be constructed from the local orbitals, for example, the linear combinations of the Gaussian orbitals from all nuclear sites. Details on how to construct the kinetic energy for the Jastrow ansatz with the Slater determinant can be found in Ceperley, Chester, and Kalos (1977).

There have been a lot of impressive variational quantum Monte Carlo simulations for electronic systems recently, for example, the work of Fahy, Wang, and Louie (1990) on carbon and silicon solids, the work of Umrigar, Wilson, and Wilkins (1988) on the improvement of variational wavefunctions; and the work of Umrigar (1993) on accelerated variational Monte Carlo simulations.

The numerical procedure for performing variational quantum Monte Carlo simulations is exactly the same as that for the Monte Carlo simulations of statistical systems. The only difference is that for the statistical system the calculations are done for a given temperature, but for the quantum system they are done for a given set of variational parameters in the variational wavefunction. One can update either the whole configuration at one Metropolis step or just the coordinates associated with one particle at a time.

9.6 Green's function Monte Carlo simulations

As we pointed out in the last section, the goal of the many-body theory is to understand the properties of a many-body Hamiltonian under given conditions. Assuming that we are dealing with an identical-particle system with an external potential $U_{ext}(\mathbf{r})$, the Hamiltonian is then given by

$$\mathcal{H} = -\frac{\hbar^2}{2m} \sum_{i=1}^{N} \nabla_i^2 + \sum_{i=1}^{N} U_{ext}(\mathbf{r}_i) + \sum_{i<j}^{N} V(\mathbf{r}_i, \mathbf{r}_j), \qquad (9.77)$$

where m is the mass of each particle and $V(\mathbf{r}_i, \mathbf{r}_j)$ is the interaction potential between the ith and jth particles. We will also assume that the interaction is spherically symmetric; that is, that $V(\mathbf{r}_i, \mathbf{r}_j) = V(r_{ij})$. As we discussed in the preceding section, the formal solution of the Schrödinger equation can be written as

$$\mathcal{H}\Psi_n(\mathbf{R}) = E_n \Psi_n(\mathbf{R}), \qquad (9.78)$$

where $\mathbf{R} = (\mathbf{r}_1, \mathbf{r}_2, \ldots, \mathbf{r}_N)$ and $\Psi_n(\mathbf{R})$ and E_n are the nth eigenstate and its corresponding eigenvalue of $\mathcal{H}$. For convenience of discussion, we will set $\hbar = m = 1$.

It is impossible to solve the above Hamiltonian analytically for most cases. Numerical simulations are the alternatives in the study of many-body Hamiltonians. In the preceding section, we discussed the use of the variational quantum Monte Carlo simulation scheme to obtain the approximate solution for the ground state. However, the result is usually limited by the variational wavefunction selected. In this section, we would like to show that one can go one step further to sample the ground-state wavefunction and for some systems can even find the exact ground state. First we will discuss in more detail a version of the Green's function Monte Carlo scheme that treats the ground state of the Schrödinger equation as the stationary solution of a diffusion equation, the so-called imaginary-time Schrödinger equation. Later we will briefly discuss the original version of the Green's function Monte Carlo scheme, which actually samples Green's function. The imaginary-time Schrödinger equation is given by

$$\frac{\partial \Psi(\mathbf{R}, t)}{\partial t} = -(\mathcal{H} - E_c)\Psi(\mathbf{R}, t), \qquad (9.79)$$

where t is the imaginary time, $\Psi(\mathbf{R}, t)$ is a time-dependent wavefunction, and E_c is an adjustable constant that is used later in the simulation to ensure that the overlap of $\Psi(\mathbf{R}, t)$ and $\Psi_0(\mathbf{R})$ is of the order of 1. One can show that at a later time the wavefunction is given by the evolution equation

$$\Psi(\mathbf{R}, t) = \int G(\mathbf{R}, \mathbf{R}'; t - t')\Psi(\mathbf{R}', t')\,d\mathbf{R}', \qquad (9.80)$$

where Green's function $G(\mathbf{R}, \mathbf{R}'; t - t')$ is given by

$$\left[\frac{\partial}{\partial t} - (\mathcal{H} - E_c)\right] G(\mathbf{R}, \mathbf{R}'; t - t') = \delta(\mathbf{R} - \mathbf{R}')\delta(t - t'), \qquad (9.81)$$

which can be expressed as

$$G(\mathbf{R}, \mathbf{R}'; t - t') = \langle \mathbf{R}| \exp[-(\mathcal{H} - E_c)(t - t')]|\mathbf{R}'\rangle. \qquad (9.82)$$

Green's function can be used to sample the time-dependent wavefunction $\Psi(\mathbf{R}, t)$ as discussed by Anderson (1975; 1976). A more efficient way of sampling the wavefunction is by constructing a probability-like function (Reynolds et al., 1982). Assume that we take a trial wavefunction for the ground state, such as the one obtained in a variational quantum Monte Carlo simulation, as the initial wavefunction $\Psi(\mathbf{R}, 0) = \Phi(\mathbf{R})$. We can easily show that the time-dependent wavefunction is given by

$$\Psi(\mathbf{R}, t) = e^{-(\mathcal{H} - E_c)t} \Phi(\mathbf{R}). \qquad (9.83)$$

Then we can construct a probability-like function

$$F(\mathbf{R}, t) = \Psi(\mathbf{R}, t)\Phi(\mathbf{R}). \qquad (9.84)$$

We can show that $F(\mathbf{R}, t)$ satisfies the diffusion equation

$$\frac{\partial F}{\partial t} = \frac{1}{2}\nabla^2 F - \nabla \cdot F\mathbf{U} - [E_c - E(\mathbf{R})]F, \qquad (9.85)$$

with

$$\mathbf{U} = \nabla \ln \Phi(\mathbf{R}) \qquad (9.86)$$

being the drifting velocity and

$$\epsilon(\mathbf{R}) = \Phi^{-1}(\mathbf{R})\mathcal{H}\Phi(\mathbf{R}) \qquad (9.87)$$

being the local energy for a given $\mathbf{R}$. If we introduce an expectation value

$$E(t) = \frac{\langle \Phi(\mathbf{R})|\mathcal{H}|\Psi(\mathbf{R}, t)\rangle}{\langle \Phi(\mathbf{R}')|\Psi(\mathbf{R}', t)\rangle} = \frac{\int F(\mathbf{R}, t)\epsilon(\mathbf{R})\,d\mathbf{R}}{\int F(\mathbf{R}', t)\,d\mathbf{R}'} \qquad (9.88)$$

at time t, we can obtain the true ground-state energy

$$E_0 = \lim_{t \to \infty} E(t). \qquad (9.89)$$

The multidimensional integral in Eq. (9.88) can be carried out by means of the Monte Carlo quadrature with $F(\mathbf{R}, t)$ treated as a time-dependent distribution function if it is positive definite. In practice, the simulation is done by rewriting the diffusion equation for $F(\mathbf{R}, t)$ into an integral form

$$F(\mathbf{R}', t + \tau) = \int F(\mathbf{R}, t) G(\mathbf{R}', \mathbf{R}; \tau) \, d\mathbf{R}, \tag{9.90}$$

where $G(\mathbf{R}', \mathbf{R}; \tau)$ is the Green's function of the diffusion equation. If τ is very small, Green's function can be approximated as

$$G(\mathbf{R}', \mathbf{R}; \tau) \simeq \mathcal{W}(\mathbf{R}', \mathbf{R}; \tau) G_0(\mathbf{R}', \mathbf{R}; \tau), \tag{9.91}$$

where

$$G_0(\mathbf{R}', \mathbf{R}; \tau) = \left(\frac{1}{2\pi\tau}\right)^{3N/2} e^{-[\mathbf{R}' - \mathbf{R} - \mathbf{U}\tau]^2/2\tau} \tag{9.92}$$

is a propagator due to the drifting and

$$\mathcal{W}(\mathbf{R}', \mathbf{R}; \tau) = e^{-\{[E(\mathbf{R}) + E(\mathbf{R}')]/2 - E_c(t)\}\tau} \tag{9.93}$$

is a branching factor.

   In order to treat $F(\mathbf{R}, t)$ as a distribution function or a probability, we have to have it positive definite. A common practice is to use the *fixed-node approximation*, which forces the function $F(\mathbf{R}, t)$ to be zero in case it becomes negative. This fixed-node approximation still provides an upper bound for the evaluation of the ground-state energy and gives good molecular energies for small systems if the trial wavefunction is properly chosen. The fixed-node approximation can be removed by releasing the fixed nodes, but this will require additional computing efforts (Ceperley and Alder, 1984). Here is a summary of the major steps in the actual Green's function Monte Carlo simulation.

   (1) We first perform variational quantum Monte Carlo simulations to optimize the variational parameters in the trial wavefunction.
   (2) Then we can generate an initial ensemble with many independent configurations from the variational simulations.
   (3) Each configuration is updated with a drifting term and a Gaussian random walk, with the new coordinate

$$\mathbf{R}' = \mathbf{R} + \mathbf{U}\tau + \chi, \tag{9.94}$$

   where χ is a $3N$-dimensional Gaussian random number generated in order to have a variance of τ on each dimension.
   (4) Not every step is accepted, and a probability

$$p = \min[1, w(\mathbf{R}', \mathbf{R}; \tau)] \tag{9.95}$$

is used to judge whether the updating move should be accepted or not. Here

$$w(\mathbf{R}', \mathbf{R}, \tau) = \frac{\Phi(\mathbf{R}')^2 G(\mathbf{R}, \mathbf{R}'; \tau)}{\Phi(\mathbf{R})^2 G(\mathbf{R}', \mathbf{R}; \tau)} \tag{9.96}$$

is necessary in order to have detailed balance between points $\mathbf{R}'$ and $\mathbf{R}$. Any move that crosses a node is rejected under the fixed-node approximation.

(5) A new ensemble is then created with branching; that is,

$$M = [\mathcal{W}(\mathbf{R}', \mathbf{R}; \tau_a) + \xi] \tag{9.97}$$

copies of the configuration $\mathbf{R}'$ are put in the new ensemble. Here ξ is a uniform random number between $[0, 1]$ that is used to make sure that the fraction of $\mathcal{W}$ is properly taken care of. τ_a is the effective diffusion time, which is proportional to τ, with the coefficient being the ratio of the mean square distance of the accepted moves to the mean square distance of the attempted moves.

(6) The average local energy at each time step $\bar{\epsilon}(\mathbf{R}')$ is then evaluated from the summation of $\epsilon(\mathbf{R}')$ of each configuration and weighted by the corresponding probability $\mathcal{W}(\mathbf{R}', \mathbf{R}; \tau_a)$. The time-dependent energy $E_c(t)$ is updated at every step with

$$E_c(t) = \frac{\bar{\epsilon}(\mathbf{R}) + \bar{\epsilon}(\mathbf{R}')}{2} \tag{9.98}$$

to ensure a smooth convergence.

Before one can take the data for the calculation, one needs to run the above steps enough times so that the error is dominated by statistics. The data itself is typically taken with an interval of ten steps. The exact size of the interval can be determined from the autocorrelation function of the physical quantities evaluated in the simulation. One can then group and average the data to the desired accuracy. Several independent runs can be carried out for a better average.

Another scheme that samples Green's function directly does not have the problems of finite time step or fixed-node approximation. This method was originally proposed by Kalos (Kalos, 1962). This method is based on the time-independent Green's function formalism. One can formally write the solution of the many-body Hamiltonian in an integral form

$$\Psi^{(n+1)}(\mathbf{R}) = E_c \int G(\mathbf{R}, \mathbf{R}') \Psi^{(n)}(\mathbf{R}') \, d\mathbf{R}', \tag{9.99}$$

with E_c being a trial energy of the system and $G(\mathbf{R}, \mathbf{R}')$ being the Green's function of the Hamiltonian given by

$$\mathcal{H} G(\mathbf{R}, \mathbf{R}') = \delta(\mathbf{R} - \mathbf{R}'), \tag{9.100}$$

where $G(\mathbf{R}, \mathbf{R}')$ also satisfies the boundary condition. One can show that the ground state of the Hamiltonian is recovered for $n = \infty$,

$$\Psi_0(\mathbf{R}) = \Psi^{(\infty)}(\mathbf{R}). \tag{9.101}$$

In order to have the desired convergence, E_c has to be a positive quantity, which can be ensured at least locally by adding a constant to the Hamiltonian. It can be shown for sufficiently large values of n that

$$\Psi^{(n+1)}(\mathbf{R}) \simeq \frac{E_c}{E_0} \Psi^{(n)}(\mathbf{R}). \tag{9.102}$$

The time-independent Green's function is related to the time-dependent Green's function by

$$G(\mathbf{R}, \mathbf{R}') = \int_0^\infty G(\mathbf{R}, \mathbf{R}'; t) \, dt, \tag{9.103}$$

which can be sampled exactly in principle. In practice, it will usually take a lot of computing time to reach a stable result. For more detail on how to sample Green's function, see Ceperley and Kalos (1986) or Lee and Schmidt (1992).

9.7 Path-integral Monte Carlo simulations

So far we have been discussing the study of the ground-state properties of quantum many-body systems. In reality, the properties of excited states are as important as those of the ground state, especially when we are interested in the temperature-dependent behavior of the system. In this section, we briefly discuss a finite-temperature quantum Monte Carlo simulation scheme introduced by Pollock and Ceperley (1984). If the system is in equilibrium, the properties of the system are generally given by the density matrix of the system,

$$\rho(\mathbf{R}, \mathbf{R}'; \beta) = \langle \mathbf{R}' | e^{-\beta \mathcal{H}} | \mathbf{R} \rangle, = \sum_n e^{-\beta E_n} \Psi_n^+(\mathbf{R}') \Psi_n(\mathbf{R}), \tag{9.104}$$

where $\beta = 1/k_B T$ is the inverse temperature and $\mathcal{H}$, E_n, and Ψ_n are related through the Schrödinger equation of the system,

$$\mathcal{H}\Psi_n(\mathbf{R}) = E_n \Psi_n(\mathbf{R}). \tag{9.105}$$

The thermodynamic quantities of the system are related to the density matrix through the partition function

$$\mathcal{Z} = \mathrm{Tr}\, \rho(\mathbf{R}, \mathbf{R}'; \beta), \tag{9.106}$$

which is the sum of all diagonal elements of the density matrix. The density matrix, as well as the partition function, satisfies a convolution integral

$$\rho(\mathbf{R}, \mathbf{R}'; \beta) = \int \rho(\mathbf{R}, \mathbf{R}_1; \tau)\rho(\mathbf{R}_1, \mathbf{R}_2; \tau) \cdots \rho(\mathbf{R}_M, \mathbf{R}'; \tau)$$

$$\times \, d\mathbf{R}_1 d\mathbf{R}_2 \cdots d\mathbf{R}_M, \qquad (9.107)$$

with $\tau = \beta/(M + 1)$. If M is a very large number, $\rho(\mathbf{R}_i, \mathbf{R}_j; \tau)$ approaches the high temperature limit with the effective temperature as $T_\tau = 1/k_B\tau \simeq MT$. One can formally show that the quantum density matrix written in the convolution integral is equivalent to a classical polymer system, which is one dimension higher in geometry. For example, the point particles of a quantum system therefore behave like classical polymer chains with the bonds and beads connected around each true particle. This can be viewed from the path-integral representation for the partition function at the limit of $M \to \infty$,

$$\mathcal{Z} = \int e^{-S[\mathbf{R}(t)]} \, \mathcal{D}\mathbf{R}(t), \qquad (9.108)$$

with

$$S[\mathbf{R}(t)] = \int_0^\beta \left[\frac{1}{2}\left(\frac{d\mathbf{R}}{dt}\right)^2 - V\mathbf{R}(t) \right] dt, \qquad (9.109)$$

and

$$\mathcal{D}\mathbf{R}(\tau) = \lim_{M \to \infty} \frac{1}{(2\pi\tau)^{3N(M+1)/2}} d\mathbf{R}_1 \, d\mathbf{R}_2 \cdots d\mathbf{R}_M. \qquad (9.110)$$

Note that the path integral in Eq. (9.108) is purely symbolic. For more details on path-integral representation, especially its relevance to polymers, see Wiegel (1986).

If one can obtain the density matrix of the system at the high temperature limit, the density matrix and other physical quantities at the given temperature are given by $3N \times M$-dimensional integrals, which can be sampled, in principle, with the Metropolis algorithm. However, in order to have a reasonable speed of convergence, one cannot simply use

$$\rho(\mathbf{R}, \mathbf{R}'; \tau) = \frac{1}{(2\pi\tau)^{3N(M+1)/2}} \exp\left[-\frac{(\mathbf{R} - \mathbf{R}')^2}{2\tau} + \tau V(\mathbf{R}) \right], \qquad (9.111)$$

which is the density matrix at the limit of $M \to \infty$ or $\tau = \beta/(M + 1) \to 0$ in the Feynman path integral representation, in the simulation. A good choice seems to be the pair-product approximation (Barker, 1979)

$$\rho(\mathbf{R}, \mathbf{R}'; \tau) \simeq \rho_0(\mathbf{R}, \mathbf{R}'; \tau) e^{-U(\mathbf{R}, \mathbf{R}'; \tau)}, \tag{9.112}$$

with

$$U(\mathbf{R}, \mathbf{R}'; \tau) = \sum_{i<j} u(r_{ij}, r'_{ij}; \tau), \tag{9.113}$$

which is the pair correction to the free particle density matrix

$$\rho_0(\mathbf{R}, \mathbf{R}'; \tau) = \frac{1}{(2\pi\tau)^{3N(M+1)/2}} \exp\left[-\frac{(\mathbf{R} - \mathbf{R}')^2}{2\tau}\right]. \tag{9.114}$$

Another important issue is that all quantum systems satisfy specific quantum statistics. Fermions and bosons behave totally differently when two particles in the system are interchanged. The general expression of the symmetrized or antisymmetrized density matrix can be written as

$$\rho_\pm(\mathbf{R}, \mathbf{R}'; \beta) = \frac{1}{N!} \sum_P (\pm 1)^P \rho(\mathbf{R}, P\mathbf{R}'; \beta), \tag{9.115}$$

where P is an indicator of permutations. This density matrix with permutations has been used in the study of liquid ^{4}He (Ceperley and Pollock, 1986), liquid ^{3}He (Ceperley, 1992), and liquid helium mixtures (Boninsegni and Ceperley, 1995). For a recent review on the subject, see Ceperley (1995).

Different Metropolis sampling schemes have also been developed for path-integral Monte Carlo simulations. Interested readers should consult Schmidt and Ceperley (1992). One can also find the discussion of a new scheme devised specifically for many-fermion systems in Lyubartsev and Vorontsov-Velyaminov (1993).

9.8 Quantum lattice models

Various quantum Monte Carlo simulation techniques discussed in the last few sections have also been applied in the study of quantum lattice models, such as the quantum Heisenberg, Hubbard, and t–J models. These studies are playing a very important role in the understanding of highly correlated materials systems that have great potential for future technology.

In this section, we will give a very brief introduction to the subject so that interested readers can get started before going into the vast literature. We will use the Hubbard model as an illustrative example to highlight the key elements in these approaches and also point out the difficulties encountered in current research. We used the Hubbard model to describe the electronic structure of H_3^+ in Eq. (4.13). The single-band Hubbard model for a general electronic system is given by the Hamiltonian

$$\mathcal{H} = -t \sum_{\langle ij \rangle}^{N} (a_{i\uparrow}^{+} a_{j\uparrow} + a_{i\downarrow}^{+} a_{j\downarrow}) + U \sum_{i=1}^{N} \left(n_{i\uparrow} - \frac{1}{2} \right)$$

$$\times \left(n_{i\downarrow} - \frac{1}{2} \right) + \mu \sum_{i=1}^{N} n_i. \qquad (9.116)$$

Here $a_{i\sigma}^{+}$ and $a_{j\sigma}$ are creation and annihilation operators of an electron with either spin up ($\sigma = \uparrow$) or down ($\sigma = \downarrow$) and $n_{i\sigma} = a_{i\sigma}^{+} a_{i\sigma}$ is the corresponding occupancy at the ith site; t is the hopping strength of an electron between the nearest sites (notated as $\langle ij \rangle$); U is the on-site interaction strength of two electrons with opposite spin orientations; and μ is the chemical potential of the system. The zero point of μ can be selected at a specific average occupancy per site $\langle n_i \rangle$ with $n_i = n_\uparrow + n_\downarrow$, for example, $\langle n_i \rangle = 2/3$ for the Hamiltonian in Eq. (4.13) and $\langle n_i \rangle = 1/2$ for the Hamiltonian in Eq. (9.116). The Hubbard model is extremely appealing, since it is very simple but contains almost all the information in the highly correlated systems. An exact solution of this model is only available in one dimension. The typical quantities to be studied are spin correlation function, charge correlation function, and excitation spectra. The spin–spin correlation function is defined as

$$C_s(l) = \langle \mathbf{s}_i \cdot \mathbf{s}_{i+l} \rangle, \qquad (9.117)$$

where the spin operator is given by

$$\mathbf{s}_i = \frac{\hbar}{2} \sum_{\sigma \nu} a_{i\sigma}^{+} \mathbf{p}_{\sigma \nu} a_{i\nu}, \qquad (9.118)$$

with $\mathbf{p}_{\sigma \nu}$ being the Pauli matrices,

$$p^x = \begin{pmatrix} 0 & 1 \\ 1 & 0 \end{pmatrix}, \quad p^y = \begin{pmatrix} 0 & -i \\ i & 0 \end{pmatrix}, \quad p^z = \begin{pmatrix} 1 & 0 \\ 0 & -1 \end{pmatrix}. \qquad (9.119)$$

The particle–particle correlation function is defined as

$$C_n(l) = \langle \Delta n_i \Delta n_{i+l} \rangle, \qquad (9.120)$$

where $\Delta n_i = n_i - n_0$ is the difference of the occupancy of the ith site and the average occupancy n_0. These and other correlation functions are very useful in understanding the physical properties of the system. For example, a solid state is formed if the density correlation function has a long-range order and the (staggered) magnetization m is given by the long distance limit of the spin correlation function $m^2 = \frac{1}{3} C_s(l \to \infty)$. These correlation functions can be evaluated up to a point by quantum Monte Carlo simulations.

Variational simulation

The variational quantum Monte Carlo simulation is a combination of the variational scheme and the Metropolis algorithm. One first needs to construct a trial wavefunction that contains as much of the relevant physics as possible. It is usually very difficult to actually come up with a state that is concise but contains all the important physics. It has happened only three times in the entire history of the many-body theory: in 1930, when Bethe proposed the Bethe ansatz for the spin-1/2 one-dimensional Heisenberg model; in 1957, when Bardeen, Cooper, and Schrieffer proposed the BCS ansatz for the ground state of superconductivity, and in 1983, when Laughlin proposed the Laughlin ansatz for the fractional quantum Hall state. It turns out that the Bethe ansatz is exact for the spin-1/2 chain and has since been applied to many other one-dimensional systems.

Since there is no such an extraordinary trial state for the two- or three-dimensional Hubbard model, we will take the best-known trial state so far, the Gutzwiller ansatz

$$|\Phi\rangle = \prod_{i=1}^{N}[1 - (1 - g)n_{i\uparrow}n_{i\downarrow}]|\Phi_0\rangle, \qquad (9.121)$$

to illustrate the scheme. Here g is the only variational parameter in the trial wavefunction and $|\Phi_0\rangle$ is a reference state, which is usually chosen as the uncorrelated state of the Hamiltonian, that is, the solution with $U = 0$, the single-band tight-binding model. This trial state is used by Vollhardt (1984) in the description of the behavior of the normal state of liquid ^{3}He. Some other variational wavefunctions have also been developed for the Hubbard model, but so far there has not been a unique description because of lack of fundamental understanding of the model in two dimensions. We can show that the Gutzwiller ansatz can be rewritten into

$$|\Phi\rangle = g^D|\Phi_0\rangle, \qquad (9.122)$$

where D is the number of sites with double occupancy. The above variational wavefunction can be generalized into a Jastrow type of ansatz

$$|\Phi\rangle = e^{-U[n_i]/2}|\Phi_0\rangle, \qquad (9.123)$$

with the correlation factor given by

$$U[n_i] = \sum_{j,k}^{N}\alpha_l(j,k)n_jn_k, \qquad (9.124)$$

where $\alpha_l(j,k)$ is a variational parameter between the jth site and the kth site. The spin index is combined into the site index for convenience, and the summation over j and k should be interpreted as also including spin degrees of freedom. The index $l = 1, 2$ is for the different spin configuration of the ith and jth sites with two distinct situations: spin parallel and spin antiparallel. The above Jastrow type of ansatz reduces back to the Gutzwiller ansatz when $\alpha_l(j,k)$ is only for the on-site occupancy with $\alpha = -\ln(1/g)$.

The simulation is performed with optimization of the energy expectation

$$E = \frac{\langle \Phi | \mathcal{H} | \Phi \rangle}{\langle \Phi | \Phi \rangle}. \tag{9.125}$$

Several aspects need special care during the simulation. First, the total spin of the system commutes with the Hamiltonian, so we cannot change the total spin of the system; that is, we can only move the particles around but not flip their spins, or flip a spin-up and spin-down pair at the same time. The optimized g is a function of t/U and n_0, so we can choose t as the unit of energy, that is, $t = 1$, and then g will be a function of U and n_0 only. Each site cannot be occupied by more than two particles with opposite spins. So we can move one particle to its nearest site if allowed, and the direction of the move can be determined by a random number generator. For example, if we label the nearest sites $1, 2, \ldots, z$, we can move the particle to the site labeled $\text{int}(\eta_i z + 1)$ if allowed by the Pauli principle. Here η_i is a uniform random number in the region $[0, 1]$. The move is accepted by comparing

$$p = \frac{|\Phi_{\text{new}}|^2}{|\Phi_{\text{old}}|^2} = \frac{\exp\left(-U\left[n_i^{\text{new}}\right]\right)}{\exp\left(-U\left[n_i^{\text{old}}\right]\right)} \tag{9.126}$$

with another uniform random number w_i. Other aspects are similar to the variational quantum Monte Carlo simulation discussed earlier in this chapter for continuous systems.

The Green's function Monte Carlo simulation

One can also use the Green's function Monte Carlo to study quantum lattice systems. As we discussed earlier, the solution of the imaginary-time Schrödinger equation can be expressed in terms of an integral equation

$$\Psi(\mathbf{R}, \mathbf{R}', t + \tau) = \int G(\mathbf{R}, \mathbf{R}', \tau)\Psi(\mathbf{R}', t)\,d\mathbf{R}', \tag{9.127}$$

where $\mathbf{R}$ or $\mathbf{R}'$ is a set of locations and spin orientations of all the particles.

So far the Green's function Monte Carlo simulation has been performed for only lattice spin problems, especially the spin-1/2 antiferromagnetic Heisenberg model on a square lattice, which is relevant to the normal state of the cuprates, which become superconducting with high transition temperatures after doping. One can show formally that this model is the limit of the Hubbard model with half-filling and infinite on-site interaction. The Heisenberg model is given by

$$\mathcal{H} = J \sum_{\langle ij \rangle}^{N} \mathbf{s}_i \cdot \mathbf{s}_j, \tag{9.128}$$

with $J = 4t^2/U$ being a positive constant for the antiferromagnetic case. This model can formally be transformed into a hard-core lattice boson model

$$\mathcal{H} = -J \sum_{\langle ij \rangle}^{N} b_i^+ b_j + J \sum_{\langle ij \rangle}^{N} n_i n_j + E_0, \tag{9.129}$$

with $b_i^+ = s_i^+ = s_i^x + i s_i^y$, $b_i = s_i^- = s_i^x - i s_i^y$, and $n_i = b_i^+ b_i = 1/2 - s_i^z$. The constant $E_0 = -Jz(N - N_b)/4$ with N being the total number of sites in the system, $N_b = N/2 - S^z$ being the number of occupied sites, and z being the number of nearest neighbors of each lattice point. S^z is the total spin along the z direction. As we discussed in Section 9.6, if we choose the time step τ to be very small, we can use the short-time approximation for Green's function

$$G(\mathbf{R}, \mathbf{R}', \tau) = \langle \mathbf{R} | e^{-\tau(\mathcal{H} - E_c)} | \mathbf{R}' \rangle \simeq \langle \mathbf{R} | [1 - \tau(\mathcal{H} - E_c)] | \mathbf{R}' \rangle, \tag{9.130}$$

which has a very simple form under the boson representation,

$$G(\mathbf{R}, \mathbf{R}', \tau) = \begin{cases} 1 - \tau[V(\mathbf{R}) - E_c] & \text{for} \quad \mathbf{R} = \mathbf{R}', \\ \tau J/2 & \text{for} \quad \mathbf{R} = \mathbf{R}' + \Delta \mathbf{R}', \\ 0 & \text{otherwise,} \end{cases} \tag{9.131}$$

where $\mathbf{R}' + \Delta \mathbf{R}'$ is a position vector with one particle moved to the nearest neighbor site in $\mathbf{R}'$ and $V(\mathbf{R})$ is the interaction part, that is, the second term in the Hamiltonian with a configuration $\mathbf{R}$. A procedure similar to that we discussed for continuous systems can be devised. Readers who are interested in the details of the Green's function Monte Carlo simulation for quantum lattice systems should consult the original articles, for example, Trivedi and Ceperley (1990).

Finite-temperature simulation

If we want to know the temperature dependence of the system, we need to simulate it at finite temperature. The starting point is the partition function

$$\mathcal{Z} = \text{Tr}\, e^{-\beta \mathcal{H}}, \tag{9.132}$$

where $\mathcal{H}$ is the lattice Hamiltonian, for example, the Hubbard model, and $\beta = 1/k_B T$ is the inverse temperature. The average of a physical quantity A is measured by

$$\langle A \rangle = \frac{1}{\mathcal{Z}} \mathrm{Tr}\, A e^{-\beta \mathcal{H}}. \tag{9.133}$$

If we divide the inverse temperature into $M + 1$ slices with $\tau = \beta/(M + 1)$, we can rewrite the partition function into a product

$$\mathcal{Z} = \mathrm{Tr}(e^{-\tau \mathcal{H}})^{M+1}. \tag{9.134}$$

In the limit of $\tau \to 0$, one can show that

$$e^{-\tau \mathcal{H}} \simeq e^{-\tau V} e^{-\tau \mathcal{H}_0}, \tag{9.135}$$

where V is the potential part of the Hamiltonian,

$$V = U \sum_{i=1}^{N} \left(n_{i\uparrow} - \frac{1}{2} \right) \left(n_{i\downarrow} - \frac{1}{2} \right) + \mu \sum_{i=1}^{N} n_i, \tag{9.136}$$

and $\mathcal{H}_0$ is the contribution of all the hopping integrals,

$$\mathcal{H}_0 = -t \sum_{\langle ij \rangle}^{N} a_{i\sigma}^{+} a_{j\sigma}. \tag{9.137}$$

Hirsch (1985a; 1985b) has showed that the discrete Hubbard–Stratonovich transformation can be used to rewrite the quartic term of the interaction into a quadratic form; that is,

$$e^{-\tau U(n_{i\uparrow} - 1/2)(n_{i\downarrow} - 1/2)} = e^{-\tau U/4} \sum_{\sigma_l = \pm 1} e^{-\tau \sigma_l \lambda (n_{i\uparrow} - n_{i\downarrow})}, \tag{9.138}$$

with λ given by

$$\cosh \tau \lambda = e^{\tau U/2}. \tag{9.139}$$

Since the kinetic energy part $\mathcal{H}_0$ has only the quadratic term, it is ready now to have the partition function simulated with the Metropolis algorithm. Readers who are interested in more details can find a good discussion of the method in Sugar (1990).

There is a fundamental problem for the Fermi systems, called the *fermion sign problem*. The sign problem appears in most Fermi systems. It is the result of the fundamental properties of the nonlocal-exchange interaction or the Pauli principle. It may appear in different forms, such as the nodal structure in the wavefunction or negative probability in the stochastic simulation. However, the challenge lies in the fundamental aspect of many-body systems with Fermi statistics. Interested readers will be able to find many discussions, but not a real solution yet, in current literature.

Exercises

9.1 Show that the Monte Carlo quadrature yields a standard deviation

$$\Delta^2 = \langle A^2 \rangle - \langle A \rangle^2 \propto \frac{1}{M},$$

where A is a physical observable and M is the number of Monte Carlo points selected.

9.2 Show that the Metropolis algorithm applied to statistical mechanics fulfills detailed balance and samples the points according to the distribution function.

9.3 Modify the program in Section 9.2 for the one-dimensional integral with the Metropolis algorithm to a three-dimensional problem and test it with the integral

$$S = \int_{-\infty}^{\infty} e^{-r^2/2} x^2 y^2 z^2 d\mathbf{r}.$$

Compare the Monte Carlo result with the exact result and analyze the Monte Carlo error bar with the number of points used.

9.4 Calculate the autocorrelation function of the integral in Exercise 9.3 numerically. From the plot of the autocorrelation function, determine how many points need to be skipped between any two data points in order to have nearly uncorrelated data points.

9.5 Develop a Monte Carlo program for the two-dimensional Ising model on a square lattice. For simplicity, the system can be chosen as an $N \times N$ square with the periodic boundary condition imposed. Update the spin configuration site by site. Study the temperature dependence of the magnetization. Is there any critical slowing down?

9.6 Implement the Swendsen–Wang algorithm in the program developed in Exercise 9.5. Is there any improvement over the single-site update scheme?

9.7 Implement the Wolff algorithm in the two-dimensional Ising model studied in Exercises 9.5 and 9.6 and compare the result with that of the Swendsen–Wang algorithm.

9.8 Study the electronic structure of the helium atom with the variational quantum Monte Carlo method. Assume that the nucleus is fixed at the origin of the coordinates. The key is to find a good parameterized variational wavefunction with proper cusp conditions built in.

9.9 Implement the Green's function Monte Carlo algorithm in the study of a hydrogen molecule. The two-proton and two-electron system should be treated as a four-body system. Calculate the ground-state energy of the system and compare the result with the best known calculation. Would the result obtained here be the best if a few fast workstations were available?

9.10 The structure of the liquid ^{4}He can be studied with the variational quantum Monte Carlo method. Develop a program to calculate the ground-state energy per particle and the pair correlation function with the selected variational wavefunction.

9.11 Modify the variational quantum Monte Carlo program developed in Exercise 9.10 with the Green's function Monte Carlo algorithm. Is there any significant improvement in the calculated ground-state energy?

10

Numerical renormalization

Renormalization is a method that was first introduced in quantum field theory to deal with far-infrared divergence in Feynman diagrams. The idea was further explored in statistical physics to tackle the problems associated with adiabatic continuity problems in phase transitions and critical phenomena. So far, the renormalization method has been applied to many problems in physics, including chaos, percolation, critical phenomena, and quantum many-body problems.

Even though many aspects of the renormalization group have been studied, it is still unclear how to apply the method to study most highly correlated systems that we are interested in now, for example, the systems described by the two-dimensional Hubbard model. However, it is believed by some physicists that the idea of renormalization will eventually be applied to solve the typical quantum many-body problems we are encountering, especially with the availability of modern computers, new numerical algorithms, and fresh ideas for setting up proper renormalization schemes.

10.1 The scaling concept

The renormalization technique in statistical physics is based on the scaling hypothesis that the competition between the long-range correlation and the fluctuation of the order parameters is the cause of all the singular behavior of the relevant physical quantities at the critical point. The longest distance of the order parameter that stays correlated at a given temperature T is defined as the correlation length of the system, which diverges when the system approaches the critical temperature T_c:

$$\xi \propto t^{-\nu}, \tag{10.1}$$

where ξ is the correlation length, $t = |T - T_c|/T_c$ is the reduced temperature, and ν is the critical exponent of the correlation length. The scaling hypothesis states that the divergence of all the other quantities at the critical temperature are the result of the divergence of the correlation length. Based on this hypothesis, one can obtain several relations among the exponents of different physical quantities. Detailed discussions on the scaling hypothesis and exponent relations can be found in Ma (1976). Here we just want to give a brief discussion of the scaling concept introduced by Widom (1965) and Kadanoff (1966). Widom's idea is based on the fact that the thermodynamic potentials, such as the free energy, are homogeneous functions. If we take a spin system as an example, the free energy satisfies

$$F(t, B) = \Lambda^{-1} F(\Lambda^x t, \Lambda^y B), \tag{10.2}$$

where Λ is a parameter that changes the length scale of the system and x and y are two exponents determining how the reduced temperature t and the magnetic field strength B scale with Λ. The critical behavior of the magnetization m is characterized by two exponents β and δ as

$$m(t, B = 0) \propto t^\beta; \quad m(t = 0, B) \propto B^{1/\delta}. \tag{10.3}$$

Now let us show how we can derive the relations of the critical exponents from the assumption that the free energy is a homogeneous function. The magnetization is defined as

$$m(t, B) = -\frac{\partial F(t, B)}{\partial B}, \tag{10.4}$$

which leads to a rescaling relation of the magnetization,

$$m(t, B) = \Lambda^{y-1} m(\Lambda^x t, \Lambda^y B), \tag{10.5}$$

which in turn gives

$$\beta = \frac{1 - y}{x}; \quad \delta = \frac{y}{1 - y}, \tag{10.6}$$

if we take $\Lambda^x t = 1$ and $B = 0$, and $t = 0$ and $\Lambda^y B = 1$. Similarly, we can obtain the exponent γ of the susceptibility from the definition

$$\chi(t, B = 0) = \left. \frac{\partial m(t, B)}{\partial B} \right|_{B=0} = \Lambda^{2y-1} \chi(\Lambda^x t, 0) \propto t^{-\gamma}, \tag{10.7}$$

which gives

$$\gamma = \frac{2y - 1}{x} = \beta(\delta - 1), \tag{10.8}$$

if we take $\Lambda^x t = 1$. Following exactly the same process, we can obtain the exponent α for the specific heat

$$C(t, B = 0) = \left. \frac{\partial^2 F(t, B)}{\partial t^2} \right|_{B=0} \propto t^{-\alpha}, \tag{10.9}$$

which leads to

$$\alpha = 2 - \frac{1}{x} = 2 - \beta(\delta + 1). \tag{10.10}$$

The above relations indicate that if we know x and y, we know β, δ, γ, and α. In other words, if we know any two exponents, we know the rest. In most cases, the behavior on both sides of the critical point is symmetric, so we do not need to distinguish the two limiting processes of approaching the critical point. However, one has to realize that this may not always be the case, so a more detailed analysis should consider them separately.

How can one relate the critical exponents of the thermodynamic quantities discussed above to the fundamental exponent of the correlation length? It was Kadanoff (1966) who first introduced the block Hamiltonian concept and then related the exponent of the correlation length to the critical exponents of the thermodynamic quantities. The idea of Kadanoff is based on the observation that the correlation length will diverge at the critical point, so the details of the behavior of the system at shorter length do not matter very much. Thus, we can imagine a process in which the spins in a block of dimension Λ point in more or less the same direction so we can treat the whole block like a single spin. The free energy then scales as

$$F(\Lambda^a t, \Lambda^b B) = \Lambda^d F(t, B), \tag{10.11}$$

with a and b the rescaling exponents for t and B, respectively, and d the number of dimensions of the lattice. The above relation shows that

$$a = xd; \quad b = yd, \tag{10.12}$$

in comparison with the Widom scaling based on the homogeneous function assumption. The spin correlation function is defined as

$$G(r_{ij}, t) = \langle s_i s_j \rangle - \langle s_i \rangle \langle s_j \rangle, \tag{10.13}$$

which satisfies a scaling relation

$$G(r_{ij}, t) = \Lambda^{2(b-d)} G(\Lambda^{-1} r_{ij}, \Lambda^a t), \tag{10.14}$$

which leads to

$$\nu = \frac{1}{xd} = \frac{2 - \alpha}{d}. \tag{10.15}$$

The exponent of the correlation function with the distance is given by

$$G(r_{ij}, t = 0) \propto \left(\frac{1}{r_{ij}}\right)^{d-2-\eta}, \tag{10.16}$$

which is related to the other exponents by

$$\eta = 2 - \frac{d(\delta - 1)}{\delta + 1}. \tag{10.17}$$

In the next section we will show how one can construct a renormalization scheme, the mathematical form of the Kadanoff block spin transformation, and then show how the exponents are related to the eigenvalues of the linear representation of the transformation around the critical point.

10.2 Renormalization transform

As argued by Wilson (1975), the physical quantities of statistical physics are controlled by fluctuations at all length scales up to the correlation length ξ. These fluctuations come into each physical observable with some influence at each different energy scale, and the renormalization scheme is aimed at finding the structure of the hierarchy. One way to obtain the structure of the hierarchy in the spin system is through the Kadanoff transform (Kadanoff, 1966), which partitions the lattice and forms a new lattice with the definition of the block variables. This procedure is repeated until a stable fixed point is reached. Intuitively, one can view this transform as a process for recovering the long-range behavior of the system. For example, when the system changes from a paramagnetic state to a ferromagnetic state, at any length scale, the magnetization is more or less the same. In other words, the spins tend to line up to form ferromagnetic domains, and the domains will continue to line up to form larger domains until the whole system becomes one big ferromagnetic domain. This transform can be formulated mathematically into a semigroup, that is, a series of operations without an inverse. To illustrate how this is done, assume that the system is the d-dimensional Ising spin system with a length L along each direction. We can first divide the system into many blocks, and a block spin s_i can then be formulated from all the lattice spins in that block:

$$s_i = g\{s_{i\sigma}\}, \tag{10.18}$$

where we have used $i = 1, 2, \ldots, (L/\Lambda)^d$, with Λ as the length of the block. $\{s_i\}$ is used to denote a configuration of s_i with $i = 1, 2, 3, \ldots$ as a generic index for the sites involved, and σ is used to label spins in a block, that is, $\sigma = 1, 2, \ldots, (\Lambda/a)^d$, with a being the lattice constant. The specific function form of $g\{s_{i\sigma}\}$ is not always simple; the choice has to reflect the physical process of the system. For the Ising model, a common practice is to use the majority rule: The block spin is assigned $+1$ (-1) if the majority of spins in the block are up (down). In case the block has the same number of up and down sites, one can assign either $+1$ or -1 to the block spin with an equal probability. The transform is continued by taking the block spins as the new lattice spins, that is,

$$s_i^{(n)} = g\{s_{i\sigma}^{(n)}\},$$ (10.19)

with $s_i^{(0)}$ being the original lattice spins. To simplify our notation, we will use $\mathbf{S} = (s_1, s_2, s_3, \ldots)$ for a total spin configuration. The so-called renormalization transform is defined with the new distribution function

$$\mathcal{W}^{(n+1)}(\mathbf{S}^{(n+1)}) = \frac{1}{\mathcal{Z}^{(n+1)}} \sum_\sigma \delta[\mathbf{S}^{(n+1)} - \mathbf{g}(\mathbf{S}_\sigma^{(n)})] e^{-\mathcal{H}^{(n)}(\mathbf{S}^{(n)})}$$ (10.20)

where the effective Hamiltonian $\mathcal{H}$ is also a function of the coupling constant and the external field. Temperature is absorbed into the coupling constant and the external field. The partition function at each iteration of the transform is given by

$$\mathcal{Z}^{(n)} = \sum_{\mathbf{S}^{(n)}} e^{-\mathcal{H}^{(n)}(\mathbf{S}^{(n)})}.$$ (10.21)

One can easily show that the above transform ensures the average of an observable $A(\mathbf{S}^{(n+1)})$, that is,

$$\langle A \rangle = \sum_{\mathbf{S}^{(n+1)}} A(\mathbf{S}^{(n+1)})\mathcal{W}(\mathbf{S}^{(n+1)}) = \sum_{\mathbf{S}^{(n)}} A(\mathbf{S}^{(n+1)})\mathcal{W}(\mathbf{S}^{(n)}),$$ (10.22)

from the definition of the transform. The block Hamiltonian $\mathcal{H}^{(n+1)}$ is completely given by $\mathcal{H}^{(n)}$ through the transform up to a constant. Symbolically, we can write all the parameters in the Hamiltonian as the components of a vector $\mathbf{H}$, and the transform can be expressed as a matrix multiplication

$$\mathbf{H}^{(n+1)} = \mathbf{T}\mathbf{H}^{(n)},$$ (10.23)

with $\mathbf{T}$ being the transform matrix defined through the distribution function with block variables. $\mathbf{T}$ is quite complex in most problems. A *fixed point* is found if the mapping goes to itself,

$$\mathbf{H}_0 = \mathbf{T}\mathbf{H}_0,$$ (10.24)

which can be one of several types. If the flow lines move toward the fixed point along all the directions, we call it a *stable fixed point*, which usually corresponds to $T = 0$ or $T = \infty$. If the flow lines move away from the fixed point, we call it an *unstable fixed point*. The critical point is a mixture of the unstable and stable behavior. Along some directions the flow lines move toward the fixed point, but along other directions the flow lines move away from the fixed point.

If we expand the above renormalization equation around the fixed point, we have

$$\mathbf{H}_0 + \delta\mathbf{H}' = \mathbf{T}(\mathbf{H}_0 + \delta\mathbf{H}).$$ (10.25)

When $\delta\mathbf{H}$ is very close to zero, we can carry out the Taylor expansion for $\mathbf{T}$ up to the linear terms,

$$\mathbf{T}(\mathbf{H}_0 + \delta\mathbf{H}) = \mathbf{T}(\mathbf{H}_0) + \mathbf{A}\delta\mathbf{H} + O(\delta\mathbf{H}^2), \tag{10.26}$$

where $\mathbf{A}$ is given by the partial derivatives

$$A_{ij} = \frac{\partial T_i(\mathbf{H}_0)}{\partial H_j}. \tag{10.27}$$

So at the neighborhood of the fixed point, we have

$$\delta\mathbf{H}' = \mathbf{A}\delta\mathbf{H}. \tag{10.28}$$

The eigenvalues λ_α of $\mathbf{A}$ from

$$\mathbf{A}\mathbf{x}_\alpha = \lambda_\alpha \mathbf{x}_\alpha \tag{10.29}$$

determine the behavior of the fixed point along the directions of the eigenvectors. If $\lambda_\alpha > 1$, then the fixed point along the direction of $\mathbf{x}_\alpha$ is unstable; otherwise, it is stable with $\lambda_\alpha < 1$ or marginal with $\lambda_\alpha = 1$.

The critical exponent can be related to the eigenvalue of the linearized transform around the critical point. For example, the critical exponent of the correlation length is given by

$$\nu = \frac{\log \Lambda}{\log \lambda_\sigma}, \tag{10.30}$$

where λ_σ is one of the eigenvalues of $\mathbf{A}$ that is greater than 1. This is derived from the fact that around the fixed point, that is, the critical point, the correlation function scales as

$$G(\delta\mathbf{H}) = \Lambda^{-d} G(\mathbf{A}\delta\mathbf{H}), \tag{10.31}$$

which can be used to relate Λ and λ_σ to the exponent x and then the exponent ν.

10.3 Critical phenomena: The Ising model

We will use the Ising model in two dimensions to illustrate several important aspects of the renormalization scheme outlined in the preceding section. The original work was carried out by Niemeijer and van Leeuwen (1974).

As we discussed earlier, the Ising model is given by the Hamiltonian

$$\mathcal{H} = -J \sum_{\langle ij \rangle}^{N} s_i s_j - B \sum_{i=1}^{N} s_i, \tag{10.32}$$

where J is the spin coupling strength, B is the external field, and i and j are nearest neighbors. For the Ising model, s_i is either 1 or -1. Assume that the lattice is a triangular lattice. Let us take the triangle of three nearest sites as the block and use the majority rule to define the block spin, that is, $s_i = +1$ if two or three spins in the ith block are up and $s_i = -1$ if two or three spins in the block are down. We will absorb the temperature into J and B. For notation simplicity, we will suppress the interaction indices when there is no confusion. The renormalization transform is given by

$$e^{-\mathcal{H}\{s_i\}} = \sum_{\sigma} e^{-\mathcal{H}\{s_{i\sigma}\}}, \tag{10.33}$$

with s_i being the block variables, which are given by

$$s_i = g\{s_{i\sigma}\} = \mathrm{sgn}(s_{i1} + s_{i2} + s_{i3}), \tag{10.34}$$

if the majority rule is adopted. s_i will be either 1 or -1 depending on whether the majority of spins are up or down. For each value of s_i there are four spin configurations of the three spins in the block: three from two spins pointing in the same direction and one from all three spins pointing in the same direction.

We can separate the lattice Hamiltonian into two parts, one is just for the intrablock spin interactions,

$$\mathcal{H}_0 = -J \sum_{i,\sigma,\mu} s_{i\sigma} s_{i\mu}, \tag{10.35}$$

and another is for the interblock interactions and the external field,

$$V = -J \sum_{\langle ij \rangle, \sigma, \mu} s_{i\sigma} s_{j\mu} - B \sum_{i,\sigma} s_{i\sigma}, \tag{10.36}$$

where $\langle ij \rangle$ indicates the summation over the nearest blocks. An expectation value under the intrablock part of the Hamiltonian with different block spin configurations,

$$\langle A \rangle_0 = \frac{1}{\mathcal{Z}_0} \sum_{\{s_i\}} A\{s_{i\sigma}\} \delta(s_i - g\{s_{i\sigma}\}) e^{-\mathcal{H}_0\{s_{i\sigma}\}}, \tag{10.37}$$

can be introduced for convenience. The partition function $\mathcal{Z}_0$ is the total contribution of the partition functions from all blocks,

$$\mathcal{Z}_0 = \sum_{\{s_i\}} \delta(s_i - g\{s_{i\sigma}\}) e^{-\mathcal{H}_0\{s_{i\sigma}\}} = z_0^M, \tag{10.38}$$

where z_0 is the partition function in a single block,

$$z_0 = \sum_{\sigma} e^{-\mathcal{H}_0\{s_{i\sigma}\}} = 3e^{-J} + e^{3J}, \tag{10.39}$$

and M is the total number of blocks.

With the definition of the average in Eq. (10.37) over the spin configurations from all the blocks, the renormalization transform can be written as

$$e^{-\mathcal{H}\{s_i\}} = \left[\sum_{\{s_i\}} \delta(s_i - g\{s_{i\sigma}\})e^{-\mathcal{H}_0\{s_{i\sigma}\}}\right]\langle e^{-V}\rangle_0, \tag{10.40}$$

where the first part is simply given by the partition function $\mathcal{Z}_0$ and the second part needs to be evaluated term by term in an expansion if one is interested in the analytic results. In the next section, we will explain how to use Monte Carlo simulations to obtain the averages needed for the renormalization transform. Equation (10.40) is the key for performing the renormalization transform, and one can easily show that it is correct by carrying out a Taylor expansion for the exponent.

We can formally write the average left as

$$\langle e^{-V}\rangle_0 = -\langle V\rangle_0 + \frac{1}{2}\langle V^2\rangle_0 - \cdots$$

$$= \exp\{-\langle V\rangle_0 - [\langle V^2\rangle_0 - \langle V\rangle_0^2]/2 + \cdots\}, \tag{10.41}$$

after the Taylor expansion of the exponential function and the resummation in terms of the moments of the statistical averages. The above equation then leads to the renormalization transform between the lattice Hamiltonian and the block Hamiltonian,

$$\mathcal{H}\{s_i\} = \ln \mathcal{Z}_0 + \langle V\rangle_0 - \frac{1}{2}[\langle V^2\rangle_0 - \langle V\rangle_0^2] + \cdots. \tag{10.42}$$

The right-hand side has an infinite number of terms, but we can truncate the series if the fixed points have small values of J and B.

Then we can work out the transform analytically as well as the exponents associated with the fixed point. We already know $\mathcal{Z}_0$ from Eqs. (10.38) and (10.39). $\langle V\rangle_0$ can then be evaluated from the definition. If we concentrate on the two nearest neighboring blocks, i and j, we have

$$\langle V_{ij}\rangle_0 = -J\langle s_{i1}s_{j2} + s_{i1}s_{j3}\rangle_0 = -2J\langle s_{i1}\rangle_0\langle s_{j2}\rangle_0, \tag{10.43}$$

due to the symmetry of each block, and the average is performed over the interactions within each individual block. The average over a specific spin can be easily computed with a fixed block spin; that is,

$$\langle s_{i\sigma}\rangle_0 = \frac{1}{z_0}\sum_{s_i} \delta[s_i - \text{sgn}(s_{i1} + s_{i2} + s_{i3})]e^{-J(s_{i1}s_{i2}+s_{i1}s_{i3}+s_{i2}s_{i3})}$$

$$= \frac{s_i}{z_0}(e^{3J} + e^{-J}). \tag{10.44}$$

If we put all this into the renormalization transform of Eq. (10.42), we have

$$
\begin{aligned}
\mathcal{H}^{(n+1)} & \{s_i^{(n+1)}\} \\
& = \ln Z_0^{(n)} - J^{(n+1)} \sum_{\langle ij \rangle} s_i^{(n+1)} s_j^{(n+1)} - B^{(n+1)} \sum_i s_i^{(n+1)},
\end{aligned}
\tag{10.45}
$$

with

$$
J^{(n+1)} \simeq 2 \left(\frac{e^{3J^{(n)}} + e^{-J^{(n)}}}{e^{3J^{(n)}} + 3e^{-J^{(n)}}} \right)^2 J^{(n)},
\tag{10.46}
$$

and

$$
B^{(n+1)} \simeq 3 \left(\frac{e^{3J^{(n)}} + e^{-J^{(n)}}}{e^{3J^{(n)}} + 3e^{-J^{(n)}}} \right) B^{(n)},
\tag{10.47}
$$

which can be used to construct the matrix for the fixed points. If we use the vector representation of the Hamiltonian parameters, we have $\mathbf{H} = (H_1, H_2) = (B, J)$. The fixed points are given by the invariants of the transform. Using Eqs. (10.46) and (10.47), we obtain $B_0 = 0$ and $J_0 = 0$, and $B_0 = 0$ and $J_0 = \ln(2\sqrt{2}+1)/4 \simeq 0.3356$. In order to study the behavior around these fixed points, we can linearize the transform as discussed in the preceding section with the matrix elements given by

$$
\mathbf{A} = \frac{\partial \mathbf{T}(\mathbf{H})}{\partial \mathbf{H}}.
\tag{10.48}
$$

For $(B_0, J_0) = (0, 0)$, we have

$$
\mathbf{A} = \begin{pmatrix} 1.5 & 0 \\ 0 & 0.5 \end{pmatrix},
\tag{10.49}
$$

which has the two eigenvalues $\lambda_B = 1.5 > 1$ and $\lambda_J = 0.5 < 1$, corresponding to the high-temperature limit, where the interaction is totally unimportant. For the other fixed point, with $(B_0, J_0) = (0, 0.3356)$, we have

$$
\mathbf{A} = \begin{pmatrix} 2.1213 & 0 \\ 0 & 1.6235 \end{pmatrix},
\tag{10.50}
$$

which has the two eigenvalues $\lambda_B = 2.1213 > 1$ and $\lambda_J = 1.6235 > 1$, corresponding to the unstable fixed point, the critical point. Since we have truncated the series at its first-order terms, these results are only approximately correct.

From the exact solution, we know that at the critical point, $J_c = J/k_B T_c = \ln \sqrt{3}/2 \simeq 0.2747$, which is smaller than the result of J_0 obtained above. We can also calculate the critical exponents, for example,

$$\nu = \frac{\log \Lambda}{\log \lambda_J} = \frac{\log \sqrt{3}}{\log 1.6235} = 1.1336 \tag{10.51}$$

and

$$yd = \frac{\log \lambda_B}{\log \Lambda} = \frac{\log 2.1213}{\log \sqrt{3}} = 1.3690, \tag{10.52}$$

which can be used further to obtain the exponents $\alpha = 2 - \nu d = -0.2672$ and $\delta = y/(1-y) = 2.1696$. These values are still quite different from the exact results with $\alpha = 0$ and $\delta = 15$. The discrepancies are due to the approximation in the evaluation of $\langle \exp(-V) \rangle_0$. There are two ways to improve the accuracy: One can either include the higher-order terms in the expansion for $\langle \exp(-V) \rangle_0$, which seems to improve the results quite slowly, or one can perform the evaluation of $\langle \exp(-V) \rangle_0$ numerically, as we will discuss in the next section. Numerical evaluation of the averages in the renormalization transform has been a very successful approach.

10.4 Renormalization with Monte Carlo simulation

As we discussed in the preceding section, we have difficulty in obtaining an accurate renormalization matrix around the fixed point because the evaluation of nontrivial averages such as $\langle \exp(-V) \rangle_0$ is needed.

However, if we recall the Monte Carlo scheme in statistical physics discussed in Chapter 9, these averages should not be very difficult to evaluate through the Metropolis algorithm. A scheme introduced by Swendsen (1979) was devised for such a purpose. Let us still take the Ising model as the illustrative example. The renormalization transform is given by

$$\mathcal{H}^{(n+1)} = -J^{(n+1)} \sum_{\langle ij \rangle} s_i^{(n+1)} s_j^{(n+1)} - B^{(n+1)} \sum_i s_i^{(n+1)}, \tag{10.53}$$

up to a constant term. We will label the Hamiltonian by a vector $\mathbf{H} = (H_1, H_2) = (B, J)$, as we did in Section 10.3. Then we have the renormalization matrix

$$A_{ij} = \frac{\partial H_i^{(n+1)}}{\partial H_j^{(n)}}. \tag{10.54}$$

The Hamiltonian can also be written in terms of the moments of the spin variables. For example, we can write

$$\mathcal{H} = \sum_{i=1}^m H_i M_i, \tag{10.55}$$

with, for example, $M_1 = \sum_j s_j$, $M_2 = \sum_{\langle jk \rangle} s_j s_k$, and so on. Now we can relate the renormalization transform matrix to the thermal averages of the moments through

$$\frac{\partial \langle M_i^{(n+1)} \rangle}{\partial H_j^{(n)}} = \sum_k \frac{\partial \langle M_i^{(n+1)} \rangle}{\partial H_k^{(n+1)}} \frac{\partial H_k^{(n+1)}}{\partial H_j^{(n)}}, \tag{10.56}$$

which is equivalent to

$$\mathbf{F} = \mathbf{GA}, \tag{10.57}$$

with

$$F_{ij} = \frac{\partial \langle M_i^{(n+1)} \rangle}{\partial H_j^{(n)}} \tag{10.58}$$

and

$$G_{ij} = \frac{\partial \langle M_i^{(n+1)} \rangle}{\partial H_j^{(n+1)}}. \tag{10.59}$$

We can express F_{ij} and G_{ij} in terms of the averages of M_i and $M_i M_j$ from the relation of H_i and M_i in the Hamiltonian and the definition of the thermal average, for example,

$$\langle M_i \rangle = \frac{\sum_{\{s_i\}} M_i e^{-\sum_j H_j M_j}}{\sum_{\{s_i\}} e^{-\sum_j H_j M_j}}. \tag{10.60}$$

If we take the derivative with respect to H_j in the above expression, we have

$$F_{ij} = \langle M_i^{(n+1)} M_j^{(n)} \rangle - \langle M_i^{(n+1)} \rangle \langle M_j^{(n)} \rangle, \tag{10.61}$$

$$G_{ij} = \langle M_i^{(n+1)} M_j^{(n+1)} \rangle - \langle M_i^{(n+1)} \rangle \langle M_j^{(n+1)} \rangle. \tag{10.62}$$

The Swendsen scheme evaluates the matrices $\mathbf{F}$ and $\mathbf{G}$ in each renormalization transform step, and then the matrix $\mathbf{A}$ is obtained through Eq. (10.57), with $\mathbf{F}$ multiplied by the inverse of $\mathbf{G}$. As we discussed in the preceding section, we can now calculate the relevant critical exponents from the eigenvalues of the matrix $\mathbf{A}$. For more details, see Swendsen (1979a; 1979b).

10.5 Crossover: The Kondo problem

The Kondo problem has been one of the most interesting problems in the history of theoretical physics. Resistance to a full solution had lasted for more than ten years. The problem was first noticed in the experimental observation of the resistivity of simple metals such as copper, embedded with very dilute magnetic impurities, for example, chromium. The resistivity first decreases when the temperature is lowered, which is common in all nonmagnetic impurity scattering cases and well understood. However, at very low temperatures, typically on the order of 1 K, the resistivity starts to increase and eventually saturates at zero temperature. Kondo argued that this was due to the spin-flip scattering of the electron and magnetic moment of the impurity, and came up with a very simple Hamiltonian for the single impurity scattering:

$$\mathcal{H} = \sum_{\mathbf{k}\sigma} \epsilon_{\mathbf{k}} c^{+}_{\mathbf{k}\sigma} c_{\mathbf{k}\sigma} + J\mathbf{S} \cdot \sum_{\mathbf{k}\sigma, \mathbf{n}\mu} c^{+}_{\mathbf{k}\sigma} \mathbf{s}_{\sigma\mu} c_{\mathbf{n}\mu}, \tag{10.63}$$

where $c^{+}_{\mathbf{k}\sigma}$ and $c_{\mathbf{k}\sigma}$ are creation and annihilation operators for the electrons, $\mathbf{s}_{\sigma\mu}$ are Pauli matrices for the electron spins, and $\mathbf{S}$ is the spin operator for the impurity. J is the spin coupling constant. For ferromagnetic coupling, that is, $J < 0$, perturbation theory provides a convergent solution of this model. The typical electron and magnetic impurity system, however, has antiferromagnetic coupling. Kondo used the Hamiltonian in Eq. (10.63) to calculate the resistivity of the electrons due to impurity scattering with a perturbation method. After summing up a specific class of infinite terms, Kondo found that the series diverges logarithmically when the temperature is approaching a characteristic temperature of the system, now known as the Kondo temperature T_K, which is typically on the order of 1 K. Theoretical work was done following the discovery of the problem by Kondo, and these efforts were summarized in Kondo (1969).

It was Wilson who devised the numerical renormalization method and solved the Kondo problem. Even though the problem later was solved exactly by the analytic method, it is still fair to say that Wilson's solution extracted all the relevant physics of the problem (the process now known as the crossover phenomenon); that is, the system changes its behavior gradually from the high temperature or the weak coupling state to the low temperature or the strong coupling state without a phase transition.

Wilson made several modifications to the original Hamiltonian in order to solve it numerically with the renormalization method. First, the free electron energy band is assumed to be linear, that is, that $\epsilon(k) \propto k$. This modification of the Hamiltonian will not significantly change the physics of the model if the relevant energy scale is much smaller that the energy bandwidth. After a proper choice of the units and the Fermi level, the modified Kondo Hamiltonian is given by

$$\mathcal{H} = \int_{-1}^{1} c^{+}_{k} c_{k} \, dk + J A^{+} \mathbf{s} A \cdot \mathbf{S}, \tag{10.64}$$

with $\mathbf{s}$ and $\mathbf{S}$ being the Pauli matrices for an electron and the impurity, respectively. The operators A and A^{+} are the collective operators defined from

$$A = \int_{-1}^{1} c_{k} \, dk. \tag{10.65}$$

Note that the spin indices have been suppressed in the above expressions for convenience of notation. Both c_k and A have spin indices and when the operators appear in pairs, the spin indices are all summed. For example,

$$A^+ \mathbf{s} A = \sum_{\sigma\mu} A_\sigma^+ \mathbf{s}_{\sigma\mu} A_\mu. \tag{10.66}$$

Then the momentum space is discretized with a logarithmic decrease of the space intervals toward the center of the band, that is, the Fermi level, with the lattice points at

$$k_m = \frac{\pm 1}{\Lambda^m}, \tag{10.67}$$

for $m = 0, 1, \ldots, \infty$, with $\Lambda > 1$. When $\Lambda \to 1$, the space approaches the continuous limit. One can then construct a set of orthogonal basis states defined in each interval $\Lambda^{-(m+1)} < k < \Lambda^{-m}$ as

$$\phi_{ml}(k) = \frac{\Lambda^{(m+1)/2}}{\sqrt{\Lambda - 1}} e^{i\omega_m kl}, \tag{10.68}$$

for $l = 0, 1, \ldots, \infty$, with

$$\omega_m = \frac{2\pi \Lambda^{m+1}}{\Lambda - 1}. \tag{10.69}$$

The state in Eq. (10.68) is nonzero only in the interval defined. Since the Fermi level is at the center of the band, $\phi_{ml}(k)$ and $\phi_{ml}(-k)$ for $k \geq 0$ form a complete basis set for the whole momentum space. The creation and annihilation operators of electrons can then be expressed in terms of these discrete states,

$$a_k = \sum_{ml} [c_{ml}\phi_{ml}(k) + d_{ml}\phi_{ml}(-k)], \tag{10.70}$$

where c_{ml} and d_{ml} satisfy the fermion anticommutation relation, for example,

$$\left[c_{kl}, c_{mn}^+\right]_+ = \delta_{km}\delta_{ln}, \tag{10.71}$$

and they can be interpreted as different fermion operators. If we only keep the $l = 0$ states in the above expansion, the Kondo Hamiltonian is further approximated and simplified into

$$\mathcal{H} = \epsilon_0 \sum_m \frac{1}{\Lambda^m} (c_m^+ c_m - d_m^+ d_m) + J A^+ \mathbf{s} A \cdot \mathbf{S}, \tag{10.72}$$

where $c_m = c_{m0}$, $d_m = d_{m0}$, and $\epsilon_0 = (1 + \Lambda^{-1})/2$. A transform can be made so that the Hamiltonian can be described by just one set of parameters,

$$\mathcal{H} = \sum_{k=0}^{\infty} \epsilon_k \left(f_k^+ f_{k+1} + f_{k+1}^+ f_k\right) + \tilde{J} f_0^+ \mathbf{s} f_0 \cdot \mathbf{S}, \tag{10.73}$$

where f_0 is defined directly from A as

$$f_0 = \frac{1}{\sqrt{2}} A, \tag{10.74}$$

and f_k for $k > 0$ are given from an orthogonal transform of c_m and d_m,

$$f_k = \sum_m (u_{km} c_m + v_{km} d_m). \tag{10.75}$$

After rescaling, one has $\epsilon_k = 1/\Lambda^{k/2}$ and $\tilde{J} = 4J\Lambda/(\Lambda + 1)$. This rescaling has some effects on the terms with small values of k but not on the terms with large values of k. One can obtain all the transform coefficients u_{km} and v_{km} and show that f_k satisfies the fermion anticommutation relation

$$\left[f_k, f_l^+\right]_+ = \delta_{kl}, \tag{10.76}$$

from the conditions on c_{kl} and d_{kl} as well as the orthogonal transform. For more details on the approximation and transform, see Wilson (1975).

Before we describe Wilson's solution of the above Hamiltonian, a technical detail is worth mentioning first. Assume that we want to obtain the spectrum of the Hamiltonian

$$\mathcal{H} = \mathcal{H}_0 + \mathcal{H}_1 + \mathcal{H}_2 + \cdots, \tag{10.77}$$

where $\mathcal{H}_0$ has a degenerate ground state and the elements in $\mathcal{H}_n$ are several orders smaller than the elements in $\mathcal{H}_{n-1}$. In order to have accurate structure of the eigenvalue spectrum, one cannot simply diagonalize $\mathcal{H}$ with all the elements in $\mathcal{H}_n$ with $n > 0$ substituted. The reason is that the numerical rounding error will kill the detailed structure of the spectrum due to the smallness of $\mathcal{H}_n$ for $n > 0$. A better way, perhaps the only way, to maintain the accuracy in the spectrum is to diagonalize $\mathcal{H}_0$ first and then treat $\mathcal{H}_n$ as perturbations term by term. Note that the accuracy we are talking about here is not the absolute values of the eigenvalues but rather the structure, that is, the correct splitting of the energy levels or the hierarchical structure of the spectrum. The structure of the energy spectrum determines the properties of the system.

The Hamiltonian of Eq. (10.73) is ready to be formulated in a recursive form that defines the renormalization transform. Let us define a Hamiltonian with k truncated at $k = n - 1$:

$$\mathcal{H}^{(n)} = \Lambda^{(n-1)/2} \sum_{k=0}^{n-1} \epsilon_k \left(f_k^+ f_{k+1} + f_{k+1}^+ f_k\right) + \tilde{J} f_0^+ \mathbf{s} f_0 \cdot \mathbf{S}, \tag{10.78}$$

where the rescaling factor $\Lambda^{(n-1)/2}$ is used to make the smallest term in the Hamiltonian on the order of 1. The results of the original Hamiltonian can be recovered by dividing the eigenvalues by $\Lambda^{(n-1)/2}$ after they are obtained. The recursion relation or the renormalization transform is then formulated as

$$\mathcal{H}^{(n+1)} = T[\mathcal{H}^{(n)}] = \frac{1}{\sqrt{\Lambda}}\mathcal{H}^{(n)} + f_n^+ f_{n+1} + f_{n+1}^+ f_n - E_0^{(n+1)}, \quad (10.79)$$

where $E_0^{(n+1)}$ is the ground-state energy of the Hamiltonian $\mathcal{H}^{(n+1)}$. The Hamiltonian can now be diagonalized as described earlier in this section with a combination of numerical diagonalization of the first part of the Hamiltonian and then degenerate perturbations of other terms. The number of states will increase exponentially with n. So one has to truncate the number of states used in the recursions at some reasonable number so it can be dealt with on current computers.

Wilson discovered that no matter how small the original coupling constant $J > 0$ is, the behavior of the system will always cross over to that of $J = \infty$ after many steps of recursion. In physical terms, this means that no matter how small the coupling between an electron and the impurity, if it is antiferromagnetic, when the temperature approaches zero, the system will eventually move to the strong coupling limit; that is, the electron spins will screen the impurity spin completely at a temperature of zero. The temperature where the crossover happens is the Kondo temperature, which is a function of the original coupling constant J.

We are not going to have more discussions here of the methods or evaluations of physical quantities such as the magnetic susceptibility or specific heat of the impurity. Interested readers can find excellent explanations in Wilson (1975).

10.6 Quantum lattice renormalization

After Wilson's celebrated work on the numerical renormalization study of the Kondo problem in the early 1970s, a series of attempts were made to generalize the idea in the study of quantum lattice models. These attempts turned out not to be as successful as the study of the Kondo problem. However, the studies did provide some qualitative understanding of these systems and have formed the basis for further development, especially the density matrix renormalization of White (1992; 1993).

In this section, we would like to discuss how to generalize the Wilson method to quantum lattice models. What we need to find is the transform

$$\mathcal{H}^{(n+1)} = T[\mathcal{H}^{(n)}], \quad (10.80)$$

which can accurately describe the energy structure of the system, at least for low-lying excited states. We will consider an infinite chain system and divide it first into small blocks. The renormalization transform then combines two nearest blocks into one new block. We focus on two blocks and assume that the left block has

L independent states and the right block has R independent states, so the total number of states of the two block system is $N = L \times R$. The Hamiltonian can be symbolically written as

$$\mathcal{H}^{(n)} = \mathcal{H}_L^{(n)} + \mathcal{H}_R^{(n)} + V_{LR}^{(n)}, \tag{10.81}$$

with $\mathcal{H}_{L,R}^{(n)}$ being the intrablock parts and $V_{LR}^{(n)}$ being the coupling between two blocks. Let us take $|l\rangle$ with $l = 1, 2, \ldots, L$ and $|r\rangle$ with $r = 1, 2, \ldots, R$ as the eigenstates of the left and right blocks, respectively, and the direct product $|i\rangle = |l\rangle \otimes |r\rangle$ with $i = 1, 2, \ldots, N$ as the basis for constructing the eigenstates of $\mathcal{H}^{(n)}$,

$$|k\rangle = \sum_{i=1}^{N} a_{ki} |i\rangle, \tag{10.82}$$

where $k = 1, 2, \ldots, N$, and then we have

$$\mathcal{H}^{(n)} |k\rangle = \epsilon_k |k\rangle. \tag{10.83}$$

The direct product of two vectors, $|i\rangle = |l\rangle \otimes |r\rangle$, will be illustrated later in this section in an example. The renormalization transform is then performed with $K \le N$ states selected from $|k\rangle$ with

$$\mathcal{H}^{(n+1)} = \mathbf{A} \mathcal{H}^{(n)} \mathbf{A}^+, \tag{10.84}$$

where $\mathbf{A}$ is a $K \times N$ matrix constructed from

$$A_{ki} = a_{ki} \tag{10.85}$$

and the matrix representation of $\mathcal{H}^{(n)}$ on the right side of Eq. (10.84) is in the original direct product state $|i\rangle$,

$$\langle i | \mathcal{H}^{(n)} | i' \rangle = \langle l | \mathcal{H}_L | l' \rangle \otimes \mathbf{I}_R + \mathbf{I}_L \otimes \langle r | \mathcal{H}_R | r' \rangle + \langle i | V_{LR} | i' \rangle, \tag{10.86}$$

where $\mathbf{I}_R$ and $\mathbf{I}_L$ are unit matrices with dimensions $R \times R$ and $L \times L$, respectively, and $\otimes$ has the same meaning as the direct product of two irreducible matrix representations of groups. One aspect needing special care is the matrix representation of the interaction term between two blocks. For example, if the interaction is of the Ising type, that is,

$$V_{LR} = J s_r^x s_l^x, \tag{10.87}$$

we have

$$\langle i | V_{LR} | i' \rangle = J \langle l | s_r^{x(n)} | l' \rangle \otimes \langle r | s_l^{x(n)} | r' \rangle, \tag{10.88}$$

where s_r and s_l mean the spin operator at the right and left boundaries, respectively, of the block. Later in this section we will demonstrate how to work out the direct products of vectors and matrices. In fact, all other variables are transformed in a way similar to the transformation of the Hamiltonian, particularly the operators at the boundaries, which are needed to construct the next iteration of the Hamiltonian. For example, the new spin operator for the right boundary is given by

$$s_r^{x(n+1)} = \mathbf{A} s_r^{x(n)} \mathbf{A}^+, \tag{10.89}$$

with again $s_r^{x(n)}$ represented under the states $|i\rangle$. Note that one also needs to expand the dimensions of $s^{x(n)}$ with the direct product of a unit matrix, for example,

$$\langle i | s_r^{x(n)} | i' \rangle = \mathbf{I}_L \otimes \langle r | s_r^{x(n)} | r' \rangle; \quad \langle i | s_l^{x(n)} | i' \rangle = \langle l | s_l^{x(n)} | l' \rangle \otimes \mathbf{I}_R. \tag{10.90}$$

There is one important aspect we have not specified: the criteria for selecting the K states out of a total number of N states to construct the renormalization transform. Traditionally, the states with lowest energies were taken for the renormalization transform. However, there is a difference between the lattice Hamiltonian and the Kondo Hamiltonian worked out by Wilson. In the Kondo problem, every term added during the transform is much smaller than the original elements, so one would not expect, for example, level crossing. The lower states will contribute more to the true ground state. However, in the lattice case there are no such small elements. The terms added in are as large as other elements. A problem arises when we perform the transform by keeping the few (for example 100) lowest states, since the higher levels will still contribute to the true ground state or lower excited states at the infinite limit. In the next section, we will discuss the density matrix formulation of the renormalization transform introduced by White (1992; 1993), which solves the problem of the level crossing, at least for one-dimensional systems.

In order to have a better understanding of the scheme outlined above, let us apply it to an extremely simple model, the spin-1/2 quantum Ising chain in a transverse magnetic field. This model was studied by Jullien et al. (1978) in great detail. If we include only the nearest-neighbor interaction, then the model Hamiltonian is given by

$$\mathcal{H} = J \sum_{i=-\infty}^{\infty} s_i^x s_{i+1}^x - B \sum_{i=-\infty}^{\infty} s_i^z, \tag{10.91}$$

with s^x and s^z being the Pauli matrices,

$$s^x = \begin{pmatrix} 0 & 1 \\ 1 & 0 \end{pmatrix}; \quad s^z = \begin{pmatrix} 1 & 0 \\ 0 & -1 \end{pmatrix}. \tag{10.92}$$

What we would like to obtain is a transform that relates the lattice Hamiltonian $\mathcal{H}^{(n)}$ to the block Hamiltonian $\mathcal{H}^{(n+1)}$,

$$\mathcal{H}^{(n+1)} = T[\mathcal{H}^{(n)}]. \tag{10.93}$$

In order to simplify our discussion, we will only take one site in each block and then combine two blocks into a new block during the transform. We will also take the two states with lowest eigenvalues for the transform, which means that $L = R = K = 2$ and $N = 4$. So the states in each block $|r\rangle$ or $|l\rangle$ are given by

$$|1\rangle = (1\ 0); \quad |2\rangle = (0\ 1), \tag{10.94}$$

Then the direct product $|i\rangle = |l\rangle \otimes |r\rangle$ is given by

$$|1\rangle = (1\ 0\ 0\ 0); \quad |2\rangle = (0\ 1\ 0\ 0);$$

$$|3\rangle = (0\ 0\ 1\ 0); \quad |4\rangle = (0\ 0\ 0\ 1), \tag{10.95}$$

which are obtained by taking the product of the first element of the first vector and the second vector to form the first two elements and then taking the second element of the first vector and the second vector to form the last two elements. The matrix direct products are performed in exactly the same manner. In general, we want to obtain the direct product $\mathbf{C} = \mathbf{A} \otimes \mathbf{B}$, where $\mathbf{A}$ is an $n \times n$ matrix and $\mathbf{B}$ is an $m \times m$ matrix. $\mathbf{C}$ is then an $nm \times nm$ matrix, which has $n \times n$ blocks with $m \times m$ elements in each block constructed from the corresponding element of $\mathbf{A}$ multiplied by the the second matrix $\mathbf{B}$. Based on this rule of direct product, we can obtain all the terms in a Hamiltonian of two blocks. And if we put all these terms together, we have

$$\mathcal{H}^{(n)} = \begin{pmatrix} -2B & 0 & 0 & J \\ 0 & 0 & J & 0 \\ 0 & J & 0 & 0 \\ J & 0 & 0 & 2B \end{pmatrix}, \tag{10.96}$$

which can be diagonalized by the following four states:

$$|1\rangle = \frac{1}{\sqrt{2}}(0\ -1\ 1\ 0); \qquad |2\rangle = \frac{1}{\sqrt{2}}(0\ 1\ 1\ 0);$$

$$|3\rangle = \frac{1}{\sqrt{1+\alpha^2}}(-\alpha\ 0\ 0\ 1); \quad |4\rangle = \frac{1}{\sqrt{1+\beta^2}}(-\beta\ 1\ 1\ 0), \tag{10.97}$$

where α and β are given by

$$\alpha = \frac{2B + \sqrt{4B^2 + J^2}}{J}; \quad \beta = \frac{2B - \sqrt{4B^2 + J^2}}{J}. \tag{10.98}$$

The corresponding eigenvalues are

$$\epsilon_k = -J,\ J,\ -\sqrt{4B^2 + J^2},\ \sqrt{4B^2 + J^2}. \tag{10.99}$$

Since the first and third states have lower energy for $J > 0$, we have our transform matrix as

$$\mathbf{A} = \begin{pmatrix} 0 & -1/\sqrt{2} & 1/\sqrt{2} & 0 \\ -\alpha/\sqrt{1+\alpha^2} & 0 & 0 & 1/\sqrt{1+\alpha^2} \end{pmatrix}, \qquad (10.100)$$

which can be used to construct the new block Hamiltonian

$$\mathcal{H}^{(n+1)} = \mathbf{A}\mathcal{H}^{(n)}\mathbf{A}^+ = \begin{pmatrix} -J & 0 \\ 0 & -\gamma \end{pmatrix}, \qquad (10.101)$$

with $\gamma = \sqrt{4B^2 + J^2}$. Similarly, one can construct new spin operators at the boundaries, for example,

$$s_r^{x(n+1)} = \mathbf{A}\begin{pmatrix} 1 & 0 \\ 0 & 1 \end{pmatrix} \otimes \begin{pmatrix} 0 & 1 \\ 1 & 0 \end{pmatrix}\mathbf{A}^+ = \frac{\alpha+1}{\sqrt{2(1+\alpha^2)}}\begin{pmatrix} 0 & 1 \\ 1 & 0 \end{pmatrix}, \quad (10.102)$$

and $s_l^{x(n+1)} = -s_r^{x(n+1)}$. The new interaction term is then given by

$$V_{LR}^{(n+1)} = J s_r^{x(n+1)} \otimes s_l^{x(n+1)}. \qquad (10.103)$$

Then the above scheme can be carried on to infinity. As we pointed out at the beginning of the section, this procedure turns out to be unsuccessful in most systems. The difficulty is due to the facts that the addition of two blocks is not a very small perturbation to the single-block Hamiltonian and that the overlap of the higher-energy states with the lower-energy states in the infinite system is very significant. In the next section, we will discuss recent work of White (1992; 1993) that has been very successful in the study of one-dimensional quantum lattice systems.

10.7 Density matrix renormalization

Recent progress in the numerical renormalization study of highly correlated systems has been made by White (1992). The basic idea comes from the observation that the reduced density matrix constructed from a set of states of a specific segment of an infinite system should have the largest eigenvalues if we want this set of states to have the largest overlap with the lower-lying states of the system. Here we will just give a brief introduction of the scheme developed by White (1992); interested readers should consult White (1993).

The density matrix is defined for a system that is in contact with the environment. Assume that the system is described by a set of orthonormal and complete states $|i\rangle$ for $i = 1, 2, \ldots, m$ and the environment by another set of orthonormal and complete states $|r\rangle$ for $r = 1, 2, \ldots, n$. Then an arbitrary state including the system and its environment can be expressed as

$$|\psi\rangle = \sum_{i,r=1}^{m,n} c_{ir}|i\rangle|r\rangle, \tag{10.104}$$

where $|i\rangle|r\rangle$ is used as a two-index notation, instead of the direct product $|i\rangle \otimes |r\rangle$ used earlier. This is an easier way of keeping track of the indices. The expectation value of an operator A of the system under the state $|\psi\rangle$ can then be written as

$$\langle A\rangle = \sum_{i,i',r} c_{i'r}^{+} c_{ir} \langle i'|A|i\rangle = \text{Tr}\,\rho A, \tag{10.105}$$

with the reduced density matrix ρ defined from

$$\rho_{ij} = \sum_{r} c_{jr}^{+} c_{ir}, \tag{10.106}$$

for the above expectation value.

One can show easily that ρ is Hermitian and therefore can be diagonalized and written in the diagonal form of its eigenstates $|\alpha\rangle$ as

$$\rho = \sum_{\alpha=1}^{m} w_{\alpha}|\alpha\rangle\langle\alpha|, \tag{10.107}$$

where w_{α} can be interpreted as the probability of the system in the state $|\alpha\rangle$ under the given system and environment. Since $|\alpha\rangle$ is a state of the system, it can be written as

$$|\alpha\rangle = \sum_{i=1}^{m} u_{\alpha i}|i\rangle, \tag{10.108}$$

with $u_{\alpha i}$ as the elements of a unitary matrix, since $|\alpha\rangle$ also forms an orthonormal basis set. The states with higher w_{α} will then contribute more than other states when an average of a physical quantity $\text{Tr}\,\rho A$ is carried out.

This concept of the reduced density matrix forms the basic idea of the renormalization scheme proposed by White (1992; 1993). The states selected to construct the new blocks are the eigenstates of the reduced matrix with the largest eigenvalues, since they represent the highest probability of a system under an environment. The system can then be viewed as a segment of an infinite system. White has also provided a more rigorous mathematical argument on the selection of the states based on the eigenvalues of the reduced density matrix (White, 1993).

Because of the fast increase in the number of states in the system and in its environment, the selection of the system and its environment becomes very important in order to have a practical and accurate scheme. As we discussed in the last section, a new block is usually constructed from two identical blocks. Numerical results show that it is more efficient in most cases to construct the new block by adding one more site to the old block. Because the system is growing only at one end if one site is added to the old block, the result corresponds to the condition of an open end. If a periodic boundary condition is imposed on the system, the added site(s) should also be in contact with the other end, as the size of a circle is increased by introducing a new segment to its circumference. For an infinite chain, imposing a periodic condition for the renormalization transform requires many more states to be maintained in order to have the same accuracy for the condition of an open end.

There are two important criteria in the selection of the environment. It has to be convenient for constructing the new block Hamiltonian, and it has to be the best representation of the rest of the actual physical system within the capacity of current computers. Here we would like to discuss a little more the algorithm of White in the context of the one-dimensional Ising model discussed in the preceding section. One first selects a block with a fixed number of sites. The block will be denoted as $B^{(n)}$ at the nth iteration of the renormalization transform. Then we add a single site to the right-hand side of the block to form the new block denoted by $B^{(n+1)} = B^{(n)}\cdot$, where the dot ($\cdot$) means the site added to the block $B^{(n)}$. The environment is selected as the reflection of $B^{(n)}\cdot$, denoted as $\cdot B_r^{(n)}$. The system $(B^{(n)}\cdot)$ and the environment $(\cdot B_r^{(n)})$ together form a superblock $B^{(n)}\cdot\cdot B_r^{(n)}$. One can solve the Hamiltonian of the superblock and construct the reduced density matrix for the system $(B^{(n)}\cdot)$ from some of the eigenstates (for example, the ground state) of the superblock.

After diagonalizing the reduced density matrix, one can select the corresponding eigenstates with the largest eigenvalues as the bases to form the states for the new block $B^{(n+1)}$ in the next iteration. Then one can add one more site to the block to repeat the whole process again. After many iterations, one obtains approximate results of an infinite system.

One can start the renormalization scheme with four sites as the first superblock $B^{(1)}\cdot\cdot B_r^{(1)}$. So each block, $B^{(1)}$ and $B_r^{(1)}$, has only one site. The matrix of the Hamiltonian of a superblock is usually very sparse and can be solved with the Lanczos method discussed in Chapter 4. Then one can construct a reduced density matrix for the system,

$$\rho_{ij} = \sum_r c_{jr}^+ c_{ir}, \tag{10.109}$$

from the ground state of the superblock,

$$|\psi_0^{(1)}\rangle = \sum_{i,r} c_{ir} |l\rangle |r\rangle, \tag{10.110}$$

where $|i\rangle$ is for the states of $B^{(1)}$· and $|r\rangle$ is for the states of $\cdot B_r^{(1)}$. The reduced density matrix can be diagonalized with

$$\rho|\alpha\rangle = w_\alpha|\alpha\rangle \tag{10.111}$$

to obtain the l states that have the largest eigenvalues. The new block Hamiltonian is obtained from $\langle\alpha|\mathcal{H}^{(2)}|\alpha'\rangle$. One more site can then be added to form the new system, and a reflection of the new system can be used as the new environment.

The above renormalization transform scheme of White has been applied to several one-dimensional systems, including the quantum spin (White, 1992; 1993) and Kondo lattice (Yu and White, 1993) systems. The method can also be applied to a finite chain (White, 1993). However, it is not clear whether the method would also be a powerful tool for two-dimensional systems. Current computing capacity has put a limit on the size of the block as well as the size of the environment. Some recent work on coupled-chain systems (White, Noack, and Scalapino, 1993) has shown some promise for the method, at least in an anisotropic two-dimensional system.

Exercises

10.1 From the scaling relations discussed in Section 10.1, if two of the exponents are given, one would be able to know the rest. For the two-dimensional Ising model, the exact results are available. If we start with $\alpha = 0$ and $\beta = 1/8$, show that $\gamma = 7/4$, $\delta = 15$, $\eta = 1/4$, and $\nu = 1$.

10.2 As discussed in Section 10.3 that the approximation, up to the first order, of the coupling between two nearest neighboring spin blocks can be improved by including more terms in the expansion. If we rewrite the Hamiltonian in the form of the nearest-neighbor interaction and another term of next-nearest-neighbor interaction (zero in the zeroth-order iteration), we can incorporate the second-order term in the renormalization transform. Derive the renormalization transform matrix for the case in which first-order and second-order terms are included in the expansion and calculate the critical exponents for the triangular Ising model. Are there any improvements over the first-order approximation?

10.3 One can also improve the calculation outlined in Section 10.3 for the triangular Ising model by choosing larger blocks. Use the hexagon blocks with seven sites as the renormalization transform units and evaluate the exponents with the expansion up to the first order. Are there any improvements over the calculations of the triangular blocks?

10.4 Develop a program with the Swendsen Monte Carlo renormalization algorithm for the two-dimensional Ising model on a square lattice. Take the

2 × 2 block as the renormalization transform unit for a system of 40 × 40 sites and apply the majority rule for the block spin.

10.5 One of the very interesting observations of Wilson is that when one has a Hamiltonian matrix with multiple energy scales, the only way to obtain the accurate structure of the energy levels is to perform degenerate perturbations level by level. Assume that we have the following Hamiltonian:

$$\mathcal{H} = H^{(0)} + H^{(1)} + H^{(2)},$$

with the zeroth order given by

$$H_0 = \begin{pmatrix} 2 & 1 \\ 1 & -2 \end{pmatrix}$$

and the first-order and second-order terms given by

$$H_{ij}^{(1)} = \begin{cases} 0.01 & \text{for} \quad i \neq j, \\ 0 & \text{for} \quad i = j, \end{cases}$$

and

$$H_{ij}^{(2)} = \begin{cases} 0.0001 & \text{for} \quad i \neq j, \\ 0 & \text{for} \quad i = j. \end{cases}$$

Find the energy level structure of the Hamiltonian through degenerate perturbation. Keep only the ground state at each stage of the calculations.

10.6 Complete the renormalization study of the quantum Ising model discussed in Section 10.6 and discuss the fixed point and the phase diagram of the system.

10.7 Develop a program with the density matrix renormalization scheme for the same model studied in Exercise 10.6. Start the scheme with a four-site superblock.

11

Symbolic computing

Even though symbolic computing was introduced at the very beginning of computer history, it did not become a potential tool until about ten years ago, when high-speed computers became available. During the last decade, several comprehensive and affordable computer software packages were developed or became mature as a result of many efforts to combine data programming, symbolic manipulation, and graphic visualization into one advanced computational system, which we will refer to as symbolic computing. They include, but are not limited to, *Mathematica* from Wolfram Research Inc., MACSYMA from MACSYMA Inc., Maple from Water-loo Maple Inc., MATLAB from MathWorks Inc., Mathcad from MathSoft Inc., Theorist from Prescience Corporation, and REDUCE from RAND Corporation. Usually these systems are equipped to handle data processing, variable manipula-tion, equation solving, and data graphics in both two and three dimensions. The initial purpose of these packages was to simplify programming efforts and to model mathematical problems at their simplest but deepest level. Many current applica-tions of symbolic computing systems still concern the presentation of the known results; that is, the emphasis is on teaching mathematics, physics, and other subjects in science and engineering. However, as we will illustrate in this chapter, symbolic computing has started to play an active role in research and is expected to be an important tool for researchers as well as students in science and engineering.

11.1 Symbolic computing systems

In this chapter, we will not be able to cover many topics associated with symbolic computing, since it is a combination of symbolic computer language, advanced programming, and associated scientific applications, each of which could be the subject of an independent book. So the goal here is just to provide a very brief but useful introduction and to illustrate the ideas of symbolic computing and its possible applications in physics. The emphasis will be on what kind of physical problems might be studied through symbolic computing. We have to choose a specific package in order to have a solid basis for our discussions, and we will therefore use the statements and conventions of *Mathematica* for the examples and discussions in this chapter. The symbolic computing systems just mentioned are all very competitive; that is, they are quite similar in dealing with symbolic computing tasks and have little variation in computing speed for different problems or on different computer systems. So if one has access to a symbolic computing system, one should be able to apply it to the tasks encountered in study or research. The basic ideas discussed here can be generalized to the specific problems one is interested in, regardless which symbolic system is available. A comparison of different symbolic computing systems can be found in Cook et al. (1992).

Here is how *Mathematica* works. The software is usually accessed with a very simple command and then a *Mathematica* window will appear on the monitor. For example, if an X-terminal or a workstation is used, one logs into the system running *Mathematica* and types in the line mathematica -display term:0, and an untitled window will show up on the screen. Here term is the address (name) of the X-terminal or workstation. Then one can type in *Mathematica* input commands which can be executed when Shift-Return is used. The output of the execution is given right after the input line if input commands are executed line by line. The *i*th input (output) is labeled automatically as In[i] (Out[i]) in the *Mathematica* window. One can also write many input commands without intermediate outputs by ending each input command line with a semicolon. The input and output of *Mathematica* can be saved into a file known as a *Mathematica notebook*. Each notebook can be used to record a complete program, which can be used or modified later if needed.

Let us here just highlight a couple of the basic features of *Mathematica*. The arithmetic operators +, −, *, / and ^ are used for addition, subtraction, multiplication, division, and exponential operation. * can also be replaced by a space; for example, x y=x*y. No space is needed if the multiplication is between a number and a variable, for example, 2y=2 y=2*y, but not between a variable and a number. For more details on the arithmetic operations in *Mathematica*, see the manual by Wolfram (1996).

All built-in *Mathematica* commands and operations start with an upper-case letter. For example, y=Sin[x] means that $y = \sin x$, MatrixExp[-a] means e^{-a} with a being a matrix, s=Series[g[x],x,a,n] is the power series of $g(x)$ around $x = a$ up to the nth order, N[x,n] means to find the numerical value of x with n digits of accuracy. So it is a good practice in *Mathematica* to begin user-defined variables or functions with a lower-case letter. A function form is defined with an underscore. For example, f[x_] = x^2 + x*Sin[x] + Log[x^2] defines the function form of $f(x)$ for later use. x here is just a dummy variable; it can be replaced later with any other variable or number. This is similar to a function routine in Fortran. Substitution in *Mathematica* is achieved with a /. operation. For example, if sl is the solution of an equation set eq, then eq/.sl instructs the system to substitute the solution into the equation set. Some constants are also reserved under *Mathematica*; for example, Pi is reserved for $\pi = 3.1415926\ldots$, and I is reserved for $i = \sqrt{-1}$. A complex variable or number can be expressed as z = x + I y, or v = 3 + 4 I. We would like to emphasize that each symbolic computing system can also be treated as an advanced language as well. One can write application packages and full programs to handle all kinds of computational tasks, including numerical computation, symbolic manipulation, and graphic illustration in a combined program. Excellent discussions on how to program with *Mathematica* are given by Maeder (1991).

11.2 Basic symbolic mathematics

The simplest use of the symbolic computing system is in algebraic manipulation. For example, if we want to expand $(1 + x)^5$, we can easily achieve it with the following statements in *Mathematica*:

```
y = (x+1)^5;
Expand[y]
```

and then press Shift-Return. The following output appears in the *Mathematica* window as a result of the expansion,

$$1 + 5 x + 10 x^2 + 10 x^3 + 5 x^4 + x^5$$

as expected. We can illustrate this further in a more sophisticated case, where the result is not as simple as the last example. If we input

```
y = 1+x;
z = (1-x)*y;
r = z^2;
f = (r-y^2*z)/(z-y^2);
ExpandAll[f]
```

and then press Shift-Return, we obtain

$$\frac{-2 x}{-2 x - 2 x^2} - \frac{2 x^2}{-2 x - 2 x^2} + \frac{2 x^3}{-2 x - 2 x^2} + \frac{2 x^4}{-2 x - 2 x^2}$$

which is no longer trivial. Now if we want to simplify the above expansion for $f(x)$, we can achieve this in *Mathematica* with the command

```
Simplify[%]
```

and *Mathematica* returns the simplified form of the expansion,

```
      2
1-x
```

Here % is used to represent the most recent statement in *Mathematica*. One can also obtain the numerical value of $f(x)$ with

```
x = 0.5;
N[f]
```

Mathematica will respond with 0.75, which is the value of the function $f(0.5) =$ 0.75. If we would like to have higher accuracy in the numerical value of the function, we can instruct the system to round up the number to have a specific number of digits with N[f,n], where n is the desired number of digits in the result.

We can also obtain the Taylor expansion of a function in *Mathematica*. For example, if we would like to have the first few terms of the Taylor expansion of the function

$$f(x) = \exp(-x^3 + x^2 - x) \tag{11.1}$$

around $x = 0$, we can simply type in

```
f[x_] = Exp[-x^3+x^2-x];
s = Series[f[x],{x,0,5}]
```

and *Mathematica* will return the Taylor expansion of the function up to the fifth-order term,

```
              2        3        4        5
          3 x      13 x     49 x     87 x              6
1 - x + ---- - ----- + ----- - ----- + O[x]
              2        6        24       40
```

which is useful if we want to know the behavior of the function around $x = 0$.

We can also carry out a summation of a series if the terms in the series are functions of the index n. For example, if we want to obtain

$$s = \sum_{n=1}^{\infty} \frac{1}{n^2},$$ (11.2)

we can obtain an approximation of the series in *Mathematica* with the commands,

```
Sum[1/n^2, {n,1,1000,1}];
N[%,5]
```

which take the summation of the first 1,000 terms in the series. The numerical value of this approximation from *Mathematica* is 1.6439; the exact result of the summation to $n = \infty$ would be $\pi^2/6 \simeq 1.6449$. Higher accuracy can be obtained by including more terms in the summation or by the extrapolation scheme discussed in Chapter 1, which is useful if a closed expression of the series is not available.

11.3 Computer calculus

One can take the derivatives of any given function with the symbolic computing system. For example, if we want to find the first-order derivative of the function given as

$$f(x) = x^2 \ln x - \cosh x,$$ (11.3)

we can simply input

```
D[x^2 Log[x]-Cosh[x],x]
```

and *Mathematica* will respond with

```
x + 2 x Log[x] - Sinh[x]
```

which is the derivative expected. If we have a logarithmic function with a base b (other than e), `Log[b,x]` will be used instead of `Log[x]`.

Integrals can be performed in exactly the same way. For example, if we would like to carry out the integral

$$s = \int \frac{dx}{\sqrt{x^2 + a^2}}$$ (11.4)

in *Mathematica*, we can simply type in

```
Integrate[1/Sqrt[x^2+a^2],x]
```

The system will return

```
             2    2
Log[x+Sqrt[a  + x ]]
```

which is exactly what one would obtain by working it out by hand in five to fifteen minutes. However, one should not expect *Mathematica* to tell more than logically allowed. For example, if we input

```
Integrate[x/Sin[x],x]
```

then the system cannot return anything but

```
Integrate[x Csc[x], x]
```

which is just a rearrangement of the original integral. The reason is that the integral does not have a closed form or does not exist. However, the symbolic system can manage some very complicated situations in many cases. For example, if we want to know the result of the integral

$$s = \int \frac{\sin x}{x} dx, \tag{11.5}$$

we can input

```
Integrate[Sin[x]/x,x]
```

Mathematica will respond with

```
SinIntegral[x]
```

in the output. Note that the integral is interpreted by *Mathematica* as a special function SinIntegral[x]. Now if we want to know the actual value of the integral in the region [0, 1,000,000], we can input

```
Integrate[Sin[x]/x,{x,0,1000000}]
```

Mathematica will still first inform us that it is a special function,

```
SinIntegral[1000000]
```

Then we can ask *Mathematica* for the numerical value of this special function with

```
N[%]
```

which requests the numerical value of the last output, and *Mathematica* prints out the numerical value of the special function,

    1.5708

as expected.

We can also obtain approximate expressions of derivatives in discrete forms in terms of truncated series in symbolic computing. For example, we know that the three-point formula for the second-order derivative is given by

$$f''(x) \simeq \frac{1}{h^2}[f(x-h) + f(x+h) - 2f(x)]. \tag{11.6}$$

Now if we would like to know the first few higher-order terms that were truncated, we can type in

    Series[(f[x-h]+f[x+h]-2f[x])/h^2, {h,0,5}]

Then *Mathematica* returns the expansion,

```
              (4)      2     (6)      4
           f    [x] h     f    [x] h            6
  f''[x] + ---------- + ---------- + O[h]
               12            360
```

which shows us the next two terms omitted in the three-point formula. One can surely argue that the power of symbolic computing is limited only by the capability of the software packages if we know how to apply them to our problems.

11.4 Linear systems

Symbolic computing can be used to solve linear algebra problems in a very concise form. Consider the unbalanced Wheatstone bridge discussed in Section 4.1. If we solve the equation set with *Mathematica*, the procedure is very simple after we have written out all the equations correctly. We can simply input the commands;

```
eq = {rv i1 + r1 i2 + r2 i3 == v0,
      -r4 i1 + (r1 + r4 + ra) i3 == 0,
      -r3 i1 - ra i2 + (r2 + r3 - ra) i3 == 0};
iv = Solve[eq, {i1, i2, i3}]
```

in the *Mathematica* window and then press Shift-Return. In *Mathematica*, a==b means to have a equal to b but c=d means to have c assigned with the value of d.

The solution of the currents in terms of the external voltage and resistances then show up in the *Mathematica* window:

```
{{i2 -> v0/r1 + (ra r2 r4 v0)/
      (r1 (-((ra r2 + r1*(-ra + r2 + r3)) r4) -
            (ra rv - r1 r3) (ra + r1 + r4))) +
      (ra rv (ra + r1 + r4) v0)/
      (r1 (-((ra r2 + r1 (-ra + r2 + r3)) r4) -
            (ra rv - r1 r3) (ra + r1 + r4))),
  i1 -> -((ra (ra + r1 + r4) v0)/
      (-((ra r2 + r1 (-ra + r2 + r3)) r4) -
         (ra*rv - r1 r3) (ra + r1 + r4))),
  i3 -> -((ra r4 v0)/
      (-((ra r2 + r1 (-ra + r2 + r3)) r4) -
         (ra rv - r1 r3) (ra + r1 + r4)))}}
```

which can be verified with the substitution and `Simplify[ ]` commands,

```
Simplify [eq/.iv]
```

Mathematica responds with

```
{{True,True,True}}
```

if the solution obtained is correct for the equation. As we mentioned in Section 11.1, `/.` means to replace the variables in the equation set `eq` on the left side with the expressions of (`i1,i2,i3`) obtained. Then the command `Simplify[ ]` rearranges the terms and checks whether or not both sides in each equation are equal. We can also obtain the numerical values of the currents if the resistances and external voltage are explicitly given. For example, typing in

```
v0 = 110;
rv =   10;
r1 = 100;
r2 = 120;
r3 = 100;
r4 = 150;
ra =1000;
{i1,i2,i3}/.iv
```

instructs *Mathematica* to substitute the values and then solve for the numerical values of the currents. The results show up as

```
    1375      1331    55
{{----,  -(----),  --}}
     63       315    21
```

and we can also instruct *Mathematica* to approximate the results with a given number of digits in the results. For example,

    N[iv,5]

gives

    {{i2 -> -4.2254, i1 -> 21.825, i3 -> 2.6190}}

The symbolic computing system can also perform all types of matrix operations in a very simple fashion. Here we will take the Hamiltonian matrix for the two-site block of the quantum spin-1/2 Ising model discussed in Chapter 10 as an example. The Hamiltonian matrix is given as

$$
\mathcal{H} = \begin{pmatrix} -2B & 0 & 0 & J \\ 0 & 0 & J & 0 \\ 0 & J & 0 & 0 \\ J & 0 & 0 & 2B \end{pmatrix},
\tag{11.7}
$$

where B and J are the magnetic field strength and the spin–spin interaction strength, respectively. We can input this matrix in *Mathematica* as

    h = {{-2b,0,0,j},{0,0,j,0},{0,j,0,0},{j,0,0,2b}};

The above command assigns the elements of the Hamiltonian matrix to the variable h. Now if we want to obtain the eigenvalues of the matrix, we can simply input the command

    Eigenvalues[h]

and *Mathematica* will return all four eigenvalues of the matrix,

```
                  2    2              2    2
{-j, j, -Sqrt[4 b  + j ], Sqrt[4 b  + j ]}
```

The corresponding eigenvectors can be obtained with the command

```
Eigenvectors[h]
```

and *Mathematica* returns all four eigenvectors:

```
{{0, -1, 1, 0}, {0, 1, 1, 0},


                 2     2
      2 b + Sqrt[4 b  + j ]
   {-(---------------------), 0, 0, 1},
               j


                 2     2
      2 b - Sqrt[4 b  + j ]
   {-(---------------------), 0, 0, 1}}
               j
```

which are sorted according to the order of the eigenvalues. Note that the eigenvectors are not normalized; we can take a dot product of each eigenvector with itself with the *Mathematica* command Dot[x,y] to obtain the normalization constant. We can also use Eigensystem[h] to obtain the list of both eigenvalues and eigenvectors in the corresponding order. The determinant and the inverse of the matrix can be obtained with Det[h] and Inverse[h].

11.5 Nonlinear systems

Mathematica also has built-in commands to handle nonlinear equations. For a polynomial equation, one can use Roots to obtain the roots of the equation. For example, if we would like to obtain the roots of $f(x) = x^4 + 3x^2 - 2x + 12 = 0$, we can simply type in

```
f[x_] = x^4 + 3x^2-2x+12;
Roots[f[x]==0,x]
```

and *Mathematica* will return the roots of the equation,

```
       2 - 2 I Sqrt[2]              2 + 2 I Sqrt[2]
x == --------------- || x == --------------- ||
            2                        2
       -2 - 2 I Sqrt[3]             -2 + 2 I Sqrt[3]
x == --------------- || x == ---------------
            2                        2
```

and of course the above expressions can be simplified with

```
Simplify[%]
```

which modifies the expressions to

```
x == 1 - I Sqrt[2]  ||  x == 1 + I Sqrt[2]  ||
x == -1 - I Sqrt[3]  ||  x == -1 + I Sqrt[3]
```

We can also use `Solve` to obtain the roots of a polynomial equation and sometimes the roots of an equation involving the inverse functions. For example, `Solve[Sin [x]==1,x]` will return the root x = Pi/2. However, if the equation is more complicated than polynomial equations or simple inverse functions, that is, if a simple expression is no longer available for the roots, we need to use the command `FindRoot` to obtain approximate solutions. For example, if we want to obtain the approximate solution of $f(x) = e^x \ln x - x^2 = 0$ around $x = 1.5$, we can simply input

```
f[x_]= Exp[x]*Log[x]-x^2;
FindRoot[f[x]==0,{x,1.5}]
```

and *Mathematica* returns the approximate solution $x = 1.694$. The `FindRoot [f[x]==0,x,x0]` uses the Newton method for root searching. The secant method is implemented in `FindRoot[f[x]==0, {x,x0,x1}]`, with x0 and x1 being the two points used to start the algorithm. These commands can also be used to obtain the roots of a set of equations, such as

```
f[x_,y_]= Exp[x^2]*Log[y]-x^2;
g[x_,y_]= Exp[y]*Log[x]-y^2;
FindRoot[{f[x,y]==0,g[x,y]==0},{x,1.5},{y,1.5}]
```

and will find the approximate solution $x = 1.55964$ and $y = 1.23815$ with the multivariable Newton method.

Similarly, we can search for the minimum or maximum of a function with the command `FindMinimum[ ]`. For example, the commands,

```
f[x_,y_] = (x-1)^2*Exp[-y^2]+y*(y+2)*Exp[-2x^2];
FindMinimum[f[x,y],{x,1},{y,0}]
```

give the minimum of the function as -1.01513 at $x = 0.0824999$ and $y = -0.970978$. For the maximum of a function $f(x, y)$, we can search for the minimum of $-f(x, y)$ with the same command.

The Fourier transform of a discrete set of data can also be performed in *Mathematica*. Assuming that we have a set of data $f_0, f_1, \ldots, f_{n-1}$, the discrete Fourier transform of $\{f_j\}$ is defined as

$$g_k = \frac{1}{\sqrt{n}} \sum_{j=0}^{n-1} f_j e^{2\pi i k j / n}, \tag{11.8}$$

with $k = 0, 1, \ldots, n - 1$, for example, if we have $f(x) = \exp(-x^2)$ and x is a set of 16 points evenly distributed in the region $[-1, 1]$. Now we can obtain the Fourier components with the *Mathematica* commands

```
f = Table[Exp[-(2*j/15.0-1.0)^2],{j,0,15}];
g = Fourier[f]
```

Mathematica returns the Fourier components of the discrete function set $\{f_j\}$ in a complex form. The inverse Fourier transform can be obtained in a similar manner with either the command `InverseFourier` or just the command of the Fourier transform, since the inverse Fourier transform and the Fourier transform are equivalent except a constant factor. The rounding error in the Fourier transform can be estimated from the inverse transform of the Fourier components of the same function, which would recover the exact original data if there were no rounding error.

11.6 Differential equations

So far we have used symbolic computing in a very simple way, more or less a couple of commands in each example. However, one should understand that almost all symbolic computing systems allow one to program for very complicated problems. In this section we would like to discuss programming in symbolic computing systems with a few more examples in the context of differential equations. We will also illustrate how to use the application packages that come with symbolic computing systems. All symbolic computing systems have built-in commands to solve differential equations. Some commands are for linear equations and some are for general equations. Here we would like to demonstrate some of them by studying several familiar examples, which were introduced earlier in the book.

The first example is the one we solved in Chapter 3 with the shooting method,

$$u''(x) = -\frac{\pi^2}{4}[u(x) + 1], \tag{11.9}$$

with the boundary conditions $u(0) = 0$ and $u(1) = 1$. Note that this is a linear equation and has an analytic solution, given in Chapter 3. With *Mathematica*, we can solve a linear differential equation symbolically with the command DSolve. For this specific example, we can input the equation and boundary conditions, and then solve them together with the command DSolve as

```
eq = {u''[x]+Pi^2*(u[x]+1)/4 == 0,
      u[0] == 0, u[1] == 1};
DSolve[eq, u[x], x]
```

Mathematica will return the solution of the equation as

$$\{\{u[x] \rightarrow -1 + (\frac{1}{2} + I)\ E^{-I/2\ Pi\ x} + (\frac{1}{2} - I)\ E^{I/2\ Pi\ x}\}\}$$

which is the exact result of $u(x)$ in exponential form. What we have done here is to treat the boundary conditions as two more equations and solve the problem under the equation and the boundary conditions simultaneously. Note that the problem is solved completely symbolically.

Just to show the variation of the programming, we can achieve the same solution by first obtaining the general solution with

```
DSolve[u''[x]+Pi^2*(u[x]+1)/4 == 0, u[x], x]
```

and *Mathematica* returns the general solution

$$\{\{u[x] \rightarrow -1 + E^{-I/2\ Pi\ x}\ C[1] + E^{I/2\ Pi\ x}\ C[2]\}\}$$

with C[1] and C[2] being the coefficients to be determined. We obtain the coefficients from the given boundary conditions $u(0) = 0$ and $u(1) = 1$ with

```
Solve[{C[1]+C[2]-1==0, C[2]-C[1]+2I==0}, {C[1],C[2]}]
```

The output of the above command is

$$\{\{C[2] \rightarrow \frac{1}{2} - I, C[1] \rightarrow \frac{1}{2} + I\}\}$$

which are the desired coefficients. Sometimes we prefer to have the solution expressed in trigonometric functions. This can be done with the application package

Trigonometry.m, which is part of the *Mathematica* distribution. We can input the package with a command at the beginning of the program. Now if we modify the earlier program to

```
<<Trigonometry.m;
eq = {u''[x]+Pi^2*(u[x]+1)/4 == 0,
      u[0] == 0, u[1] == 1};
sl = DSolve[eq, u[x],x];
ut = ComplexToTrig[u[x]/.sl]
```

Mathematica will return the solution in the trigonometric form as

```
         1                 Pi x            Pi x
{-1 + (- + I) (Cos[----] - I Sin[----]) +
         2                  2               2


         I          Pi x           Pi x
    (1 + -) (-I Cos[----] + Sin[----])}
         2           2              2
```

which can be further simplified with

```
us = Simplify[ut]
```

Then *Mathematica* returns the desired result

```
           Pi x            Pi x
{-1 + Cos[----] + 2 Sin[----]}
            2               2
```

which is exactly what we had in Chapter 3.

Let us now turn to a more complex initial-value problem with nonlinear terms. A general initial-value problem is defined by

$$\frac{d\mathbf{y}}{dt} = \mathbf{f}(\mathbf{y}, t), \tag{11.10}$$

where both $\mathbf{y}(t)$ and $\mathbf{f}(\mathbf{y}, t)$ are vector arrays. The problem is solved for a specific given initial condition $\mathbf{y}(0) = \mathbf{y}_0$.

As we discussed in Chapter 3, the problem can be solved by the fourth-order Runge–Kutta algorithm. This algorithm has been written into a package called RungeKutta.m in *Mathematica* (Maeder, 1991), and it is now part of the *Mathematica* distribution. To use the package, one can simply list the velocity field **f**, first-order derivative of variable **y**, initial value **y**$_0$, and time variable with starting and ending values. Let us take the driven pendulum with damping as an example. The equation set for the system is given by

$$\frac{dy_1}{dt} = y_2, \tag{11.11}$$

$$\frac{dy_2}{dt} = -qy_2 - \sin y_1 + b \cos \omega_0 t, \tag{11.12}$$

if we take the same values used in Fig. 3.2 in Chapter 3, that is, $\omega_0 = 2/3$, $q = 0.5$, and $b = 0.9$. The time interval is taken as $\tau = 0.03\pi$. The initial conditions are taken as $y_{10} = 0$ and $y_{20} = 2.0$. The following *Mathematica* program uses the application package RungeKutta.m and shows the result in a two-dimensional plot:

```
<<RungeKutta.m;
      RungeKutta[{y, -0.5y-Sin[x]+0.9Cos[2t/3]},
                  {x, y}, {0.0, 2.0},
                  {t, 0, 30Pi, 3Pi/100}];
      Show[Graphics[{Line[%]}],
            Axes->Automatic, AspectRatio->Automatic,
            Frame->True, AxesLabel->{theta,omega}]
```

which produces a figure almost identical to Fig. 3.2 with a doubled period of the driving force. For comparison, let us show the output from *Mathematica* in Fig. 11.1. Of course, we can also use the differential equation solver NDSolve from *Mathematica* to solve this initial value problem in a similar fashion. For example, the above problem can be solved with the following *Mathematica* program:

```
eq = {x'[t]  == y[t],
       y'[t]  == -0.5y[t]-Sin[x[t]]+0.9Cos[2t/3],
       x[0]   == 0.0,
       y[0]   == 2.0};
r   = NDSolve[eq,{x, y}, {t,0,30Pi}, MaxSteps->1500];
ParametricPlot[Evaluate[{x[t],y[t]}/.r], {t,0,30Pi},
                Frame->True, AxesLabel->{theta,omega}]
```

which generates a figure similar to the output of the earlier Runge–Kutta program. We can modify the parameters in the above programs to study the chaotic behavior of the pendulum system. In practice, programming in symbolic computing systems

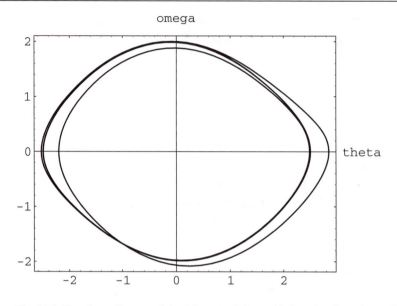

Fig. 11.1. The phase diagram of the driven pendulum with damping from the application of the Runge–Kutta package in *Mathematica*.

is similar to programming in other advanced languages. One has a choice in style, algorithm, and format. The goal, however, is always the same, that is, to solve the problem first and then to improve the efficiency if possible.

11.7 Computer graphics

Visualization of a process is extremely important in the understanding of many problems in science and engineering, especially when the problems are so complex that we may never have an analytical solution, not even an asymptotic solution in some cases. Then the understanding of the problem will rely mainly on our analysis of the numerical results or on graphic visualization of the approximate solution. This means that we need graphic description of the solution in order to establish a thorough understanding of complex problems. We will not cover many aspects of computer graphics, because a detailed discussion can be very tedious. However, the subject is extremely important and useful in practice and should be given some attention. For example, we can visualize the dynamics of atoms and molecules at the surface of a liquid by creating a "movie" within a molecular dynamics simulation of the surface with the visualization software programs now available. To create a movie of a many-body system is quite a task in itself. Ordinarily we would not need such complicated tools to create figures and plots of the solution of a problem. Most figures will still be plain plots in two or three dimensions.

Computer graphics is also a very important aspect of any symbolic computing system, since in most cases the outputs have to be in the form of graphics in order to be visualized. Almost all symbolic computing systems come with a complete graphics system interfaced. The graphics in a symbolic computing system is implemented with many options, ranging from the simplest *xy* curves to color-coded three-dimensional surface plots.

In this section, we will just show a simple example rather than discuss all the details of computer graphics. We will still use *Mathematica* plot commands in our illustrative example. Let us analyze the two plot commands used in Section 11.6, Show[Graphics[ ]] and ParametricPlot[ ]. We would like to study the chaotic behavior of the driven pendulum with damping, as discussed in Section 11.6, and would like to have the output in graphics with a point plot. We can achieve it with the following *Mathematica* program:

```
<<RungeKutta.m;
    RungeKutta[{y, -0.5y-Sin[x]+1.15Cos[2t/3]},
               {x, y}, {0.0, 2.0},
               {t, 0, 30Pi, 3Pi/100}];
    data = Table[Mod[%+Pi,2Pi]-Pi];
    ListPlot[%, Axes->Automatic, AspectRatio->Automatic,
             Frame->True, AxesLabel->{theta,omega}]
```

The command Mod[ ] is used to move the data back to the region of one period of the angle $\theta \in [-\pi, \pi]$ and the command Table[ ] is used to create a list plotted with points with the command ListPlot[ ]. Note that the command Mod[ ] also acts on ω but does not modify the data, since ω is in the region $[-\pi, \pi]$ for the given conditions. The output of the above program from *Mathematica* is shown in Fig. 11.2. The data points are less dispersed, since we only have 1,000 time steps instead of 10,000 steps used in Chapter 3 to generate Fig. 3.4. Note that if we use the same plot commands used in the Section 11.6, we will have a problem with the lines between the points wrapped back to the region $[-\pi, \pi]$ and the unwrapped data points, because the behavior of the variables is no longer periodic. To convince oneself, if *Mathematica* is available, try to plot the above result with the following commands,

```
    data = Graphics[{Line[Mod[%+Pi,2Pi]-Pi]}];
    Show[%, Axes->Automatic, AspectRatio->Automatic,
         Frame->True, AxesLabel->{theta,omega}]
```

One can in general control the output of the figures in style, shape, color, and intensity to have the data visualized in any desired way. Readers are strongly encouraged to try all the different options with the available symbolic computing systems and examine different features in computer graphics.

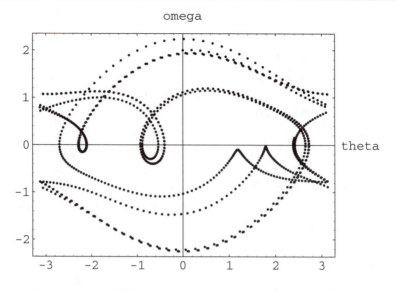

Fig. 11.2. The phase diagram of the driven pendulum with damping in a chaotic state. The data are from the first 1,000 time steps with the Runge–Kutta package in *Mathematica*, translated back to the region $\theta \in [-\pi, \pi]$.

11.8 Dynamics of a flying sphere

We have discussed several important aspects of symbolic computing. These methods and techniques can be generalized in many applications in computational physics problems. In this section, we would like go into more details of the problem-solving aspects in symbolic computing with an illustrative example. We would like to demonstrate how to combine the analysis of the physics of a specific problem with the symbolic manipulation, numerical computation, and graphic visualization available from a symbolic computing system.

The dynamics of a flying sphere is of general interest in sports, for example, the dynamics of a flying baseball, tennis ball, or golf ball. When a sphere is flying in the air, it experiences three major forces. The force of gravity acting on the object is

$$\mathbf{f}_g = m\mathbf{g}, \tag{11.13}$$

with $g = 9.8 \, \text{m/s}^2$ being the gravitational acceleration. Gravity pulls an object toward the ground. The drag force

$$\mathbf{f}_d = -\frac{c_d}{2} A \rho v \mathbf{v} \tag{11.14}$$

results from the friction between the object and the air, and it tends to slow the object down. The Magnus force

$$\mathbf{f}_m = -\frac{c_m}{2} A \rho r \mathbf{v} \times \boldsymbol{\omega} \tag{11.15}$$

is due to the rotational motion of the object, which creates an uneven relative motion on the opposite sides of the rotation and tends to move the object away from its current direction. Here $A = \pi r^2$ is the cross-sectional area of a sphere with radius r and ρ is the density of the air. In general, the coefficients c_d and c_m depend on the magnitude of the velocity $\mathbf{v}$ (that is, speed $v = |\mathbf{v}|$) and the angular velocity $\boldsymbol{\omega}$ as well as on the specifics of the object; they are nearly constant at the low-velocity limit. For a baseball the low-velocity limit is about 25 m/s. Various experiments have been conducted in order to determine the velocity and angular velocity dependences of the coefficients c_d and c_m for different objects and under different conditions. Here we will not go into a discussion of how the experiments are conducted for different objects but rather assume that the relations of the coefficients to other physical quantities are given and concentrate on the computational aspects of the model. Newton's equation for the sphere is given by

$$m\frac{d^2\mathbf{r}}{dt^2} = m\mathbf{g} - \frac{c_d}{2} A \rho v \mathbf{v} - \frac{c_m}{2} A \rho r \mathbf{v} \times \boldsymbol{\omega}, \tag{11.16}$$

with the initial conditions $\mathbf{r}(0) = \mathbf{r}_0$ and $\mathbf{v}(0) = \mathbf{v}_0$ given. The time dependence of the angular velocity can be obtained from the equation

$$I\frac{d\boldsymbol{\omega}}{dt} = \boldsymbol{\tau}, \tag{11.17}$$

if necessary. Here I is the moment of inertia and $\boldsymbol{\tau}$ is the torque around the center of the sphere.

Now let us first examine the trivial case, that is, extremely low velocity with no spinning. The object is simply a projectile that is confined in a vertical plane (for example, the xz-plane) with the initial position at the origin of the coordinate system. Then we have

$$\ddot{x} = 0, \tag{11.18}$$

$$\ddot{z} = -g, \tag{11.19}$$

with $x(0) = 0$, $z(0) = 0$, $v_x(0) = v_0 \cos\theta_0$, and $v_z(0) = v_0 \sin\theta_0$. Now we will show how to solve the equation set and show the result graphically in *Mathematica*. We will take the motion of a baseball as an example here. Let us assume that the initial height is about 2.0 m, the initial velocity is 36.8 m/s, and the total traveling time is 0.5 s. We can use the following program to perform the computation and to output the result graphically:

```
g   =  9.8;
x0  =  0.0;
z0  =  2.0;
vx0 = 36.8;
vz0 =  0.0;
t0  =  0.0;
tf  =  0.5;
eq  = {x''[t]  == 0.0,
       z''[t]  == -g,
       x[0]  == x0, x'[0]  == vx0,
       z[0]  == z0, z'[0]  == vz0};
sl =  NDSolve[eq,{x, z}, {t, t0, tf}, MaxSteps->1000];
ParametricPlot[Evaluate[{x[t],z[t]}/.sl], {t,t0,tf},
               Frame->True, FrameLabel->{x,z}]
```

The graphic output from *Mathematica* is shown in Fig. 11.3, with distance and height as the x- and z-coordinates. As one can see, the ball has dropped about 1.24 m when it reaches the base, which is about 18.4 m away. The above example is trivial in the sense that one can write down the analytic solution very quickly. However, the procedure set up here can easily be implemented in more complicated cases.

Consider now that we have drag force but no Magnus force; that is, the ball is not spinning. The trajectory will now descend faster than before. We can make some very small modifications to this program to study this case easily. Let us take

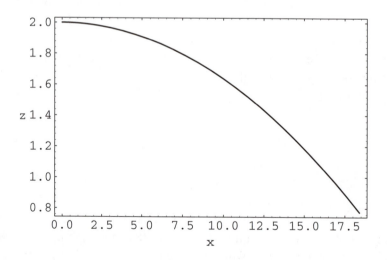

Fig. 11.3. The relation between distance and height for a forward-moving object without any resistance forces. Both height and distance are measured in meters.

$m = 0.145$ kg, $r = 0.037$ m, $\rho = 1.21$ kg/m^3, and $c_d = 0.3$. The modified *Mathematica* program for a baseball traveling under this drag force is

```
cd  =   0.3;
m   =   0.145;
rho =   1.21;
r   =   0.037;
c   =   cd*Pi*r^2*rho/(2*m);
g   =   9.8;
x0  =   0.0;
z0  =   2.0;
vx0 =   36.8;
vz0 =   0.0;
t0  =   0.0;
tf  =   0.51;
eq  =   {x'[t]  ==  vx[t],  z'[t]  ==  vz[t],
            x''[t]  ==  - c*Sqrt[x'[t]*x'[t]
                    + z'[t]*z'[t]]*x'[t],
            z''[t]  ==  - g-c*Sqrt[x'[t]*x'[t]
                    + z'[t]*z'[t]]*z'[t],
            x[0]  ==  x0,  x'[0]  ==  vx0,
            z[0]  ==  z0,  z'[0]  ==  vz0};
sl  =   NDSolve[eq,{x, z}, {t, t0, tf}, MaxSteps->1000];
ParametricPlot[Evaluate[{x[t],z[t]}/.sl], {t,t0,tf},
                Frame->True, FrameLabel->{x,z}]
```

which generates a trajectory bent a little further at the end. Similarly, the effect of the Magnus force can be included if we assume that the angular velocity is a constant. This generates the effect of a curveball. Note that if the angular velocity is downward, the ball will move toward the right. If the angular velocity is upward, the ball will move toward the left. We will also need the y-component of the equation now, since the motion of the ball is three-dimensional. We will leave it as an exercise for the reader to program and study the curveball case. For more discussions on the physics of baseball, see Adair (1995) and references therein.

Exercises

11.1 Use a symbolic computing system to show that

$$a^2 + b^2 \geq 2ab$$

for any real variables a and b.

11.2 Determine the coefficients in each term of the power series expansion of $f(x, y) = (x + y)^n$ for a given integer n. x and y are real variables.

11.3 Find the first twelve nonzero terms in the Taylor expansion of $f(x) = \exp(-x^2 - \cos x - 3 \sin x)$ around $x = 0$.

11.4 Explore the integration command of a symbolic computing system with the following integrals:

$$J_0(x) = \frac{1}{2\pi} \int_{-\pi}^{\pi} e^{ix \cos \theta} d\theta,$$

$$J_1(x) = \frac{1}{x} \int_0^x y J_0(y) \, dy,$$

$$S_F(k) = 1 - 2k_F^2 \int_0^\infty \frac{J_1^2(k_F r) J_0(kr)}{(k_F r)^2} r \, dr,$$

where $J_0(x)$ and $J_1(x)$ are the Bessel functions of the first kind of order zero and order one, respectively, and $S_F(k)$ is known as the structure factor of a two-dimensional free fermion gas at the wavevector k with a Fermi wavevector k_F. The expression of $S_F(k)$ is known as

$$S_F(k) = \begin{cases} \frac{2}{\pi} \sin^{-1}\left(\frac{k}{2k_F}\right) + \frac{k}{\pi k_F} \left[1 - \left(\frac{k}{2k_F}\right)^2\right]^{1/2} & \text{for} \quad k \le 2k_F, \\ 1 & \text{for} \quad k \ge 2k_F. \end{cases}$$

11.5 Use a symbolic computing system to show that the fourth-order Runge–Kutta algorithm satisfies the required conditions for the desired accuracy. Find another set of coefficients that can provide the same order of accuracy.

11.6 Derive the Numerov algorithm for the Sturm–Liouville equation outlined in Section 3.8 with an accuracy on the order of up to $O(h^6)$ on a symbolic computing system.

11.7 Apply the differential equation solver on a symbolic system for the Sturm–Liouville equation

$$[p(x)y'(x)]' + q(x)y(x) = s(x),$$

with
(a) $p(x) = 1 - x^2$, $q(x) = l(l + 1)$, and $s(x) = 0$, with $l = 0, 1, \ldots, \infty$;
(b) $p(x) = x$, $q(x) = 1 - m^2/x^2$, and $s(x) = 0$, with $m = 0, \pm 1, \ldots, \pm \infty$.

11.8 Use the numerical differential equation solver on a symbolic computing system to solve the Duffing model defined in Exercise 3.4. Illustrate the phase diagram of $(x, \dot{x})$ and the power spectrum of x in the graphic output of the program. Find the parameter region where the system is chaotic.

11.9 Solve the one-dimensional Schrödinger equation on a symbolic computing system with the potential defined in Exercise 3.6. Discuss the behavior of the wavefunction for both bounded states and scattering states.

11.10 Study the curveball in a baseball pitch. If the pitcher throws out the spinning ball at 75 mph (33.5 m/s) horizontally, what is the deflection distance of the baseball from its intended direction when it is caught by the catcher? Assume that the distance between the pitcher and the catcher is 18.4 m, the mass and radius of the baseball are $m = 0.145$ kg and $r = 0.037$ m, respectively, the drag force and Magnus force coefficients are $c_d = 0.30$ and $c_m = 0.40$, the density of the air is $\rho = 1.21$ kg/m^3, and the angular velocity of the baseball is $\omega = 31$ rad/s along the vertical axis.

12

High-performance computing

In the past twenty years we have experienced the rapid development of super-computing: Many grand challenges have been conceived and studied since vector processor supercomputers and parallel computing systems became available. This development is still a rapid ongoing process. However, we are at a turning point with the concept of concurrent computation implemented in scientific computing, or to use a more general term, heterogeneous computing, which combines all the available facilities with multiprocessor and/or multiplatform systems into a sole giant resource to solve large-scale scientific computing problems. This concept, if successfully implemented, will increase the capacity of the available resources by orders of magnitude.

For example, if a single task requires a huge amount of computing power for a few nights, one can construct a virtual computing system that can have conventional vector processor supercomputers, the new scalable multiprocessor computers, and individual workstations linked through a network. For just a few nights, one can perform a single task on this virtual system. After these few nights, or even just during the daytime, each of the networked computers can act as a single resource to deal with the daily tasks of other users.

At the same time, one can also partition a multiprocessor system into segments, with each working on a different individual task. In reality, there are hundreds more tasks that may require only the moderate computing power of a single microprocessor. So the same resource can be used to deal with many small computing problems and with the grand challenges at the same time.

12.1 The basic concept

What exactly is meant by high-performance computing is rather a difficult question to answer. The answer is partially dependent on the availability of technology. So the definition of high-performance computing evolves with time. Today's high-performance computing may become an ordinary task in five years. But one thing that is certain about high-performance computing is that it will always represent the cutting edge of scientific computing.

In this chapter, we will not be able to cover this fast-developing subject in great detail. However, the transition from a vector supercomputing model to a distributed or parallel computing model forces us to reformulate our algorithms, and we do need to know some basic aspects of parallel and distributed computing in order to take advantage of the intrinsic parallelism of physics problems in the new computing environments. So we will give a brief discussion on the subject here and cite some references for readers who want to know more.

The vector processors introduced in the middle of the 1970s have brought a huge improvement in the computing capacity of a single system. The concept of vector processing is rather simple. Imagine that we are dealing with operations for vectors in a vector space with a large number of components, for example, $\mathbf{a} = (a_1, a_2, \ldots, a_n)$, with n typically greater than 100. If we want to compute the addition of one vector to another multiplied by a scalar coefficient,

$$\mathbf{c} = \mathbf{a} + \alpha\mathbf{b}, \tag{12.1}$$

on a single scalar processor, it will take an amount of computing time that is proportional to n, the number of dimensions of the vector, to complete the operation. However, on a vector processor computer, it will take considerably less time. Assume that a vector processor can deal with m pairwise additions concurrently; then the time to complete the above vector addition is proportional to n/m on the vector processor computer, which is a significant increase in speed over the scalar machine with comparable processor speed. Many important simulations have been performed on vector supercomputers during the last decade. There are lots of vector problems in physics that can easily be implemented in the vector processor environment.

However, there are limitations on vector processor supercomputers. There are problems in physics that cannot be vectorized. The cost of the maintenance of these giant systems is extremely high. The speed of a single processor is also limited by the fundamental laws of physics; signals cannot be passed faster than the speed of light. The high cost is still a problem even for some massively parallel computers with sequential processors.

The development of scientific computing in the last decade has forced us to rethink what it means when information and data are handled on computers. *High-performance computing* is an integrated term, that represents all the latest advances in hardware, software, networks, and algorithms. However, several key elements will remain within the definition of high-performance computing in the future even though the units of measurements may change: the maximum speed of a single processor, memory access, and interprocessor communication; the scalability of the multiprocessor system; the ratio of the computing power to the cost of the system; and above all, performance.

12.2 High-performance computing systems

Even though the definition of high-performance computing evolves with time due to the development of technology, the concept is always associated with the measure of the performance of a computing system and the best technology available.

The speed of a single processor plays an important role in achieving high performance in a system. The speed of a processor is measured by how many cycles can be completed in one second. For example, if a processor can complete 100,000,000 cycles per second, we say the processor has a *clock speed* of 100 MHz (megahertz). A cycle itself is the completion of a task corresponding to one of the basic operations of computation, for example, an OR or AND operation in logic computing. Most current microprocessors can deliver more than one instruction in each cycle. Here an instruction is associated with typical calculations; for example, the addition of two integers,

$$I_3 = I_1 + I_2, \tag{12.2}$$

is an integer addition instruction, and the subtraction of one floating-point number from another,

$$F_3 = F_1 - F_2, \tag{12.3}$$

is a floating-point subtraction instruction. A floating-point operation usually takes longer that an integer operation to complete with a single computing unit. The average number of instructions carried out in one clock cycle is determined by the number of units in the processor. There are separate integer units for integer operations, floating-point units for floating-point operations, and other units, such as branch units for data branching. Most computing time in physics problems is spent on performing floating-point operations. However, if a program is written purely in integers, only the integer units on the processor will be used and the performance is then limited by the integer performance of the processor. If a program requires a lot of load and storage operations, the speed of the computation can be dominated by the load/storage speed of the processor.

One way to measure the performance of a system is by the number of floating-point operations per second (flop) delivered by the processor. The common units used for current computing systems range from megaflops (10^6 flops) to gigaflops (10^9 flops). We may have to use teraflops (10^{12} flops) as the unit of measurement of the computing power of a system in a few years. A computing system can be measured according to its maximum performance (peak performance) or its sustained performance (average performance). Specially designed computing systems may deliver near-peak performance for a specific problem but be of no use for other problems. A general-purpose system cannot deliver its peak performance for a variety of users and tasks. So average performance is the correct measure of a general-purpose system.

Another important issue regarding a computing system is the ratio of its computing power to its cost. This is a very important measure, since financial resources are always limited.

The suitability of a computing task being performed on a multiprocessor system is determined from the scalability of the task, which depends on the problem, the algorithm applied, and the architecture of the system. The factor used to measure the scalability is called the *speedup*. The actual speedup is limited by the ideal speedup, that is, the greatest speedup that can be achieved for a given problem. Assuming that a fraction of a task, $0 < q < 1$, is intrinsically sequential, the ideal speedup on an n-processor system is given by

$$s_n = \frac{n}{1 + q(n-1)},$$
(12.4)

which is known as the Amdahl law. Usually the fraction q decreases with the size of the task.

Traditional supercomputers use vector processors or attached processor arrays to achieve parallelism in vector and matrix computations. A vector processor is based on a pipeline concept, which is similar to an automobile assembly line: Each unit completes a designated part of the task and passes it to the next unit in line to complete another part of the task. This type of architecture is further developed and simplified in the new microprocessors that form the core of current workstations and massively parallel systems. The microprocessors used in the current parallel or distributed computing systems are mainly based on the RISC (Reduced Instruction Set Computer) architecture. The RISC idea was a very important breakthrough in manufacturing fast and powerful microprocessors now used in almost all major computing systems. The RISC architecture came from an observation that almost all computing tasks can be reduced to four basic steps: fetch, decode, execute, and store. Thus one can produce microchips with only these four basic steps and pipeline all the tasks. Three aspects associated with RISC architecture are used in microprocessors to speed up computation. Instructions are executed in pipeline order with four separate basic steps in one clock cycle. A fast memory unit (called cache memory) is attached to the processor on a chip. Multiple instruction units are built into a single processor. For example, a processor can have two integer units, four floating-point units, a load/store unit, and a branch unit for the operations that do not have exact four-step decompositions. Another trend in microtechnology is to implement 64-bit processors, which will automatically give double the precision for floating-point calculations and allow data to be moved in eight byte segments instead of four byte segments within 32-bit processors.

The performance of a single processor is limited by two factors of the chip: the clock cycle and the average number of instructions the processor can deliver in each clock cycle. The clock speed can be increased. For example, the fastest microprocessors can now deliver a clock speed of 500 MHz and it may still go up in the near future. The average number of instructions per cycle has moved up to eight.

Most current high-performance computing systems can be roughly divided into two categories: parallel computers and distributed computing platforms.

Parallel computers are the ones with multiple processors tightly connected through a fast system bus or network to pass the information among the processors and their fast memory units. The system can have a single memory unit shared by all the processors or many memory units with each associated with a processor. There are several important issues related to a parallel system. The most important one is how the processors are physically connected. The topology of the connections determines the possible speedup of the system for a specific computing task. While most current parallel systems deliver good speedup with a small number of processors involved, saturation is eventually reached because of the limits of the capacity of the system bus or network. The capacity of the system bus or network is measured in terms of its bandwidth, the total amount of the data that can be transferred in a unit of time, and the data is measured in bytes. For example, the fastest connection would be one with all processors connected directly. However, it would be impossible to do this if we have, for example, 60,000 processors in the system. So while a direct connection would work well for a small parallel system, it is simply impossible to achieve for a massively parallel system. An alternative is to construct some hierarchical network to connect a few processors at the first level and then a few clusters at the second level and so on. For each cluster of processors, a multiple connection switch can be devised and used. The average time it takes to pass a message from one processor to another processor increases with the total number of processors in the system. If all processors are connected by a hierarchical switchboard, the average communication time between any two processors increases logarithmically with the total number of processors in the system.

A more promising high-performance system is a cluster of workstations loosely connected through a network. Workstations are considerably less expensive than a massively parallel system with the same number of processors and require very little maintenance. Thousands of workstations can perform daily tasks individually during the daytime and then act as a concurrent resource at night to deal with grand challenge problems. Microprocessor speed is so high now that the bottleneck is purely in the network bandwidth if a task requires significant message-passing among all the participating processors. The ongoing transition from the traditional STM (synchronous transfer mode) network to the ATM (asynchronous transfer mode) network will increase the usage of a network drastically. The bandwidth of a network is divided into many equal and fixed segments. An STM network allows messages passed between a pair of processors on a designated segment. So a processor on the network can communicate with another only through one segment of the bandwidth. When the message is larger than the segment can hold, part of the message has to wait for the same segment in the next cycle of communication. This way of passing information wastes a large percentage of network capacity, since most segments are empty most of the time but cannot be used by processors other than the designated pair. ATM allocates the segments to the processors according to availability rather than simply for the designated pair. This allows a processor to pass large messages quickly if segments are available on the network.

There are also issues having to do with the speed of the memory units, largely because the increase of the clock speed of the microprocessors. This memory bottleneck is currently handled with a memory hierarchy, that is, with a small, fast memory unit (cache memory) very close to each processor but a larger, slower memory unit at the bottom of the hierarchy. The speed of the storage unit is also important if a task requires a lot of data input and output. There are also new technologies used to increase the reliability and performance of the disk drives. For example, most current computers use mirroring to improve reliability and striping to improve performance.

The ultimate goal of high-performance computing, of course, is to utilize the available resources with maximum flexibility. The current practice in some super-computing centers has indicated the trend of development: A future high-performance system will be a system with different architectures and even different physical locations networked at different levels; but to a user, it will be as transparent as a single machine is now, and the user will be able to request the allocation of computing power. One will be able to request that machines on different continents be connected to form a giant system to model global environmental change for a couple of days and then use a single vector processor computer with a huge memory unit to model the detail of the climate in San Francisco Bay. The concept of scientific computing may not be too different from what it is now, but the systems will surely be different.

12.3 Parallelism and parallel computing

The need for parallel computing is due to the fact that many problems in scientific computing have inherent parallelism in nature. A trivial example is the calculation of a very similar set of physical quantities with different given conditions, such as the evaluation of the statistical quantities of a collection of particles with different temperature, pressure, or external field. This is trivial, because there is no correlation between the two different calculations.

A typical physical system in a large computer simulation is either a continuous system with many particles or a discrete lattice system with many lattice sites. For the calculation of a specific quantity at a given simulation time step, one can divide the system into many small cells geometrically. The behavior of the particles in a specific cell is mainly influenced by the other particles in the same cell. The behavior of the particles in each cell can then be calculated by a single processor on a multiprocessor computing system. The effects due to the interaction between the particles in different cells, or to the motion of the particles from one cell to another, can be ascertained by means of communication among the processors. When communication is needed, synchronization of the processors becomes important, since a physical system cannot violate the *causality* requirement; that is, the behavior of the current event can be affected only by past events. The causality requirement is the intrinsic limitation of a physical problem on the speedup of the parallel system. This sets up the fraction of the task that cannot be parallelized. As we pointed out in the preceding section, this fraction usually decreases with the increase of the size of the system to be simulated.

The programming model of a parallel environment can be as simple as having a single instruction on all the processors for a large set of data, the so-called SIMD (single instruction on multiple data) model. The vector processor computers and early parallel array computers are built for this programming model. The simplicity of the model makes the programming task straightforward but puts a limit on flexibility and applications. Ultimately, one would like to have an environment that can deal with multiple instructions and multiple tasks on large sets of data. The MIMD (multiple instructions on multiple data) model is highly desirable to maximize the utilization of a parallel computing system. Almost all current parallel computers are built on this programming model.

There are several paradigms for achieving parallel computing. Control parallelism is achieved with different units of the system working on different parts of a correlated task. Pipelining is one of these types of parallelism. A more practical aspect for users is data parallelism, which is accomplished with each processor in the system working on part of a set of uncorrelated data. For example, if we are performing a molecular dynamics simulation for a system of one billion particles on one hundred thousand processors of a massively parallel computing system, we can divide the system into one hundred thousand cells and dedicate one processor to deal with the approximately ten thousand particles in each cell. Assume that the cutoff distance of the interaction is $r_{ij} = r_0$. Data parallelism is achieved for the movements of the particles that are affected only by the particles in the same cell; that is, the particles are not in the layer of thickness r_0 at the boundary of the cells. The particles in the boundary layer and the particles that will move into another cell in one time step can then be dealt with under control parallelism with communication among the processors for different cells.

Now let us turn to a simple model of how parallelization can be developed on an ideal parallel system, which has an unlimited number of processors and both private and global memories accessible randomly. This system is usually called a PRAM (parallel random access machine) system.

The conventional single processor sequential computers are based on the RAM (random access machine) system. The RAM computing model consists of a memory with an unlimited number of registers, a read-only device, a write-only device, and an unchangeable program. This RAM model captures the essence of the von Neumann computer concept. We can assume that the input is a series of integers and that a read or write operation is advanced to the next number of the data automatically after each operation until the end of the data is reached. The numerical tasks performed on the RAM model can only be purely sequential.

The PRAM system is instead for the ideal parallel computing model, which has a control unit and an unlimited number of processors that can operate synchronously and communicate with each other through a shared global memory. An index is assigned to each processor as a switch, and a private memory is attached to each processor for all local operations. In each clock cycle, each processor can read from or write to a memory unit if it is allowed. The conflicts between concurrent read or write operations issued by more than one processor for a shared memory unit need further restriction. The strongest model of the PRAM system is the *priority* PRAM model, which allows concurrent read operations from a single memory unit and allows only the processor with the lowest index to write to a unit if concurrent write operations are attempted.

The measure of the computing speed of a specific task on a given system comes from its *time complexity* and/or its *space complexity*. Time complexity is the relation between the computing time and the size of the task, and space complexity is the relation between the memory needed and the task size. For example, if a one-dimensional array with n elements is read on a RAM system, the time complexity is linear in n; that is, $t_r \simeq O(n)$. The space complexity is also linear in n in this case. To simplify the analysis, the time for any single operation, for example, an addition or a multiplication, can be assumed to be the same as for any other and taken as the unit of time complexity. The unit of space complexity is assumed to be the memory size of one register. Both time complexity and space complexity are critical for the development of the actual parallel algorithms for a specified problem on a specified system. Each can have its worst case, which is termed the maximum time or space complexity.

Let us take the molecular dynamics simulation of a many-particle system as an illustrated example again. The Hamiltonian is given by

$$\mathcal{H} = \sum_{i=1}^{N} \left[\frac{\mathbf{p}_i^2}{2m_i} + U(\mathbf{r}_i) \right] + \sum_{i<j}^{N} V(\mathbf{r}_{ij}), \tag{12.5}$$

where $U(\mathbf{r}_i)$ is the external potential and $V(\mathbf{r}_{ij})$ is the interaction potential truncated at $r_{ij} = r_0$. If the density of the system is $\rho = N/\Omega$, the average number of particles within the truncated radius r_0 around a particle is

$$n_0 = \frac{4\pi r_0^3 N}{3\Omega} \simeq O(1) \tag{12.6}$$

for a very large system. Since the evaluation of the interaction is the major effort in a molecular dynamics simulation, the time complexity for a single processor is $O(N)$, provided N is very large and r_0 is just a few times the average distance between two nearest particles in the system. Of course, if the simulation is performed with a full interaction potential, the time complexity is modified to $O(N^2)$, because there are $N(N-1)/2$ pairs of interactions to be evaluated. The time complexity of the molecular dynamics simulation with the truncated interaction potential on an n-processor parallel computer is then

$$t_d = O[\log_a(N/n)], \tag{12.7}$$

where a is an integer that depends on how the processors are connected.

One can simulate the priority PRAM model with a double exclusion PRAM model, that is, a model that does not allow any concurrent read or write operations on a single unit of the shared global memory. The time complexity of the simulated priority PRAM model is given by

$$t_p = \alpha t_d, \tag{12.8}$$

with $\alpha \geq 1$. One should have noticed that the PRAM models are very close to the SIMD environment of an actual parallel computer system.

PRAM algorithms can then be designed for the PRAM models. Let us here examine a very simple example. Imagine that we want to find the convergence and the asymptotic value of a series

$$s = \sum_{k=1}^{\infty} a_k, \tag{12.9}$$

where a_k is a function of the index k. Assume that we have n processors in the system. The algorithm can be divided into two steps. First, we can sum the n segments of the series with each segment performed on one processor,

$$b_i = \sum_{k=1}^{m} a_{k+m(i-1)},$$ (12.10)

for $i = 1, 2, \ldots, n$, where m is a selected integer and can be modified in the test of convergence. The second step of the algorithm is to perform the summation

$$c_j = \sum_{i=1}^{j} b_i,$$ (12.11)

for $i = 1, 2, \ldots, n$, and then store c_j back to the memory by the ith processor. This type of summation is called the *prefix sum*. The convergence of the series can now be examined with the extrapolation scheme discussed at the very beginning of this book for the approximate evaluation of π. An alternative algorithm can be designed to study the convergence and asymptotic behavior of the series. For more discussions on PRAM algorithms, see Quinn (1994, pp. 25–51).

12.4 Data parallel programming

In order to take the advantage of high-performance parallel computing systems, several advanced languages have been developed to help users accomplish parallel computing without too much modification of their existing programs or programming habits. Most parallel programming languages are either Fortran-based or C-based, for example, Fortran 90, HPF (High-Performance Fortran), C*, and C**. In this section, we will just describe some basic features of Fortran 90, which is available on almost all high-performance computing systems now, and HPF, which is partially implemented on most systems.

Fortran 90

Fortran 90 is a new version of the Fortran language with the traditional Fortran 77 as a subset. It includes all the standard statements of Fortran 77 but is extended to have features that are extremely suitable for parallel architectures. Even though Fortran 90 is designed to preserve all the features of Fortran 77, there are some significant changes in its style, array structure, and data control.

The new style of Fortran 90 reflects the flexibility of current computer systems. The column convention in traditional Fortran is no longer required; one can start a statement from any place on a line. No line indices are needed. More than one statement can appear on a single line. Comments can start anywhere after the symbol !. The continuation of a statement is achieved with the symbol & at the end of the line. & also appears at the beginning of the continuation line if the breaking point is in the middle of a word string. A line can be as long as 132 characters. All procedures, for example, programs, subroutines, and loops, are ended with the END statement. Empty spaces are meaningful now. The data type and value and other related characteristics can be all assigned at the data declaration. Common blocks are replaced by modules. The format is constructed inside the READ and PRINT or WRITE statements. For details of Fortran 90 specifications, see Adams et al. (1992), and for discussions on programming with Fortran 90, see Brainerd et al. (1990) and Metcalf and Reid (1990). We will highlight some basic features of Fortran 90.

To illustrate the new style of Fortran 90, the following Fortran 90 subroutine converts polar coordinates into rectangular coordinates.

```
SUBROUTINE RPHI_TO_XY (R,PHI,X,Y)
  IMPLICIT NONE
  REAL, INTENT (IN) :: R,PHI
  REAL, INTENT (OUT) :: X,Y
  X = R*COS(PHI); Y = R*SIN(PHI) ! Assign x and y
END SUBROUTINE RPHI_TO_XY
```

This subroutine contains some features that are not in Fortran 77. For example, there is no RETURN statement, the intentions (input, output, or both) of the variables are explicitly given, and multiple statements exist on the same line. The IMPLICIT NONE statement is an added feature in Fortran 90. With this statement, all the variables have to be defined. This avoids wrong results due to any typographic errors. Some Fortran 77 compilers have already had this statement as a machine extension.

The most significant new features in Fortran 90 are in its array structures and data manipulations. An array dimension can now be dynamically allocated; the indices can be implicit, and segments of the array can be extracted in the computation. Built-in intrinsic functions can find all the information about the array, for example, its maximum or minimum elements and its size and shape, and perform many array operations, for example, the summation of the array elements, the product of the array elements, the product of two matrices, the transpose of a matrix, and so on. Let us see a simple example to show the array operations and structures.

```
PROGRAM ARRAY_EXAMPLE
  IMPLICIT NONE
  REAL :: TWO_PI
  REAL, ALLOCATABLE :: A(:,:), B(:,:), C(:,:), D(:)
```

```
      INTEGER :: N,M,L,I
 !
      TWO_PI = 8.0*ATAN(1.0)
      READ "(3I4)", N, M, L
      ALLOCATE (A(N,M)); ALLOCATE (B(L,N))
      ALLOCATE (C(L,M)); ALLOCATE (D(M))
      CALL RANDOM_NUMBER (A); CALL RANDOM_NUMBER (B)
      A = SIN(TWO_PI*A); B = COS(TWO_PI*B)
      C = MATMUL(B,A)
      DO    I = 1, M
         D(I) = DOT_PRODUCT(A(:,I),B(I,:))
      END DO
      PRINT "(8F10.6)", D
   END PROGRAM ARRAY_EXAMPLE
```

We have used in the above example the built-in random number generator
RANDOM_NUMBER to assign all the elements in the matrices in the region [0, 1].
We have also used the IMPLICIT NONE to require that every variable be defined in
advance. This is a highly recommended practice. MATMUL is the intrinsic matrix
multiplication function, and DOT_PRODUCT is the intrinsic function for the dot product of two vectors. The arrays are defined as dynamical arrays, and the dimensions
are assigned later with the ALLOCATE statement. The intrinsic functions such as
SIN and COS are extended to handle the array arguments and return the values of
elements under the given function operation. Note that the format of the data is in
the READ and PRINT statements.

Fortran 90 also allows procedures, generic functions, and derived types to be
introduced through a MODULE that can be used by any other part of the program
with the command USE. The variables in a module are public unless declared private.
Subroutines no longer need to be placed at the end of the main program; they can
appear anywhere in the program with the CONTAINS command. Let us here illustrate
the use of the module in the following program, which generates an array of uniform
random numbers and then arranges them in incremental order.

```
   MODULE ORDER_AN_ARRAY
     PRIVATE EXCHANGE
 !
     CONTAINS
 !
     SUBROUTINE REARRANGE (A)
       IMPLICIT NONE
```

```
          REAL, INTENT(INOUT) :: A(:)
          LOGICAL, ALLOCATABLE :: MASK(:)
          INTEGER :: I, N
          INTEGER, DIMENSION(1) :: K
          N = SIZE (A)
          ALLOCATE (MASK(N))
          MASK = .TRUE.
          DO I = 0, N-1
            MASK(N-I) = .FALSE.
             K  = MAXLOC(A,MASK)
            CALL EXCHANGE(A(K(1)),A(N-I))
          END DO
        END SUBROUTINE REARRANGE
    !
        SUBROUTINE EXCHANGE (X,Y)
          IMPLICIT NONE
          REAL, INTENT(INOUT):: X,Y
          REAL TX
          TX = X; X = Y; Y = TX
        END SUBROUTINE EXCHANGE
    !
    END MODULE ORDER_AN_ARRAY
    !
    PROGRAM RANDOM_ARRAY_ORDERED
      USE ORDER_AN_ARRAY
    !
      IMPLICIT NONE
      INTEGER, PARAMETER :: N = 100
      REAL, DIMENSION(N) :: A
      INTEGER :: I
    !
      CALL RANDOM_NUMBER (A)
      CALL REARRANGE (A)
      WRITE(6, "(F10.8)") (A(I),I=1,N)
    END PROGRAM RANDOM_ARRAY_ORDERED
```

This example program shows how a module plays its role in a program and can be repeatedly used by other parts of a program. The variables that are not needed by any other parts can be declared private variables.

Another useful feature of Fortran 90 is the interface construct. An interface is used in a program when there are external functions or subroutines that will be linked to the program after compiling. The following is an example of a simple interface construct for two external functions.

```
INTERFACE
  FUNCTION FXY(X,Y) RESULT (F)
    REAL :: F
    REAL, INTENT(IN) :: X,Y
  END FUNCTION FXY
!
  FUNCTION GNX(N,X) RESULT (G)
    REAL :: G
    INTEGER, INTENT(IN) :: N
    REAL, INTENT(IN) :: X
  END FUNCTION GNX
END INTERFACE
```

One can have more than one item in an interface construct that contains the necessary information for the compiler to check the logical connection of the program and the external parts that will be linked later. The interface can also be assigned a generic name but does not need to end with the name.

Fortran 90 has also introduced a pointer feature, which can symbolically construct a link or a list to the target it points to. This is useful feature in other languages and has finally entered Fortran.

One does not need to write any loops now for a recursive function or subroutine. The subroutine or function is allowed to call itself in the new provision. For example, the Legendre polynomial introduced in Section 5.7,

$$(l + 1)P_{l+1}(x) = (2l + 1)x P_l(x) - l P_{l-1}(x),$$ (12.12)

with $P_0(x) = 1$ and $P_1(x) = x$, can be generated simply with

```
RECURSIVE FUNCTION PL(N,X) RESULT (P)
  REAL :: P
  REAL, INTENT(IN):: X
  INTEGER, INTENT (IN):: N
  IF (N==0) THEN
```

```
      P = 1.0
    ELSE IF (N==1) THEN
      P = X
    ELSE
      P = ((2.0*N-1.0)*X*PL(N-1,X)+(N-1)*PL(N-2,X))/N
    END IF
  END FUNCTION PL
```

This recursive provision has simplified the programming significantly. All these features in Fortran 90 make it a good choice for high-performance computing. Most current parallel computers use the Fortran 90 statements and some extensions in the directives that are similar to HPF directives, for example, CM Fortran for CM-2 and CM-5 computers. Interested readers can find some very interesting discussions on these aspects in the book edited by Sabot (1995).

High-Performance Fortran

As mentioned, the Fortran 90 programming language is a good start for developing a data parallel programming environment. The efforts of the HPFF (High-Performance Fortran Forum) are an important step toward standardization of data parallel programming specifications. The HPFF is a group of people from industries, universities, and national laboratories with the mission of developing a set of extensions of Fortran 90, now known as the HPF (High-Performance Fortran), aimed at setting up a set of standards for data parallel programming. Its main goal is to have an architecture-independent high-level language for all parallel computers. The idea behind HPF is to maintain traditional programming style and structure and let the computer compiler perform the task of data distribution and communication among different processors.

The major extensions of HPF that are not in Fortran 90 are the FORALL statement, which is the key statement to complete data parallel programming instructions, a set of HPF directives that are used to allocate data in a specified manner, and new library and intrinsic function routines for process monitoring and processor communication. Some restrictions on the data storage structures are introduced in order to achieve the desired parallelism.

Just to show the practical side of HPF, we here give an example of a program for solving the Laplace equation

$$\nabla^2 u = 0 \tag{12.13}$$

in two dimensions, with the boundary conditions $u(0, y) = y/2$, $u(1, y) = y^2/4$, $u(x, 0) = 0$, and $u(x, 2) = 1$.

```
PROGRAM LAPLACE_2D
!
! Example program for the 2D Laplace equation.
!
  IMPLICIT NONE
  INTEGER, PARAMETER :: NX=101,NY=201,IMAX=20
  INTEGER I
  REAL, DIMENSION (NX,NY) :: BOUND,FIELD,U,UTMP
  REAL DU,TL,Q,P,A,B
  LOGICAL, DIMENSION (NX,NY) :: MASK
  LOGICAL IOUT
!
  IOUT = .TRUE.
  TL = 1.0E-08
  A  = 1.0/(NY-1)
  B  = 1.0/(NY-1)**2
  P  = 0.5
  Q  = 1.0-P
!
! Align the data accordingly
!
!HPF$ DISTRIBUTE U(BLOCK,BLOCK)
!
!HPF$ ALIGN BOUND(:,:) WITH U(:,:)
!HPF$ ALIGN FIELD(:,:) WITH U(:,:)
!HPF$ ALIGN UTMP(:,:)  WITH U(:,:)
!HPF$ ALIGN MASK(:,:)  WITH U(:,:)
!
  MASK  = .FALSE.
  MASK(2:NX-1,2:NY-1) = .TRUE.
!
  BOUND = 0.0
  FIELD = 0.0
  WHERE (MASK)
    FIELD = 0.25
  ELSEWHERE
    BOUND = 0.25
  END WHERE
!
! Assign the initial value for iteration
```

```
!
      FORALL (I=1:NX,J=1:NY) U(I,J) = A*(J-1)*(NX-I)/(NX-1)&
                              + B*(J-1)**2*(I-1)/(NX-1)

!
! The main iteration loop of the relaxation scheme
! applied to the 2D Laplace equation
!
      DO  I = 1, IMAX
        UTMP = FIELD*(CSHIFT(U,SHIFT= 1,DIM=1) &
                     +CSHIFT(U,SHIFT= 1,DIM=2) &
                     +CSHIFT(U,SHIFT=-1,DIM=1) &
                     +CSHIFT(U,SHIFT=-1,DIM=2))&
             +BOUND
        U    = Q*U+P*UTMP
        DU   = MAXVAL(ABS(U-UTMP))
        IF (DU.LE.TL) EXIT
      END DO
      IF (I.NE.IMAX) THEN
        PRINT*, 'The number of iterations is ', I, &
' and the largest error is ', DU
        IF(IOUT) WRITE (15,"(10F10.8)") U
      ELSE
        PRINT*, 'The required convergence is not reached.'
      END IF
END PROGRAM LAPLACE_2D
```

The above program is quite simple, and most of its statements are from Fortran 90. The only new statements from HPF are the FORALL statement, which is used to assign the initial guess of the solution that satisfies the given boundary conditions. The other HPF statements are the HPF directives, which start with !HPF$ as an indicator. There are two ways to distribute an array on a multiprocessor system. The BLOCK assignment requires the data to be distributed block by block on the processors. For example,

```
!HPF$ DISTRIBUTE A(BLOCK)
```

would divide the array $a(i)$ for $i = 1, 2, \ldots, n$ into p segments for a p-processor system with the first n/p elements on the first processor, $n/p + 1$ to $2n/p$ elements on the second processor, and so on. Another way to distribute the data with HPF is cyclically, with `BLOCK` replaced by `CYCLIC`. The `CYCLIC` assignment will assign the first element of the array to the first processor, the second element to the second processor, ..., the pth element to the pth processor, and then the $(p+1)$th element to the first processor, $(p+2)$th element to the second processor, and so on. The `FORALL` statement can also have a logic condition attached. For example,

```
FORALL (I=1:N:2,A(I)>0) C(I) = SQRT(A(I))-1.0/A(I)
```

works only on the odd and positive elements of the array. An associated HPF directive is the `INDEPENDENCE` directive, which indicates that the loop followed can be evaluated term by term on different processors. For a complete description of HPF, see Koelbel et al. (1994).

12.5 Distributed computing and message passing

Another model of high-performance computing utilizes a set of distributed processors connected through either a network or a fast switch as a concurrent computing resource to achieve computing tasks such as those encountered in grand challenges. This model of computing is based on the assumption that the memory is distributed, that is, that each processor has its own memory that cannot be accessed by other processors. The key to distributed computing is the capacity to perform a computing task concurrently across a set of loosely connected processors, which can be either identical or totally different in their architectures.

This model of high-performance computing also works with systems with parallel processors connected through a fast system bus and with a shared memory unit, since most parallel systems currently available can have the global memory partitioned. For such a shared memory system, as long as one can partition the global memory to have one domain dedicated to one processor, the system can act as a distributed memory system.

The advantage of distributed computing is that users have enough freedom to fine-tune their programs to take full advantage of the available resources. Usually, the distributed model can be achieved with less expensive hardware, and has little communication overhead in many applications, in comparison to the data parallel programming paradigm used with high-performance computing languages such as HPF.

However, a distributed computational program is more involved and requires more skill and effort in programming. It usually takes more human time to develop and debug a distributed computing program than a program with a high-performance language. The data parallel programming languages, discussed in the preceding section, simplify the task of users, so they can concentrate on the scientific aspect of the problem. It is still not clear that we can eventually be able to rely on the compilers of high-performance languages, since the complexity of some grand challenges can be so high that any well-conceived compiler may still have difficulties in fully utilizing its resources. It is likely that the overall performance of the data distributed model will always be ahead of that of the data parallel model.

There have been rapid developments in the area of networked, or distributed, computing. Several software packages have emerged as a result of these efforts, for example, PVM (Geist et al., 1994), P4 (Butler and Lust, 1994), and more. The desire for standardization of these systems is extremely high. Message Passing Interface, commonly known as MPI (Snir et al., 1996), is a set of specifications that resulted from the efforts of many scientists and organizations to standardize the various message-passing protocols.

MPI has been implemented through PVM, P4, CHIMP, and LAM, and it is expected that all vendors will implement MPI on their systems directly within a few years. MPI is designed to be language independent – for example, it can work with C and Fortran – and free of a specific computer architecture. It is designed to deal with all heterogeneities encountered in distributed computing in a transparent manner.

The major statements in MPI are designed for comprehensive point-to-point communication and collective communication among the participating processors. MPI works with both Fortran and C. In this section, we will consider only how to use MPI in a Fortran environment. For an extensive discussion of using MPI with C, see Pacheco (1995).

Here is how MPI works. A user first needs to define an MPI world, the collection of all processors to be used for a specified computing task. This is usually carried out with an option in running MPI. The option links the MPI initiation to a file which contains addresses of the processors to be used in the MPI world. One can have more than one process on a single processor. Each process is assigned a unique index called *rank* by MPI, and each message is labeled with a *tag*. The data type of the message is explicitly given in the protocols. Each MPI Fortran program has an INCLUDE 'mpif.h' statement at the beginning to link it to the MPI libraries. Each MPI statement is achieved with a call to an MPI subroutine. An MPI program is always initiated with a call to the MPI_INIT subroutine and ended with a call to MPI_FINALIZE. The most important statements in MPI are the *send* and *receive* statements. Let us illustrate the structure of an MPI program with a simple example of sending a message from process 0 to process 1.

```
PROGRAM TALK_0_TO_1
INCLUDE 'mpif.h'
INTEGER IRANK,IPROC,ITAG,ISEND,IRECV,IERR,IM,ID,
```

```
   &              ISTAT(MPI_STATUS_SIZE),IFINISH
          CHARACTER*60 HELLO
   C
          ITAG  = 730
          ID    = 60
          ISEND = 0
          IRECV = 1
          CALL MPI_INIT (IERR)
          CALL MPI_COMM_RANK (MPI_COMM_WORLD,IRANK,IERR)
          CALL MPI_COMM_SIZE (MPI_COMM_WORLD,IPROC,IERR)
          PRINT*, IRANK, IPROC
          CALL MPI_BARRIER (MPI_COMM_WORLD,IERR)
          IF (IRANK.EQ.ISEND) THEN
            HELLO = 'Hello. It is nice to be connected.'
            IM = 33
            CALL MPI_SEND (HELLO,IM,MPI_CHARACTER,IRECV,
   &                       ITAG,MPI_COMM_WORLD,IERR)
            PRINT*, 'I sent the message: ', HELLO
          ELSE IF (IRANK.EQ.IRECV) THEN
            CALL MPI_RECV (HELLO,ID,MPI_CHARACTER,ISEND,
   &                       ITAG,MPI_COMM_WORLD,ISTAT,IERR)
            PRINT*, 'I got your message, which is: ', HELLO
          ELSE
          END IF
          CALL MPI_FINALIZE(IFINISH)
          STOP
          END
```

The call to MPI_COMM_RANK will find the rank (0 to $p-1$) of each process and assign it to the variable IRANK in that process. Then the call to MPI_COMM_SIZE will find out the total number of processes (p) and assign it to the variable IPROC. Note that IRANK is different on each process but IPROC is exactly the same on all the processes. The send statement involves the message, the size (not necessarily the dimension) of the message, the type of the message, the rank of the target process, the tag of the message, and the communicator. All items starting with MPI_ are intrinsic to the MPI libraries. For example, MPI_COMM_WORLD is defined as a communicator that allows communication among all the participating processes. The receive statement has one more item, ISTAT, to record the status of the communication. Also, the dimension of a message rather than its actual size is used in the receive statement. This is designed to specify the maximum size of the message. We have used the call of the barrier statement to synchronize all the processes to reach the point where MPI_BARRIER is called.

There is a total of 128 MPI statements, but only about a dozen are really needed frequently. We have used several of them already in the program TALK_0_TO_1. The send and receive statements in the above example use the blocking mechanism; that is, the process is blocked before the communication is completed. This is a synchronous message-passing model. MPI also allows asynchronous message passing. For more details, see Snir et al. (1996).

Now let us turn to a simple application. We would like to carry out a computation that requires an extremely high degree of accuracy on a cluster of processors. The Euler constant is defined as

$$\gamma = \lim_{n \to \infty} \left[\sum_{k=1}^{n} \frac{1}{k} - \ln n \right], \tag{12.14}$$

which requires a huge number of summations in order to obtain a high accuracy. Assuming we have p processors in our system, we can divide the task into parts, with each of the first $p-1$ processes to sum up $\text{int}(n/p)$ terms and the last process to sum up whatever is left. Here is an implementation of the evaluation of the Euler constant in MPI with Fortran 77.

```
      PROGRAM EULER_CONST
      INCLUDE 'mpif.h'
      DOUBLE PRECISION SUMI, SUM, SUM25
      PARAMETER (SUM25 = 0.5772156649015328606065122D0)
      INTEGER N,K,IERR,IRANK,IPROC,IFINISH
C
      CALL MPI_INIT (IERR)
      CALL MPI_COMM_RANK (MPI_COMM_WORLD,IRANK,IERR)
      CALL MPI_COMM_SIZE (MPI_COMM_WORLD,IPROC,IERR)
C
      IF (IRANK.EQ.0) THEN
        PRINT*, 'Enter number of terms in the series: '
        READ (5,*) N
      ELSE
      END IF
C
C Broadcast the total number of terms to every process
C
```

```
C
      CALL MPI_BCAST (N,1,MPI_INTEGER,0,
     *                 MPI_COMM_WORLD,IERR)
      K = (N/IPROC)
      SUMI = 0.D0
      IF (IRANK.NE.(IPROC-1)) THEN
         DO    100 I = IRANK*K+1, (IRANK+1)*K
           SUMI = SUMI+1.D0/DFLOAT(I)
  100    CONTINUE
      ELSE
         DO    200 I = IRANK*K+1, N
           SUMI = SUMI+1.D0/DFLOAT(I)
  200    CONTINUE
      END IF
C
C Collect the sums from every process
C
      CALL MPI_REDUCE (SUMI,SUM,1,MPI_DOUBLE_PRECISION,
     *                 MPI_SUM,0,MPI_COMM_WORLD,IERR)
      IF (IRANK.EQ.0) THEN
         SUM = SUM-DLOG(DFLOAT(N))
         PRINT*, 'The evaluated Euler constant is ', SUM,
     *'with the estimated error of ', DABS(SUM-SUM25)
      END IF
C
      CALL MPI_FINALIZE(IFINISH)
      STOP
      END
```

This program uses the MPI *broadcast* statement and *reduce* statement, the other two key statements in MPI. The broadcast statement is used to pass the information gained by process 0 to all processes in the MPI world, including process 0. Each process then performs the required segment of summations. The reduce statement is used to collect the segments and sum them up. The number of summations performed by process $p - 1$ is determined by the maximum number of terms included. The subtraction of the logarithmic term is accomplished by process 0 after receiving the segments of the series from all processes. The partitioning of the task can also be made with each process to sum up the terms that are p terms apart. The loop in the above program can be replaced with

```
          DO 100 I = IRANK+1, N, IPROC
            SUMI = SUMI+1.D0/DFLOAT(I)
  100   CONTINUE
```

to achieve such a distribution. This is faster if the remainder of n/p is large.

Of course, more applications of MPI are expected in the next few years. Another forum has been formed to extend what the current version of MPI covers, and the new edition of MPI will have more protocols, such as remote executions and shared memory operations, in a year or two. We cannot show all the strengths or aspects of MPI in this short section, but interested readers should consult the books mentioned in this section for a complete and up-to-date description of MPI. Descriptions of other message-passing systems can also be obtained via the references in these books.

12.6 Some current applications

There have been rapid developments in computing resources and techniques in the last decade. As a matter of fact, the rate of these developments still shows no indication of slowing down. Scientists from different disciplines are taking advantage of new developments and are trying to push the applications to the very limit of what a state-of-the-art system can deliver. In this section, we would like to give a few examples of such applications, mainly in computational physics. It is not our goal here to survey the possibilities but rather to highlight a few cases.

In Chapter 8 we discussed simulation techniques for the fluid dynamics. Scientists have implemented these simulations on both massively parallel computers and distributed computers. A recent scheme for performing simulations of the Navier–Stokes equation in a three-dimensional system on a massively parallel computing system is described by Chen and Shan (1992). They performed the simulation on a system with 65,536 processors with a total of eight gigabytes of memory. The work is very impressive, with simulations performed with high Reynolds numbers, providing a direct view of turbulent flows on a computer.

Another simulation is described by Punzo, Massaioli, and Succi (1994) using the Boltzmann lattice-gas model described in Chapter 8. They have implemented the scheme on a cluster of high-speed processors interconnected through a fast switchboard. They have performed the simulations on systems of up to 128 processors and gained about four gigaflops in performance with about 800 million variables. All these implementations are significant in the sense that they have shown the potential of high-performance computing systems in large-scale fluid dynamics simulations that may eventually be applied to the simulation of global environmental change and oceanic dynamics.

Monte Carlo simulations have also been implemented in parallel and distributed computing environments. For example, one of the quantum Monte Carlo simulation techniques has been implemented on several parallel systems (Ding, 1993). The system being studied is a two-dimensional quantum spin system, which is important in the understanding of the magnetic properties of exotic systems such as high-temperature superconductors. Two parallel schemes are implemented, one for the different Monte Carlo updating processes and another for the measurements related to the spin variables. The algorithm is designed to require only exchange and broadcast message-passing operations, implementation on some systems reached more than 90% parallelization. This type of simulation will surely have an impact on quantum Monte Carlo simulations of other highly correlated systems in the future.

Classical Monte Carlo simulations of crystal growth have also been implemented in parallel computing environments (Haider et al., 1995). The system is partitioned geometrically, and each cell is put on a processor. The update inside each cell is completed locally if the particle is not at the boundary. Communication between particles at the boundaries of the neighboring cells is achieved with a broadcast model. For a more detailed discussion of the algorithm and its implementation, see Haider et al. (1995).

There have also been many implementations of molecular dynamics simulation on parallel and distributed computing environments. For example, a recent molecular dynamics simulation has 400 million particles in the system (Deng et al., 1995).

More practical molecular dynamics simulations on parallel computing systems have recently been performed by Vashishta et al. (1995). They studied the structure and dynamics of covalent amorphous insulators with realistic particle–particle interaction potentials. There are 41,472 particles in their simulation box, and the interaction includes a long-range Coulomb term and a three-body term. They have obtained some very interesting results on the structure and dynamics of real materials such as amorphous silica and silicon nitride. Interested readers should read their article. There are also parallel programs developed for ab initio molecular dynamics simulation (Yeh, Kim, and Khan, 1995). This work uses semiconductor clusters as examples and has about 500 silicon atoms in the simulation box.

The development of the applications of high-performance computing systems is progressing rapidly, and it is still too early to say just which direction is the right one. Researchers in high-performance computing are now exploring all the possible directions, models, and paradigms and we will have a better understanding about where it is going and how far it can actually reach in just a few years.

Exercises

12.1 Write a Fortran 90 subroutine that generates the Bessel functions of the first and second kinds.

12.2 Write a Fortran 90 function routine that can generate the Chebyshev polynomials of the first kind recursively.

12.3 Rewrite the groundwater dynamics program given in Section 6.6 into an HPF program. Discuss the advantages and disadvantages of the new program.

12.4 One can calculate π from the relation

$$\frac{\pi^2}{6} = \sum_{k=1}^{\infty} \frac{1}{k^2}.$$

Develop an MPI program that can calculate π with the above formula up to any desired accuracy on a distributed multiprocessor computing system.

12.5 Rewrite the program for the time-dependent temperature distribution around a nuclear waste rod given in Section 6.8 into an MPI program on a multiprocessor computing system with each given time performed on a processor.

12.6 Develop a program that performs the molecular dynamics simulation for two-dimensional Lennard–Jones clusters discussed in Section 7.4 on a multiprocessor computing system. Is there any temperature region at which liquid and solid states coexist? Use Fortran 90 as your primary programming language first and then incorporate it into the MPI environment. Compare your results for isolated clusters with the ones having periodic boundary conditions.

REFERENCES

Adair, R.K. (1995). The physics of baseball, *Physics Today* **48**(5), 26–31.

Adams, J.C., Brainerd, W.S., Martin, J.T., Smith, B.T., and Wagener, J.L. (1992). *Fortran 90 Handbook* (McGraw–Hill, New York).

Aitken, A.C. (1932). An interpolation by iteration of proportional parts, without the use of differences, *Proceedings of Edinburgh Mathematical Society* **3**, Series 2, 56–76.

Akulin, V.M., Bréchignac, C., and Sarfati, A. (1995). Quantum shell effect on dissociation energies, shapes, and thermal properties of metallic clusters from random matrix model, *Physical Review Letters* **75**, 220–3.

Allen, M.P. and Tildesley, D.J. (1987). *Computer Simulation of Liquids* (Clarendon, Oxford).

Altschuler, E.L, Williams, T.J., Ratner, E.R., Dowla, F., and Wooten, F. (1994). Method of constrained global optimization, *Physical Review Letters* **72**, 2671–4.

Altschuler, E.L, Williams, T.J., Ratner, E.R., Dowla, F., and Wooten, F. (1995). Altschuler et al. reply, *Physical Review Letters* **74**, 1483.

Andersen, H.C. (1980). Molecular dynamics simulations at constant pressure and/or temperature, *Journal of Chemical Physics* **72**, 2384–93.

Anderson, J.B. (1975). A random walk simulation of the Schrödinger equation: H_3^+, *Journal of Chemical Physics* **63**, 1499–503.

Anderson, J.B. (1976). Quantum chemistry by random walk. $H\ ^2P$, $H_3^+\ D_{3h}\ ^1A_1'$, $H_2\ ^3\Sigma_u^+$, $H_4\ ^1\Sigma_g^+$, Be ^1S, *Journal of Chemical Physics* **65**, 4121–7.

Anderson, S.L. (1990). Random number generators on vector super computers and other advanced architectures, *SIAM Review* **32**, 221–51.

Baker, G.L. and Gollub, J.P. (1990). *Chaotic Dynamics: An Introduction* (Cambridge University, Cambridge).

Barker, J.A. (1979). A quantum-statistical Monte Carlo method; path integrals with boundary conditions, *Journal of Chemical Physics* **70**, 2914–18.

Benzi, R., Succi, S., and Vergassola, M. (1992). The lattice Boltzmann equation: Theory and applications, *Physics Reports* **222**, 145–97.

Berendsen, H.J.C., Postma, J.P.M., van Gunsteren, W.F., DiNola, A., and Haak, J.R. (1984). Molecular dynamics with coupling to an external bath, *Journal of Chemical Physics* **81**, 3684–90.

Binder, K. (ed.) (1986). *Monte Carlo Methods in Statistical Physics* (Springer–Verlag, Berlin).

Binder, K. (ed.) (1987). *Monte Carlo Methods in Statistical Physics II* (Springer–Verlag, Berlin).

Binder, K. (ed.) (1992). *Monte Carlo Method in Condensed Matter Physics* (Springer–Verlag, Berlin).

Binder, K. and Heermann, D.W. (1988). *Monte Carlo Simulation in Statistical Physics: An Introduction* (Springer–Verlag, Berlin).

Bohigas, O. (1991). Random matrix theories and chaotic dynamics, in *Chaos and Quantum Physics*, ed. by M.-J. Giannoni, A. Voros, and J. Zinn-Justin (North-Holland, Amsterdam), pp. 87–199.

Boninsegni, M. and Ceperley, D.M. (1995). Path integral Monte Carlo simulation of isotopic liquid helium mixtures, *Physical Review Letters* **74**, 2288–91.

Bouchaud, J.P. and Lhuillier, C. (1989). New microscopic description of liquid ^{3}He, *Zeitschrift für Physik B* **75**, 283–9.

Brainerd, W.S., Goldberg, C.H., and Adams, J.C. (1990). *Programmer's Guide to Fortran 90* (McGraw–Hill, New York).

Bristeau, M.O., Glowinski, R., Mantel, B., Périaux, J., and Perrier, P. (1985). Numerical methods for incompressible and compressible Navier–Stokes problems, in *Finite Elements in Fluids, Volume 6*, ed. by R.H. Gallagher, G. Carey, J.T. Oden, and O.C. Zienkiewicz (Wiley, Chichester, New York), pp. 1–40.

Brody, T.A., Flores, J., French, J.B., Mello, P.A., Pandey, A., and Wong, S.S.M. (1981). Random-matrix physics: Spectrum and strength fluctuations, *Reviews of Modern Physics* **53**, 383–479.

Broyden, C.G. (1970). The convergence of a class of double-rank minimization algorithms, Part I and II, *Journal of Institute of Mathematics and Its Applications* **6**, 76–90; 222–36.

Burks, A.R. and Burks, A.W. (1988). *The First Electronic Computer: The Atanasoff Story* (University of Michigan, Ann Arbor, Michigan).

Burnett, D.S. (1987). *Finite Element Analysis: From Concepts to Applications* (Addison–Wesley, Reading, Massachusetts).

Burrus, C.S. and Parks, T.W. (1985). *DFT/FFT and Convolution Algorithms* (Wiley, New York).

Butler, R. and Lust, E. (1994). Monitors, messages, and clusters: The P4 parallel programming system, *Parallel Computing* **20**, 547–64.

Car, R. and Parrinello, M. (1985). Unified approach for molecular dynamics and density-functional theory, *Physical Review Letters* **55**, 2471–4.

Ceperley, D.M. (1992). Path-integral calculations of normal liquid ^{3}He, *Physical Review Letters* **69**, 331–4.

Ceperley, D.M. (1995). Path integrals in theory of condensed helium, *Reviews of Modern Physics* **67**, 279–356.

Ceperley, D.M. and Alder, B.J. (1984). Quantum Monte Carlo for molecules: Green's function and nodal release, *Journal of Chemical Physics* **81**, 5833–44.

Ceperley, D.M. Chester, G.V., and Kalos, M.H. (1977). Monte Carlo simulations of a many-fermion study, *Physical Review B* **16**, 3081–99.

Ceperley, D.M. and Kalos, M.H. (1986). Quantum many-body problems, in *Monte Carlo Methods in Statistical Physics*, ed. by K. Binder (Springer–Verlag, Berlin), pp. 145–94.

Ceperley, D.M. and Pollock, E.L. (1986). Path-integral computation of the low-temperature properties of liquid ^{4}He, *Physical Review Letters* **56**, 351–4.

Chen, H., Chen, S., and Matthaeus, W. (1992). Recovery of the Navier–Stokes equations using a lattice-gas Boltzmann method, *Physical Review A* **45**, R5339–42.

Chen, S. and Shan, X. (1992). High-resolution turbulent simulations using the Connection Machine-2, *Computers in Physics* **6**, 643–6.

Chui, C.K. (1992). *An Introduction to Wavelets* (Academic, San Diego, California).

Chui, C.K., Montefusco, L., and Puccio, L. (eds.) (1994). *Wavelets: Theory, Algorithms, and Applications* (Academic, San Diego, California).

Combes, J.M., Grossmann, A., and Tchanmitchian, P. (eds.) (1990). *Wavelets* (Springer–Verlag, Berlin).

Cook, D.M., Dubisch, R., Sowell, G., Tam, P., and Donnelly, D. (1992). A comparison of several symbol-manipulating programs, Part I and II, *Computers in Physics* **6**, 411–19; 530–40.

Cooley, J.W., Lewis, P.A.W., and Welch, P.D. (1969). The fast Fourier transform and its applications, *IEEE Transactions on Education* **E-12**, 27–34.

Cooley, J.W. and Tukey, J.W. (1965). An algorithm for the machine calculation of complex Fourier series, *Mathematics of Computation* **19**, 297–301.

Courant, R. and Hilbert, D. (1989). *Methods of Mathematical Physics, Volume I and II* (Wiley, New York).

Cullum, J.K. and Willoughby, R.A. (1985). *Lanczos Algorithms for Large Symmetric Eigenvalue Computations, Volume 1 and Volume 2* (Birkhauser, Boston, Massachusetts).

Dagotto, E. and Moreo, A. (1985). Improved Hamiltonian variational technique for lattice models, *Physical Review D* **31**, 865–70.

Dagotto, E. (1994). Correlated electrons in high temperature super conductors, *Reviews of Modern Physics* **66**, 763–840.

Daubechies, I. (1988). Orthonormal bases of compactly supported wavelets, *Communications on Pure and Applied Mathematics* **41**, 909–96.

Daubechies, I. (1992). *Ten Lectures on Wavelets* (SIAM, Philadelphia, Pennsylvania).

Davis, P.J. and Polonsk, I. (1965). Numerical integration, differentiation and integration, in *Handbook of Mathematical Functions*, ed. by M. Abramowitz and I.A. Stegun (Dover, New York), pp. 875–924.

Deng, Y., McCoy, R.A., Marr, R.B., and Peierls, R.F. (1995). Molecular dynamics for 400 million particles with short-range interactions, in *High Performance Computing 1995: Grand Challenges in Computer Simulation*, ed. by A. Tentner (Society for Computer Simulation, San Diego, California), pp. 95–100.

Dennis, J.E. Jr. and Schnabel, R.B. (1983). *Numerical Methods for Unconstrained Optimization and Nonlinear Equations* (Prentice–Hall, Englewood Cliffs, New Jersey).

Ding, H.Q. (1993). Monte Carlo simulations of quantum systems on massively parallel super computers, *Computers in Physics* **7**, 687–95.

Doolen, G.D. (ed.) (1991). *Lattice Gas Methods: Theory, Applications, and Hardware* (MIT, Cambridge, Massachusetts).

Edgar, S.L. (1992). *Fortran for the '90s* (Freeman, New York).

Erber, T. and Hockney, G.M. (1995). Comment on "Method of constrained global optimization," *Physical Review Letters* **74**, 1482.

Ercolessi, F., Tosatti, E., and Parrinello, M. (1986). Au(100) surface reconstruction, *Physical Review Letters* **57**, 719–22.

Evans, D.J., Hoover, W.G., Failor, B.H., Moran, B., and Ladd, A.J.C. (1983). Nonequilibrium molecular dynamics via Gauss's principle of least constraint, *Physical Review A* **28**, 1016–21.

Evans, M.W. and Harlow, F.H. (1957). The particle-in-cell method for hydrodynamic calculations, *Los Alamos Scientific Laboratory Report* No. LA-2139.

Faddeev, D.K. and Faddeeva, W.N. (1963). *Computational Methods of Linear Algebra* (Freeman, San Francisco).

Fahy, S., Wang, X.W., and Louie, S.G. (1990). Variational quantum Monte Carlo nonlocal pseudopotential approach to solids: Formulation and application to diamond, graphite, and silicon, *Physical Review B* **42**, 3503–22.

Fletcher, R. (1970). A new approach to variable metric algorithms, *Computer Journal* **13**, 317–22.

Fortin, M. and Thomasset, F. (1983). Application to the Stokes and Navier–Stokes equations, in *Argumented Lagrangian Methods: Applications to the Numerical Solution of Boundary-Value Problems*, ed. by M. Fortin and R. Glowinski (North-Holland, Amsterdam), pp. 47–95.

Foufoula-Georgiou, E. and Kumar, P. (eds.) (1994). *Wavelets in Geophysics* (Academic, San Diego, California).

Frisch, U., d'Humières, D., Hasslacher, B., Lallemand, P., and Pomeau,Y. (1987). Lattice gas hydrodynamics in two and three dimensions, *Complex Systems* **1**, 649–701.

Frisch, U., Hasslacher, B., and Pomeau, Y. (1986). Lattice gas automata for the Navier–Stokes equation, *Physical Review Letters* **56**, 1505–8.

Gauss, C.F. (1866). Nachlass: Theoria interpolation is methodo nova tractata, in *Carl Friedrich Gauss, Werke, Band 3*, ed. by E. Schering, F. Klein, M. Brendel, and L. Schlesinger (Königlichen Gesellschaft derWissenschaften, Gïtingen), pp. 265–303.

Gear, C.W. (1971). *Numerical Initial Value Problems in Ordinary Differential Equations* (Prentice–Hall, Englewood Cliffs, New Jersey).

Geist, A., Beguelin, A., Dongarra, J., Jiang, W., Machek, R., and Sunderam, V. (1994). *PVM: User's Guide and Tutorial for Networked Parallel Computing* (MIT, Cambridge, Massachusetts).

Goldfarb, D. (1970). A family of variable metric methods derived by variational means, *Mathematics of Computation* **24**, 23–6.

Goldstine, H.H. (1977). *A History of Numerical Analysis from the 16th through the 19th Century* (Springer–Verlag, New York).

Grossmann, A., Kronland-Martinet, R., and Morlet, J. (1989). Reading and understanding continuous wavelet transforms, in *Wavelets*, ed. by J.M. Combes, A. Grossmann, and P. Tchanmitchian (Springer–Verlag, Berlin), pp. 2–20.

Haar, A. (1910). Zur theorie der orthogonalen funktionensysteme, *Mathematische Annalen* **69**, 331–71.

Hackbusch, W. (1994). *Iterative Solution of Large Sparse Systems of Equations* (Springer–Verlag, New York).

Hackenbroich, G. and Weidenmüller, H.A. (1995). Universality of random-matrix results for non-Gaussian ensembles, *Physical Review Letters* **75**, 4118–21.

Haider, N., Khaddaj, S.A., Wilby, M.R., and Vvedensky, D. (1995). Parallel Monte Carlo simulations of epitaxial growth, *Computers in Physics* **9**, 85–96.

Haile, J.M. (1992). *Molecular Dynamics Simulation* (Wiley, New York).

Harlow, F.H. (1964). The particle-in-cell computing method for fluid dynamics, in *Methods in Computational Physics, Volume 3, Fundamental Methtods in Hydrodynamics*, ed. by B. Alder, S. Fernbach, and M. Rotenberg (Academic, New York), pp. 319–43.

Heermann, D.W. (1986). *Computer Simulation Methods in Theoretical Physics* (Springer–Verlag, Berlin).

Higuera, F. and Jiménez, J. (1989). Boltzmann approach to lattice gas simulations, *Europhysics Letters* **9**, 663–8.

Hirsch, J.E. (1985a). Attractive interaction and pairing in fermion systems with strong on-site repulsion, *Physical Review Letters* **54**, 1317–20.

Hirsch, J.E. (1985b). Two-dimensional Hubbard model: Numerical simulation study, *Physical Review B* **31**, 4403–19.

Hochstrasser, U.W. (1965). Orthogonal polynomials, in *Handbook of Mathematical Functions*, ed. by M. Abramowitz and I.A. Stegun (Dover, New York), pp. 771–802.

Hockney, R.W. and Eastwood, J.W. (1988). *Computer Simulation Using Particles* (McGraw–Hill, London).

Hohenberg, P. and Kohn, W. (1964). Inhomogeneous electron gas, *Physical Review* **136**, B864–71.

Holschneider, M. (1995). *Wavelets: An Analysis Tool* (Clarendon, Oxford).

Hoover, W.G. (1985). Canonical dynamics: Equilibrium phase-space distributions, *Physical Review A* **31**, 1695–7.

Hostetter, G.H., Santina, M.S., and D'Carpio-Montalvo, P. (1991). *Analytical, Numerical, and Computational Methods for Science and Engineering* (Prentice–Hall, Englewood Cliffs, New Jersey).

Hulthén, L. (1938). Uber das austauschproblem eines kristalles, *Arkiv för Matematik, Astronomi och Fysik* **26** (11), 1–105.

Jones, R.O. and Gunnarsson, O. (1989). The density functional formalism, its applications and prospects, *Reviews of Modern Physics* **61**, 689–746.

Jullien, R., Pfeuty, P., Fields, J.N., and Doniach, S. (1978). Zero-temperature renormalization method for quantum systems. I. Ising model in a transverse field in one dimension, *Physical Review B* **18**, 3568–78.

Kadanoff, L.P. (1966). Scaling laws for Ising models near T_c, *Physics* **2**, 263–72.

Kalos, M.H. (1962). Monte Carlo calculations of the ground state of three- and four-body nuclei, *Physical Review* **128**, 1791–5.

Kernighan, B.W. and Pike, R. (1984). *The UNIX Programming Environment* (Prentice–Hall, Englewood Cliffs, New Jersey).

Kernighan, B.W. and Ritchie, D.M. (1988). *The C Programming Language* (Prentice–Hall, Englewood Cliffs, New Jersey).

Kittel, C. (1986). *Introduction to Solid State Physics* (Wiley, New York).

Knuth, D.E. (1981). *The Art of Computer Programming, Volume 2, Seminumerical Algorithms* (Addison–Wesley, Reading, Massachusetts).

Koelbel, C.H., Loveman, D.B., Schreiber, R.S., Steel, G.L. Jr., and Zosel, M.E. (1994). *High Performance Fortran Handbook* (MIT, Cambridge, Massachusetts).

Kohn, W. and Sham, L. (1965). Self-consistent equations including exchange and correlation effects, *Physical Review* **140**, A1133–8.

Kohn, W. and Vashishta, P. (1983). General functional density theory, in *Theory of the Inhomogeneous Electron Gas*, ed. by S. Lundqvist and N.H. March (Plenum, New York), pp. 79–148.

Kondo, J. (1969). Theory of dilute magnetic alloys, in *Solid State Physics, Volume 23*, ed. by F. Seitz, D. Turnbull, and H. Ehrenreich (Academic, New York), pp. 183–281.

Koonin, S.E. (1986). *Computational Physics* (Benjamin, Menlo Park, California).

Kunz, R.E. and Berry, R.S. (1993). Coexistence of multiple phases in finite systems, *Physical Review Letters* **71**, 3987–90.

Kunz, R.E. and Berry, R.S. (1994). Multiple phase coexistence in finite systems, *Physical Review E* **49**, 1895–908.

Landau, L.D. and Lifshitz, E.M. (1987). *Fluid Dynamics* (Pergamon, Oxford).

Lee, M.A. and Schmidt, K.E. (1992). Green's function Monte Carlo, *Computers in Physics* **6**, 192–7.

Lewis, D.W. (1991). *Matrix Theory* (World Scientific, Singapore).

Lifshitz, E.M. and Pitaevskii, L.P. (1981). *Physical Kinetics* (Pergamon, Oxford).

Liu, K., Brown, M.G., Carter, C., Saykally, R.J., Gregory, J.K., and Clary, D.C. (1996). Characterization of a cage form of the water hexamer, *Nature* **381**, 501–3.

Loshin, D. (1994). *High Performance Computing Demystified* (Academic, Boston, Massachusetts).

Lyubartsev, A.P. and Vorontsov-Velyaminov, P.N. (1993). Path-integral Monte Carlo method in quantum statistics for a system of N identical fermions, *Physical Review A* **48**, 4075–83.

Ma, S.-K. (1976). *Modern Theory of Critical Phenomena* (Benjamin, Reading, Massachusetts).

Mackintosh, A.R. (1987). The first electronic computer, *Physics Today* **40** (3), 25–32.

Madelung, O. (1978). *Introduction to Solid-State Theory* (Springer–Verlag, Berlin).

Maeder, R. (1991). *Programming in Mathematica* (Addison–Wesley, Redwood City, California).

Mallat, S. (1989). A theory for multiresolution signal decomposition: The wavelet representation, *IEEE Transactions on Pattern Analysis and Machine Intelligence* **11**, 674–93.

Matsuoka, H., Hirokawa, T., Matsui, M., and Doyama, M. (1992). Solid–liquid transitions in argon clusters, *Physics Review Letters* **69**, 297–300.

McCammon, J.A. and Harvey, S.C. (1987). *Dynamics of Proteins and Nucleic Acids* (Cambridge University, Cambridge).

McNamara, G.R. and Zanetti, G. (1988). Use of the Boltzmann equation to simulate lattice-gas automata, *Physical Review Letters* **61**, 2232–5.

Mehta, M.L. (1991). *Random Matrices* (Academic, Boston, Massachusetts).

Metcalf, M. and Reid, J. (1990). *Fortran 90 Explained* (Oxford University, New York).

Metropolis, N. and Frankel, S. (1947). Calculations in the liquid drop model of fission, *Physical Review* **72**, 186.

Metropolis, N., Rosenbluth, A.W., Rosenbluth, M.N.,Teller, A.H., and Teller, E. (1953). Equation of state calculations by fast computing machines, *Journal of Chemical Physics* **21**, 1087–92.

Meyer, Y. (1993). *Wavelets: Algorithms & Applications*, translated and revised by R.D. Ryan (SIAM, Philadelphia, Pennsylvania).

Millikan, R.A. (1910). The isolation of an ion, a precision measurement of its charge, and the correction of Stokes's law, *Science* **32**, 436–48.

Mollenhoff, C.R. (1988). *Atanasoff: Forgotten Father of the Computer* (Iowa State University, Ames, Iowa).

Monaghan, J.J. (1985). Particle methods for hydrodynamics, *Computer Physics Reports* **3**, 71–124.

Moreau, R. (1984). *The Computer Comes of Age: The People, the Hardware, and the Software* (MIT, Cambridge, Massachusetts).

Morlet, J., Arens, G., Fourgeau, E., and Giard, D. (1982a). Wave propagation and sampling theory–Part I: Complex signal and scattering in multilayered media, *Geophysics* **47**, 203–21.

Morlet, J., Arens, G., Fourgeau, E., and Giard, D. (1982b). Wave propagation and sampling theory–Part II: Sampling theory and complex waves, *Geophysics* **47**, 222–36.

Nash, S.G. (ed.) (1990). *A History of Scientific Computing* (Addison–Wesley, Reading, Massachusetts).

Newland, D.E. (1993). *An Introduction to Random Vibrations, Spectral and Wavelet Analysis* (Longman, Essex, UK).

Niemeijer, T. and van Leeuwen, J.M.J. (1974). Wilson theory for 2-dimensional Ising spin systems, *Physica* **71**, 17–40.

Nosé, S. (1984a). A molecular dynamics method for simulations in the canonical ensemble, *Molecular Physics* **52**, 255–68.

Nosé, S. (1984b). A unified formulation of the constant temperature molecular dynamics, *Journal of Chemical Physics* **81**, 511–19.

Nosé, S. (1991). Constant temperature molecular dynamics methods, *Progress of Theoretical Physics Supplement* **103**, 1–46.

Oguchi, T. and Sasaki, T. (1991). Density-functional molecular-dynamics method, *Progress of Theoretical Physics Supplement* **103**, 93–117.

Onodera, Y. (1994). Numerov integration for radial wave function in cylindrical symmetry, *Computers in Physics* **8**, 352–4.

Pacheco, P.S. (1995). *Programming Parallel Processors Using MPI* (Kaufmann, San Francisco).

Pang, T. (1995). A numerical method for quantum tunneling, *Computers in Physics* **9**, 602–5.

Parisi, G. (1988). *Statistical Field Theory* (Addison–Wesley, Redwood City, California).

Park, S.K. and Miller, K.W. (1988). Random number generators: Good ones are hard to find, *Communications of the ACM* **31**, 1192–201.

Parrinello, M. and Rahman, A. (1980). Crystal structure and pair potentials: A molecular-dynamics study, *Physical Review Letters* **45**, 1196–9.

Parrinello, M. and Rahman, A. (1981). Polymorphic transition in single crystal: A new molecular dynamics method, *Journal of Applied Physics* **52**, 7182–90.

Pollock, E.L. and Ceperley, D.M. (1984). Simulation of quantum many-body systems by path-integral methods, *Physical Review B* **30**, 2555–68.

Potter, D. (1977). *Computational Physics* (Wiley, London).

Press, W.H., Teukolsky, S.A., Vetterling, W.T., and Flannery, B.R. (1992). *Numerical Recipes* (Cambridge University, Cambridge).

Pryce, J.D. (1993). *Numerical Solutions of Sturm–Liouville Problems* (Clarendon, Oxford).

Punzo, G., Massaioli, F., and Succi, S. (1994). High-resolution lattice-Boltzmann computing on the IBM SP1 scalable parallel computer, *Computers in Physics* **8**, 705–11.

Quinn, M.J. (1994). *Parallel Computing: Theory and Practice* (McGraw–Hill, New York).

Rasetti, M. (ed.) (1991). *The Hubbard Model* (World Scientific, Singapore).

Reynolds, P.J., Ceperley, D.M., Alder, B.J., and Lester, W.A. Jr. (1982). Fixed-node quantum Monte Carlo for molecules, *Journal of Chemical Physics* **77**, 5593–603.

Rowe, A.C.H. and Abbott, P.C. (1995). Daubechies wavelets and Mathematica, *Computers in Physics* **9**, 635–48.

Ryckaert, J.P., Ciccotti, G., and Berendsen, H.J.C. (1977). Numerical integration of the Cartesian equations of motion of a system with constraints: Molecular dynamics of *n*-alkanes, *Journal of Computational Physics* **23**, 327–41.

Sabot, G.W. (ed.) (1995). *High Performance Computing: Problem Solving with Parallel and Vector Architectures* (Addison–Wesley, Reading, Massachusetts).

Schmidt, K.E. and Ceperley, D.M. (1992). Monte Carlo techniques for quantum fluids, solids, and droplets, in *The Monte Carlo Method in Condensed Matter Physics*, ed. by K. Binder (Springer–Verlag, Berlin), pp. 205–48.

Shanno, F. (1970). Conditioning of quasi-Newton methods for function minimization, *Mathematics of Computation* **24**, 647–57.

Simos, T.E. (1993). A variable-step procedure for the numerical integration of the one-dimensional Schrödinger equation, *Computers in Physics* **7**, 460–4.

Snir, M., Otto, S., Huss-Lederman, S., Walker, D., and Dongarra, J. (1996). *MPI: The Complete Reference* (MIT, Cambridge, Massachusetts).

Stauffer, D. and Aharony, A. (1992) *Introduction to Percolation Theory* (Taylor & Francis, London).

Strandburg, K.J. (ed.) (1992). *Bond-Orientational Order in Condensed Matter Systems* (Springer–Verlag, Berlin).

Strang, G. (1989). Wavelets and dilation equations: A brief introduction, *SIAM Review* **31**, 614–27.

Strang, W.G. and Fix, G.J. (1973). *An Analysis of the Finite Element Method* (Prentice–Hall, Englewood Cliffs, New Jersey).

Stroustrup, B. (1992). *The C++ Programming Language* (Addison–Wesley, Reading, Massachusetts).

Succi, S., Amati, G., and Benzi, R. (1995). Challenges in lattice Boltzmann computing, *Journal of Statistical Physics* **81**, 5–16.

Sugar, R.L. (1990). Monte Carlo studies of many-electron systems, in *Computer Simulation Studies in Condensed Matter Physics II: New Directions*, ed. by D.P. Landau, K.K. Mon, and H.-B. Schüttler (Springer–Verlag, Berlin), pp. 116–36.

Swendsen, R.H. (1979a). Monte Carlo renormalization group, *Physical Review Letters* **42**, 859–61.

Swendsen, R.H. (1979b). Monte Carlo renormalization-group studies of the $d = 2$ Ising model, *Physical Review B* **20**, 2080–7.

Swendsen, R.H., Wang, J.-S., and Ferrenberg, A.M. (1992). New Monte Carlo methods for improved efficiency of computer simulation in statistical mechanics, in *The Monte Carlo Methods in Condensed Matter Physics*, ed. by K. Binder (Springer–Verlag, Berlin), pp. 75–91.

Swendsen, R.H. and Wang, J.-S. (1987). Nonuniversal critical dynamics in Monte Carlo simulations, *Physical Review Letters* **58**, 86–8.

Tassone, F., Mauri, F., and Car, R. (1994). Accelerated schemes for ab initio molecular-dynamics simulations and electronic-structure calculations, *Physical Review B* **50**, 10561–73.

Trivedi, N. and Ceperley, D.M. (1990). Ground-state correlations of quantum antiferromagnets: A Green-function Monte Carlo study, *Physical Review B* **41**, 4552–69.

Umrigar, C.J. (1993). Accelerated Metropolis method, *Physical Review Letters* **71**, 408–11.

Umrigar, C.J., Wilson, K.G., and Wilkins, J.W. (1988). Optimized trial wave functions for quantum Monte Carlo calculations, *Physical Review Letters* **60**, 1719–22.

van Gunsteren, W.F. and Berendsen, H.J.C. (1977). Algorithms for micromolecular dynamics and constraint dynamics, *Molecular Physics* **34**, 1311–27.

Vichnevetsky, R. (1981). *Computer Methods for Partial Differential Equations, Volume I, Elliptic Equations and the Finite-Element Method* (Prentice–Hall, Englewood Cliffs, New Jersey).

van Hove, L. (1954). Correlations in space and time and Born approximation scattering in systems of interacting particles, *Physical Review* **95**, 249–62.

Vashishta, P., Nakano, A., Kalia, R.K., and Ebbsjö, I. (1995). Molecular dynamics simulations of covalent amorphous insulators on parallel computers, *Journal of Non Crystalline Solids* **182**, 59–67.

Vollhardt, D. (1984). Normal ^{3}He: An almost localized Fermi liquid, *Reviews of Modern Physics* **56**, 99–120.

Wang, C.Z., Chan, C.T., and Ho, K.M. (1989). Empirical tight-binding force model for molecular-dynamics simulation of Si, *Physical Review B* **39**, 8586–92.

White, S.R. (1992). Density matrix formulation for quantum renormalization groups, *Physical Review Letters* **69**, 2863–6.

White, S.R. (1993). Density-matrix algorithms for quantum renormalization groups, *Physical Review B* **48**, 10345–56.

White, S.R., Noack, R.M., and Scalapino, D.J. (1993). Resonating valence bond theory of coupled Heisenberg chains, *Physical Review Letters* **73**, 886–9.

Widom, B. (1965a). Surface tension and molecular correlations near the critical point, *Journal of Chemical Physics* **43**, 3892–7.

Widom, B. (1965b). Equation of state in the neighborhood of the critical point, *Journal of Chemical Physics* **43**, 3898–905.

Wiegel, F.W. (1986). *Introduction to Path-Integral Methods in Physics and Polymer Science* (World Scientific, Singapore).

Wilkinson, J.H. (1963). *Rounding Errors in Algebraic Processes* (Prentice–Hall, Englewood Cliffs, New Jersey).

Wilkinson, J.H. (1965). *Algebraic Eigenvalue Problems* (Clarendon, Oxford).

Wilson, K.G. (1975). The renormalization group: Critical phenomena and the Kondo problem, *Reviews of Modern Physics* **47**, 773–840.

Wolff, U. (1989). Collective Monte Carlo updating for spin systems, *Physical Review Letters* **62**, 361–4.

Wolfram, S. (1986). Cellular automaton fluids 1: Basic theory, *Journal of Statistical Physics* **45**, 471–526.

Wolfram, S. (1996). *The Mathematica Book* (Cambridge University, Cambridge).

Wong, S.S.M. (1992). *Computational Methods in Physics and Engineering* (Prentice–Hall, Englewood Cliffs, New Jersey).

Yeh, M.-L., Kim, J., and Khan, F.S. (1995). Parallel decomposition of the tight-binding fictitious Lagrangian algorithm for molecular dynamics simulations of semiconductors, *Computers in Physics* **9**, 108–20.

Yonezawa, F. (1991). Glass transition and relaxation of disordered structures, in *Solid State Physics, Volume 45*, ed. by H. Ehrenreich and D. Turnbull (Academic, Boston, Massachusetts), pp. 179–254.

Young, D.M. and Gregory, R.T. (1988). *A Survey of Numerical Mathematics, Volumes I and II* (Dover, New York).

Young, R. (1993). *Wavelet Theory and Its Applications* (Kluwer Academic, Boston, Massachusetts).

Yu, C.C. and White, S.R. (1993). Numerical renormalization group study of the one-dimensional Kondo insulator, *Physical Review Letters* **71**, 3866–9.

INDEX